Springer Texts in Statistics

Advisors:
George Casella Stephen Fienberg Ingram Olkin

Springer
New York
Berlin
Heidelberg
Barcelona
Hong Kong
London
Milan
Paris
Singapore
Tokyo

Springer Texts in Statistics

Alfred: Elements of Statistics for the Life and Social Sciences
Berger: An Introduction to Probability and Stochastic Processes
Bilodeau and Brenner: Theory of Multivariate Statistics
Blom: Probability and Statistics: Theory and Applications
Brockwell and Davis: Introduction to Times Series and Forecasting, Second Edition
Carmona: Statistical Analysis of Financial Data in S-Plus
Chow and Teicher: Probability Theory: Independence, Interchangeability, Martingales, Third Edition
Christensen: Advanced Linear Modeling: Multivariate, Time Series, and Spatial Data; Nonparametric Regression and Response Surface Maximization, Second Edition
Christensen: Log-Linear Models and Logistic Regression, Second Edition
Christensen: Plane Answers to Complex Questions: The Theory of Linear Models, Third Edition
Creighton: A First Course in Probability Models and Statistical Inference
Davis: Statistical Methods for the Analysis of Repeated Measurements
Dean and Voss: Design and Analysis of Experiments
du Toit, Steyn, and Stumpf: Graphical Exploratory Data Analysis
Durrett: Essentials of Stochastic Processes
Edwards: Introduction to Graphical Modelling, Second Edition
Finkelstein and Levin: Statistics for Lawyers
Flury: A First Course in Multivariate Statistics
Heiberger and Holland: Statistical Analysis and Data Display: An Intermediate Course with Examples in S-PLUS, R, and SAS
Jobson: Applied Multivariate Data Analysis, Volume I: Regression and Experimental Design
Jobson: Applied Multivariate Data Analysis, Volume II: Categorical and Multivariate Methods
Kalbfleisch: Probability and Statistical Inference, Volume I: Probability, Second Edition
Kalbfleisch: Probability and Statistical Inference, Volume II: Statistical Inference, Second Edition
Karr: Probability
Keyfitz: Applied Mathematical Demography, Second Edition
Kiefer: Introduction to Statistical Inference
Kokoska and Nevison: Statistical Tables and Formulae
Kulkarni: Modeling, Analysis, Design, and Control of Stochastic Systems
Lange: Applied Probability
Lange: Optimization
Lehmann: Elements of Large-Sample Theory

(continued after index)

Deborah Nolan Terry Speed

Stat Labs

Mathematical Statistics Through Applications

With 45 Illustrations

Deborah Nolan
Department of Statistics
University of California, Berkeley
Berkeley, CA 94720
USA
nolan@stat.berkeley.edu

Terry Speed
Department of Statistics
University of California, Berkeley
Berkeley, CA 94720
USA
terry@stat.berkeley.edu

Cover illustrations: Clockwise from top left are paired ribbons of DNA forming the double helix structure; a scatter plot of heights and weights of mothers in the Child Health and Development Studies; and box-and-whisker plots of weights of male mice with genetically altered DNA. These figures overlay a population map of Minnesota.

Library of Congress Cataloging-in-Publication Data
Nolan, Deborah Ann.
Stat labs: mathematical statistics through applications/Deborah Nolan, Terry Speed.
p. cm. — (Springer texts in statistics)
Includes bibliographical references and index.
ISBN 0-387-98974-9 (softcover : alk. paper)
1. Mathematical statistics. I. Speed, T.P. II. Title. III. Series.
QA276.12.N68 2000
519.5—dc21 99-057464

Printed on acid-free paper.

Production managed by Frank McGuckin; manufacturing supervised by Jeffrey Taub.
Typeset by the Bartlett Press, Inc., Marietta, GA.
Printed and bound by R.R. Donnelley and Sons, Harrisonburg, VA.
Printed in the United States of America.

9 8 7 6 5 4 3

ISBN 0-387-98974-9

Springer-Verlag New York Berlin Heidelberg
A member of BertelsmannSpringer Science+Business Media GmbH

To Ben and Sammy

—D.N.

To Sally

—T.P.S

Preface

This book uses a model we have developed for teaching mathematical statistics through in-depth case studies. Traditional statistics texts have many small numerical examples in each chapter to illustrate a topic in statistical theory. Here, we instead make a case study the centerpiece of each chapter. The case studies, which we call labs, raise interesting scientific questions, and figuring out how to answer a question is the starting point for developing statistical theory. The labs are substantial exercises; they have nontrivial solutions that leave room for different analyses of the data. In addition to providing the framework and motivation for studying topics in mathematical statistics, the labs help students develop statistical thinking. We feel that this approach integrates theoretical and applied statistics in a way not commonly encountered in an undergraduate text.

The Student

The book is intended for a course in mathematical statistics for juniors and seniors. We assume that students have had one year of calculus, including Taylor series, and a course in probability. We do not assume students have experience with statistical software so we incorporate lessons into our course on how to use the software.

Theoretical Content

The topics common to most mathematical statistics texts can be found in this book, including: descriptive statistics, experimental design, sampling, estimation, testing, contingency tables, regression, simple linear least squares, analysis of variance, and multiple linear least squares. Also found here are many selected topics, such as quantile plots, the bootstrap, replicate measurements, inverse regression, ecological regression, unbalanced designs, and response surface analysis.

This book differs from a standard mathematical statistics text in three essential ways. The first way is in how the topic of testing is covered. Although we address testing in several chapters and discuss the z, t, and F tests as well as chi-square tests of independence and homogeneity, Fisher's exact test, Mantel-Haenszel test, and the chi-square goodness of fit test, we do not cover all of the special cases of t tests that are typically found in mathematical statistics texts. We also do not cover nonparametric tests. Instead we cover more topics related to linear models.

The second main difference is the depth of the coverage. We are purposefully brief in the treatment of most of the theoretical topics. The essential material is there, but details of derivations are often left to the exercises.

Finally this book differs from a traditional mathematical statistics text in its layout. The first four sections of each chapter provide the lab's introduction, data description, background, and suggestions for investigating the problem. The theoretical material comes last, after the problem has been fully developed. Because of this novel approach, we have included an Instructor's Guide to *Stat Labs*, where we describe the layout of the chapters, the statistical content of each chapter, and ideas for how to use the book in a course.

The design of *Stat Labs* is versatile enough to be used as the main text for a course, or as a supplement to a more theoretical text. In a typical semester, we cover about 10 chapters. The core chapters that we usually cover are Chapter 1 on descriptive statistics, Chapter 2 on simple random sampling, Chapter 4 on estimation and testing, and Chapter 7 on regression. Other chapters are chosen according to the interests of students. We give examples of semester courses for engineering, social science, and life science students in the Instructor's Guide.

Acknowledgments

This book has been many years in the making, and has changed shape dramatically in its development. We are indebted to those who participated in this project. It began with the creation of a modest set of instruction sheets for using the computer for simulation and data analysis. Ed Chow helped us develop these first labs. Chad Heilig helped reorganize them to look more like the labs found here. In the final stages of preparation, Gill Ward helped clarify and improve the presentation of the material. Others have assisted us in the preparation of the manuscript: Yoram Gat prepared many of the figures; Liza Levina wrote the probability appendix; Chyng-Lan Liang prepared the news articles and Gang Liang made countless corrections to the manuscript. Thanks also go to Frank McGuckin, the production editor at Springer, for turning the many idiosyncrasies of our manuscript into a book.

Several undergraduates assisted in the preparation of the labs. Christine Cheng researched the topics of AIDS and hemophilia for Chapter 6, Tiffany Chiang pilot-tested several experiments for Chapter 5, Cheryl Gay prepared the background material for Chapter 11, and Jean-Paul Sursock helped formulate the investigations of the data for many of the labs.

We are very grateful to those who shared with us their data and subject-matter knowledge for the labs. Without them, this book would not exist. Thanks go to David Azuma, Brenda Eskanazi, David Freedman, James Goedert, Bill Kahn, Stephen Klein, David Lein, Ming-Ying Leung, Mike Mohr, Tony Nero, Phillip Price, Roger Purves, Philip Rosenberg, Eddie Rubin, and Desmond Smith.

We also thank those who have tried out some of our labs in their courses: Geir Eide, Marjorie Hahn and Pham Quan, and our colleagues Ani Adhikari, Ching-Shui Cheng, Kjell Doksum, John Rice, Philip Stark, Mark van der Laan, and Bin Yu. Similar thanks go to all the graduate students who taught from the ever evolving *Stat Labs*, including Ed Chow, Francois Collin, Imola Fodor, Bill Forrest, Yoram Gat, Ben Hansen, Chad Heilig, Hank Ibser, Jiming Jiang, Ann Kalinowski, Namhyun Kim, Max Lainer, Vlada Limic, Mike Moser, Jason Schweinsberg, and Duncan Temple Lang. Apologies to anyone left off this list of acknowledgments.

This project was partially supported by an educational mini-grant from the University of California, and by a POWRE grant from the National Science Foundation.

Finally, we are indebted to Joe Hodges, who allowed us to adopt the title from his book with Krech and Crutchfield, which was an inspiration to us. We also thank John Kimmel, our editor. Throughout this long evolution John patiently waited and supported our project.

Berkeley, California — Deborah Nolan
Berkeley, California — Terry Speed
March 2000

Instructor's Guide to *Stat Labs*

The labs you find here in this text are case studies that serve to integrate the practice and theory of statistics. The instructor and students are expected to analyze the data provided with each lab in order to answer a scientific question posed by the original researchers who collected the data. To answer a question, statistical methods are introduced, and the mathematical statistics underlying these methods are developed.

The Design of a Chapter

Each chapter is organized into five sections: Introduction, Data, Background, Investigations, and Theory. Sometimes we include a section called Extensions for more advanced topics.

Introduction

Here a clear scientific question is stated, and motivation for answering it is given. The question is presented in the context of the scientific problem, and not as a request to perform a particular statistical method. We avoid questions suggested by the data, and attempt to orient the lab around the original questions raised by the researchers who collected the data.

The excerpt found at the beginning of a chapter relates the subject under investigation to a current news story, which helps convey the relevance of the question at hand.

Data

Documentation for the data collected to address the question is provided in the Data section. Also, this section includes a description of the study protocol. The data can be found at the *Stat Labs* website:
`www.stat.berkeley.edu/users/statlabs/`

Background

The Background section contains scientific material that helps put the problem in context. The information comes from a variety of sources, and is presented in nontechnical language.

Investigations

Suggestions for answering the question posed in the Introduction appear in the Investigations section. These suggestions are written in the language of the lab's subject matter, using very little statistical terminology. They can be used as an assignment for students to work on outside of the classroom, or as a guide for the instructor for discussing and presenting analyses to answer the question in class.

The suggestions vary in difficulty, and are grouped to enable the assignment of subsets of investigations. Also included are suggestions on how to write up the results. Appendix A gives tips on how to write a good lab report.

Theory

The theoretical development appears at the end of the chapter in the Theory section. It includes both material on general statistical topics, such as hypothesis testing and parameter estimation, and on topics specific to the lab, such as goodness-of-fit tests for the Poisson distribution and parameter estimation for the log-normal distribution. The exercises at the end of the Theory section are designed to give practice with the theoretical material introduced in the section. Some also extend ideas introduced in the section. The exercises can be used for paper-and-pencil homework assignments.

Statistical Topics

The table below lists the main statistical topics covered in each chapter. All of the basic topics found in most mathematical statistics texts are included here: descriptive statistics, experimental design, sampling, estimation, testing, contingency tables, regression, simple linear least squares, analysis of variance, and multiple linear least squares. We also list some of the additional specialized topics covered in each chapter.

Chapter	Main Topic	Some Additional Topics
1	descriptive statistics	quantile plots, normal approximation
2	simple random sampling	confidence intervals
3	stratified sampling	parametric bootstrap allocation
4	estimation and testing	goodness-of-fit tests, information, asymptotic variance
5	contingency tables	experimental design
6	Poisson counts and rates	Mantel-Haenszel test
7	regression	prediction
8	simple linear model	replicate measurements, transformations, inverse regression
9	ecological regression	weighted regression
10	multiple linear regression	model checking, projections
11	analysis of variance	unbalanced designs, indicator variables
12	response surface analysis	factorial design

Sample Courses

This book can be used as the main text for a course, or as a supplement to a more theoretical text. In a typical semester, we cover eight to ten chapters. We spend about one week on each of the chapters, with the exception of Chapters 4, 10, and 11, which require up to two weeks to cover.

The core chapters that we usually include in a course are Chapter 1 on descriptive statistics, Chapter 2 on simple random sampling, Chapter 4 on estimation and testing, and Chapter 7 on regression. Other chapters are chosen according to the interests of students. In a one semester course for engineers we may include Chapter 3 on stratified sampling, Chapter 5 on experimental design, Chapter 8 on calibration and inverse regression, Chapter 11 on analysis of variance, and Chapter 12 on response surface analysis. In a course designed for social and life science majors we tend to include Chapter 3 on stratified sampling, Chapter 6 on estimating mortality rates, Chapter 9 on ecological regression, Chapter 10 on multiple regression, and Chapter 11 on analysis of variance.

Lab Assignments

We have found that our course is most successful when we incorporate the labs into the class room, and not leave them as exercises for students to work on solely outside of class. Often, we have students work on four or five of the ten labs covered in course. They have as their assignment, to address the suggestions in the Investigations section. Students analyze the data and write a report on their findings. We give two to three weeks for them to complete the assignment, and we sometimes allow them to work in groups of two or three. An alternative to this approach has students work on final projects of their own choosing. In this case, we make fewer lab assignments.

Software

For most of the labs, statistical software is needed to analyze the data. The exceptions are Chapters 2, 5, and 12. For these three chapters a statistical calculator is sufficient.

We have had success in using the software S-plus and R in the course. For those unfamiliar with R, the syntax of the language is similar to that of S-plus, and it is free. Students can easily down load a copy from the worldwide web (`lib.stat.cmu.edu/R/`) to run on their PCs, and so be able to work on assignments at home.

We advise students to consult an introductory text on how to use the software. For S-plus, we recommend *An Introduction to S and S-plus*, P. Spector, Wadsworth, 1994. For R we recommend *An Introduction to R*, the R Development Core Team, which can be found at the R website (`lib.stat.cmu.edu/R/`) and copied at no cost.

In our experience, we have found it important to provide assistance outside of class time on how to use the statistical software. One place where we do this is in section, where we sometimes meet in a computer laboratory room to work on the assignment and provide advice as needed. We also build a Frequently Asked Questions (FAQ) web page for each lab assignment. The page contains sample code and answers to questions students have asked in office hours, class, and section. These FAQs are available at the *Stat Labs* website (`www.stat.berkeley.edu/users/statlabs/`).

Grading

It can be difficult to grade the lab reports, because the investigations allow students to be creative in their solutions to the problem. We often base our grading on four aspects of the report: composition and presentation, basic analyses, graphs and tables, and advanced analyses. Sometimes we also request an appendix to the report for technical material.

Class time

This book is ideal for teaching statistics in a format that breaks away from the traditional lecture style. We have used it in classes where the enrollment ranges from 20 to 60 students, and where classes meet three hours a week with the instructor and one to two hours a week with a teaching assistant. In all of these classes we have found that interactive teaching is both possible and desirable.

In smaller classes, we run the class in a seminar style with time spent brainstorming on how to solve the problems in the Investigations section. Solving these problems leads to new statistical methods, and class time then alternates to lectures in order to cover these topics in mathematical statistics.

In the larger classes, we rely on group work in class to facilitate the discussion and analysis. We often supply handouts with an abbreviated list of investigations, and ask students to come up with a plan of attack for how to begin to address the questions. After they discuss their plan, we present results from analyses we have prepared in advance in anticipation of their suggestions. Other times, students are given a set of charts and graphs and they are asked to further summarize and interpret the output in order to answer questions from the Investigations section. Groups present their solutions and the instructor leads the class in a discussion of the analysis.

Another alternative that we have tried has groups of students prepare in advance to discuss a lab in class. They make a presentation to the class, and supply their own handouts and materials.

Whatever way you decide to use this book, we hope you will find it useful in augmenting your mathematical statistics course with substantial applications.

Contents

1

Maternal Smoking and Infant Health

WEDNESDAY, MARCH 1, 1995 ★★★★★ New York Times[1]

Infant Deaths Tied to Premature Births

Low weights not solely to blame

A new study of more than 7.5 million births has challenged the assumption that low birth weights per se are the cause of the high infant mortality rate in the United States. Rather, the new findings indicate, prematurity is the principal culprit.

Being born too soon, rather than too small, is the main underlying cause of stillbirth and infant deaths within four weeks of birth.

Each year in the United States about 31,000 fetuses die before delivery and 22,000 newborns die during the first 27 days of life.

The United States has a higher infant mortality rate than those in 19 other countries, and this poor standing has long been attributed mainly to the large number of babies born too small, including a large proportion who are born "small for date," or weighing less than they should for the length of time they were in the womb.

The researchers found that American-born babies, on average, weigh less than babies born in Norway, even when the length of the pregnancy is the same. But for a given length of pregnancy, the lighter American babies are no more likely to die than are the slightly heavier Norwegian babies.

The researchers, directed by Dr. Allen Wilcox of the National Institute of Environmental Health Sciences in Research Triangle Park, N.C., concluded that improving the nation's infant mortality rate would depend on preventing preterm births, not on increasing the average weight of newborns.

Furthermore, he cited an earlier study in which he compared survival rates among low-birth-weight babies of women who smoked during pregnancy.

Ounce for ounce, he said, "the babies of smoking mothers had a higher survival rate". As he explained this paradoxical finding, although smoking interferes with weight gain, it does not shorten pregnancy.

[1]Reprinted by permission.

Introduction

One of the U.S. Surgeon General's health warnings placed on the side panel of cigarette packages reads:

> Smoking by pregnant women may result in fetal injury, premature birth, and low birth weight.

In this lab, you will have the opportunity to compare the birth weights of babies born to smokers and nonsmokers in order to determine whether they corroborate the Surgeon General's warning. The data provided here are part of the Child Health and Development Studies (CHDS)—a comprehensive investigation of all pregnancies that occurred between 1960 and 1967 among women in the Kaiser Foundation Health Plan in the San Francisco–East Bay area (Yerushalmy [Yer71]). This study is noted for its unexpected findings that ounce for ounce the babies of smokers did not have a higher death rate than the babies of nonsmokers.

Despite the warnings of the Surgeon General, the American Cancer Society, and health care practitioners, many pregnant women smoke. For example, the National Center for Health Statistics found that 15% of the women who gave birth in 1996 smoked during their pregnancy.

Epidemiological studies (e.g., Merkatz and Thompson [MT90]) indicate that smoking is responsible for a 150 to 250 gram reduction in birth weight and that smoking mothers are about twice as likely as nonsmoking mothers to have a low-birth-weight baby (under 2500 grams). Birth weight is a measure of the baby's maturity. Another measure of maturity is the baby's gestational age, or the time spent in the womb. Typically, smaller babies and babies born early have lower survival rates than larger babies who are born at term. For example, in the CHDS group, the rate at which babies died within the first 28 days after birth was 150 per thousand births for infants weighing under 2500 grams, as compared to 5 per thousand for babies weighing more than 2500 grams.

The Data

The data available for this lab are a subset of a much larger study — the Child Health and Development Studies (Yerushalmy [Yer64]). The entire CHDS database includes all pregnancies that occurred between 1960 and 1967 among women in the Kaiser Foundation Health Plan in Oakland, California. The Kaiser Health Plan is a prepaid medical care program. The women in the study were all those enrolled in the Kaiser Plan who had obtained prenatal care in the San Francisco–East Bay area and who delivered at any of the Kaiser hospitals in Northern California.

In describing the 15,000 families that participated in the study, Yerushalmy states ([Yer64]) that

> The women seek medical care at Kaiser relatively early in pregnancy. Two-thirds report in the first trimester; nearly one-half when they are pregnant for

> 2 months or less. The study families represent a broad range in economic, social and educational characteristics. Nearly two-thirds are white, one-fifth negro, 3 to 4 percent oriental, and the remaining are members of other races and of mixed marriages. Some 30 percent of the husbands are in professional occupations. A large number are members of various unions. Nearly 10 percent are employed by the University of California at Berkeley in academic and administrative posts, and 20 percent are in government service. The educational level is somewhat higher than that of California as a whole, as is the average income. Thus, the study population is broadly based and is not atypical of an employed population. It is deficient in the indigent and the very affluent segments of the population since these groups are not likely to be represented in a prepaid medical program.

At birth, measurements on the baby were recorded. They included the baby's length, weight, and head circumference. Provided here is a subset of this information collected for 1236 babies — those baby boys born during one year of the study who lived at least 28 days and who were single births (i.e., not one of a twin or triplet). The information available for each baby is birth weight and whether or not the mother smoked during her pregnancy. These variables and sample observations are provided in Table 1.1.

Background

Fetal Development

The typical gestation period for a baby is 40 weeks. Those born earlier than 37 weeks are considered preterm. Few babies are allowed to remain in utero for more than 42 weeks because brain damage may occur due to deterioration of the placenta. The placenta is a special organ that develops during pregnancy. It lines the wall of the uterus, and the fetus is attached to the placenta by its umbilical cord (Figure 1.1). The umbilical cord contains blood vessels that nourish the fetus and remove its waste.

TABLE 1.1. Sample observations and data description for the 1236 babies in the Child Health and Development Studies subset.

Birth weight	120	113	128	123	108	136	138	132
Smoking status	0	0	1	0	1	0	0	0

Variable	Description
Birth weight	Baby's weight at birth in ounces. (0.035 ounces = 1 gram)
Smoking status	Indicator for whether the mother smoked (1) or not (0) during her pregnancy.

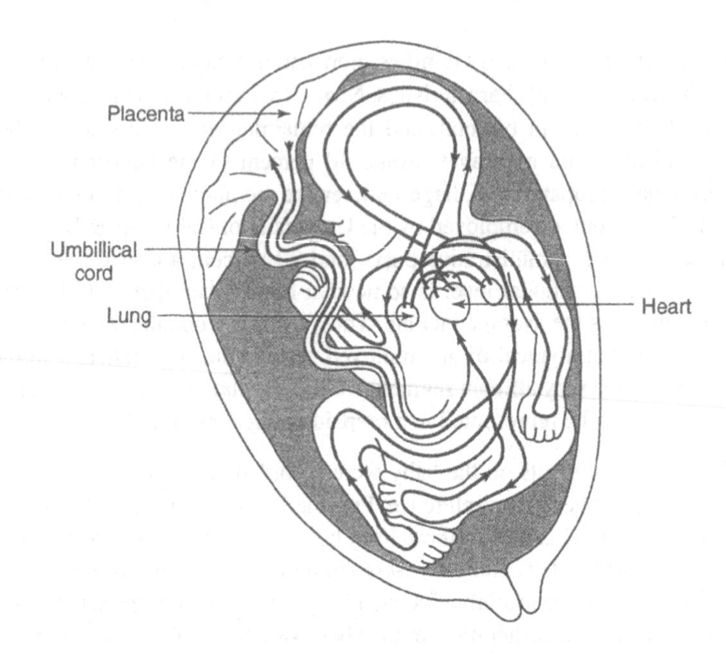

FIGURE 1.1. Fetus and placenta.

At 28 weeks of age, the fetus weighs about 4 to 5 pounds (1800 to 2300 grams) and is about 40 centimeters (cm) long. At 32 weeks, it typically weighs 5 to 5.5 pounds (2300 to 2500 grams) and is about 45 cm long. In the final weeks prior to delivery, babies gain about 0.2 pounds (90 grams) a week. Most newborns range from 45 to 55 cm in length and from 5.5 to 8.8 pounds (2500 to 4000 grams). Babies born at term that weigh under 5.5 pounds are considered small for their gestational age.

Rubella

Before the 1940s, it was widely believed that the baby was in a protected state while in the uterus, and any disease the mother contracted or any chemical that she used would not be transmitted to the fetus. This theory was attacked in 1941 when Dr. Norman Gregg, an Australian ophthalmologist, observed an unusually large number of infants with congenital cataracts. Gregg checked the medical history of the mothers' pregnancies and found that all of them had contracted rubella in the first or second month of their pregnancy. (There had been a widespread and severe rubella epidemic in 1940.) In a presentation of his findings to the Opthalmological Society of Australia, Gregg ([Gre41]) replied to comments on his work saying that

> ... he did not want to be dogmatic by claiming that it had been established the cataracts were due solely to the "German measles." However, the evidence afforded by the cases under review was so striking that he was convinced that

> there was a very close relationship between the two conditions, particularly because in the very large majority of cases the pregnancy had been normal except for the "German measles" infection. He considered that it was quite likely that similar cases may have been missed in previous years either from casual history-taking or from failure to ascribe any importance to an exanthem [skin eruption] affecting the mother so early in her pregnancy.

Gregg was quite right. Oliver Lancaster, an Australian medical statistician, checked census records and found a concordance between rubella epidemics and later increase in registration at schools for the deaf. Further, Swan, a pediatrician in Australia, undertook a series of epidemiological studies on the subject and found a connection between babies born to mothers who contracted rubella during the epidemic while in their first trimester of pregnancy and heart, eye, and ear defects in the infant.

A Physical Model

There are many chemical agents in cigarette smoke. We focus on one: carbon monoxide. It is commonly thought that the carbon monoxide in cigarette smoke reduces the oxygen supplied to the fetus. When a cigarette is smoked, the carbon monoxide in the inhaled smoke binds with the hemoglobin in the blood to form carboxyhemoglobin. Hemoglobin has a much greater affinity for carbon monoxide than oxygen. Increased levels of carboxyhemoglobin restrict the amount of oxygen that can be carried by the blood and decrease the partial pressure of oxygen in blood flowing out of the lungs. For the fetus, the normal partial pressure in the blood is only 20 to 30 percent that of an adult. This is because the oxygen supplied to the fetus from the mother must first pass through the placenta to be taken up by the fetus' blood. Each transfer reduces the pressure, which decreases the oxygen supply.

The physiological effects of a decreased oxygen supply on fetal development are not completely understood. Medical research into the effect of smoking on fetal lambs (Longo [Lon76]) provides insight into the problem. This research has shown that slight decreases in the oxygen supply to the fetus result in severe oxygen deficiency in the fetus' vital tissues.

A steady supply of oxygen is critical for the developing baby. It is hypothesized that, to compensate for the decreased supply of oxygen, the placenta increases in surface area and number of blood vessels; the fetus increases the level of hemoglobin in its blood; and it redistributes the blood flow to favor its vital parts. These same survival mechanisms are observed in high-altitude pregnancies, where the air contains less oxygen than at sea level. The placenta at high altitude is larger in diameter and thinner than a placenta at sea level. This difference is thought to explain the greater frequency in high-altitude pregnancies of abruptia placenta, where the placenta breaks away from the uterine wall, resulting in preterm delivery and fetal death (Meyer and Tonascia [MT77]).

Is the Difference Important?

If a difference is found between the birth weights of babies born to smokers and those born to nonsmokers, the question of whether the difference is important to the health and development of the babies needs to be addressed.

Four different death rates — fetal, neonatal, perinatal, and infant — are used by researchers in investigating babies' health and development. Each rate refers to a different period in a baby's life. The first is the fetal stage. It is the time before birth, and "fetal death" refers to babies who die at birth or before they are born. The term "neonatal" denotes the first 28 days after birth, and "perinatal" is used for the combined fetal and neonatal periods. Finally, the term "infant" refers to a baby's first year, including the first 28 days from birth.

In analyzing the pregnancy outcomes from the CHDS, Yerushalmy ([Yer71]) found that although low birth weight is associated with an increase in the number of babies who die shortly after birth, the babies of smokers tended to have much lower death rates than the babies of nonsmokers. His calculations appear in Table 1.2. Rather than compare the overall mortality rate of babies born to smokers against the rate for babies born to nonsmokers, he made comparisons for smaller groups of babies. The babies were grouped according to their birth weight; then, within each group, the numbers of babies that died in the first 28 days after birth for smokers and nonsmokers were compared. To accommodate the different numbers of babies in the groups, rates instead of counts are used in making the comparisons.

The rates in Table 1.2 are not adjusted for the mother's age and other factors that could potentially misrepresent the results. That is, if the mothers who smoke tend to be younger than those who do not smoke, then the comparison could be unfair to the nonsmokers because older women, whether they smoke or not, have more problems in pregnancy. However, the results agree with those from a Missouri study (see the left plot in Figure 1.2), which did adjust for many of these factors (Malloy et al. [MKLS88]). Also, an Ontario study (Meyer and Tonascia [MT77]) corroborates the CHDS results. This study found that the risk of neonatal death for babies who were born at 32+ weeks gestation is roughly the same for smokers and

TABLE 1.2. Neonatal mortality rates per 1000 births by birth weight (grams) for live-born infants of white mothers, according to smoking status (Yerushalmy [Yer71]).

Weight category	Nonsmoker	Smoker
≤ 1500	792	565
1500–2000	406	346
2000–2500	78	27
2500–3000	11.6	6.1
3000–3500	2.2	4.5
3500+	3.8	2.6

Note: 1500 to 2000 grams is roughly 53 to 71 ounces.

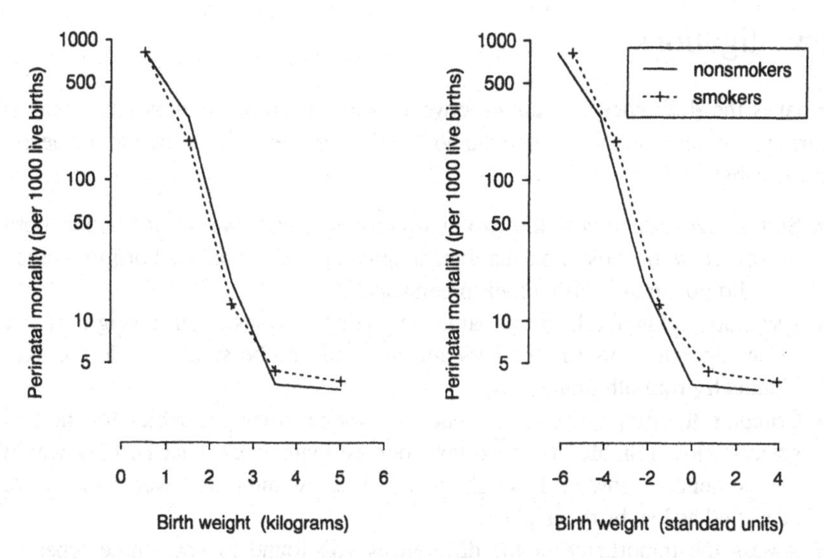

FIGURE 1.2. Mortality curves for smokers and nonsmokers by kilograms (left plot) and by standard units (right plot) of birth weight for the Missouri study (Wilcox [Wil93]).

nonsmokers. It was also found that the smokers had a higher rate of very premature deliveries (20–32 weeks gestation), and so a higher rate of early fetal death.

As in the comparison of Norwegian and American babies (New York Times, Mar. 1, 1995), in order to compare the mortality rates of babies born to smokers and those born to nonsmokers, Wilcox and Russell ([WR86]) and Wilcox ([Wil93]) advocate grouping babies according to their relative birth weights. A baby's relative birth weight is the difference between its birth weight and the average birth weight for its group as measured in standard deviations(SDs); it is also called the standardized birth weight. For a baby born to a smoker, we would subtract from its weight the average birth weight of babies born to smokers (3180 grams) and divide this difference by 500 grams, the SD for babies born to smokers. Similarly, for babies born to nonsmokers, we standardize the birth weights using the average and SD for their group, 3500 grams and 500 grams, respectively. Then, for example, the mortality rate of babies born to smokers who weigh 2680 grams is compared to the rate for babies born to nonsmokers who weigh 3000 grams, because these weights are both 1 SD below their respective averages. The right plot in Figure 1.2 displays in standard units the mortality rates from the left plot. Because the babies born to smokers tend to be smaller, the mortality curve is shifted to the right relative to the nonsmokers' curve. If the babies born to smokers are smaller but otherwise as healthy as babies born to nonsmokers, then the two curves in standard units should roughly coincide. Wilcox and Russell found instead that the mortality curve for smokers was higher than that for nonsmokers; that is, for babies born at term, smokers have higher rates of perinatal mortality in every standard unit category.

Investigations

What is the difference in weight between babies born to mothers who smoked during pregnancy and those who did not? Is this difference important to the health of the baby?

- Summarize numerically the two distributions of birth weight for babies born to women who smoked during their pregnancy and for babies born to women who did not smoke during their pregnancy.
- Use graphical methods to compare the two distributions of birth weight. If you make separate plots for smokers and nonsmokers, be sure to scale the axes identically for both graphs.
- Compare the frequency, or incidence, of low-birth-weight babies for the two groups. How reliable do you think your estimates are? That is, how would the incidence of low birth weight change if a few more or fewer babies were classified as low birth weight?
- Assess the importance of the differences you found in your three types of comparisons (numerical, graphical, incidence).

Summarize your investigations for the CHDS babies. Include the most relevant graphical output from your analysis. Relate your findings to those from other studies.

Theory

In this section, several kinds of summary statistics are briefly described. When analyzing a set of data, simple summaries of the list of numbers can bring insight about the data. For example, the mean and the standard deviation are frequently used as numerical summaries for the location and spread of the data. A graphical summary such as a histogram often provides information on the shape of the data distribution, such as symmetry, modality, and the size of tails.

We illustrate these statistics with data from the 1236 families selected for this lab from the Child Health and Development Study (CHDS). The data used here are described in detail in the Data section of the continuation of this lab in Chapter 10. For each statistic presented, any missing data are ignored, and the number of families responding is reported.

The Histogram

Figure 1.3 displays a histogram for the heights of mothers in the CHDS. The histogram is unimodal and symmetric. That is, the distribution has one mode (peak), around 64 inches, and the shape of the histogram to the left of the peak looks roughly like the mirror image of the part of the histogram to the right of the

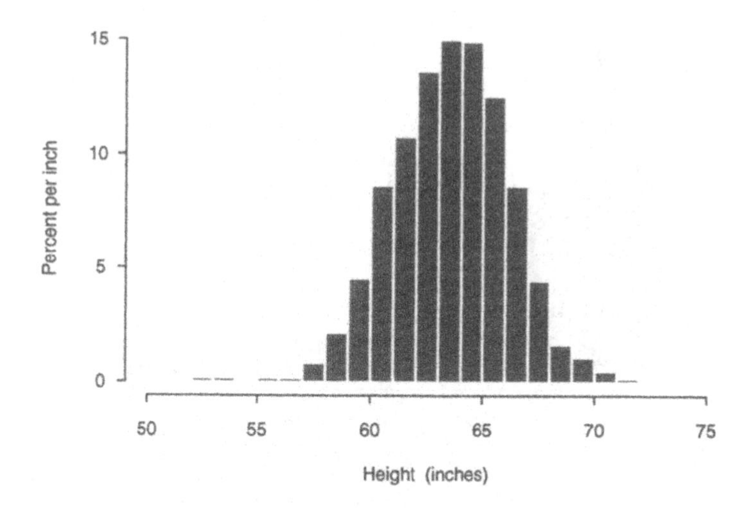

FIGURE 1.3. Histogram of mother's height for 1214 mothers in the CHDS subset.

peak. Outliers can be detected via histograms as well. They are observations that fall well outside the main range of the data. There appear to be a few very short mothers in the study.

In contrast to the height distribution, the histogram of the number of cigarettes smoked per day for those mothers who smoked during their pregnancy has a very different appearance (Figure 1.4). It shows two modes, one at 5–10 cigarettes and the other at 20–30 cigarettes. The distribution is asymmetric; that is it is right-skewed with the mode around 20–30 cigarettes less peaked than the mode at 0–5 cigarettes and with a long right tail. For unimodal histograms, a right-skewed distribution has more area to the right of the mode in comparison with that to the left; a left-skewed distribution has more area to the left.

A histogram is a graphical representation of a distribution table. For example, Table 1.3 is a distribution table for the number of cigarettes smoked a day by mothers who smoked during their pregnancy. The intervals include the left endpoint but not the right endpoint; for example the first interval contains those mothers who smoke up to but not including 5 cigarettes a day. In the histogram in Figure 1.4, the area of each bar is proportional to the percentage (or count) of mothers in the corresponding interval. This means that the vertical scale is percent per unit of measurement (or count per unit). The bar over the interval from 0 to 5 cigarettes is 3.2% per cigarette in height and 5 cigarettes in width: it includes all women who reported smoking up to an average of 5 cigarettes a day. Hence the area of the bar is

$$5 \text{ cigarettes} \times 3.2\%/\text{cigarette} = 16\%.$$

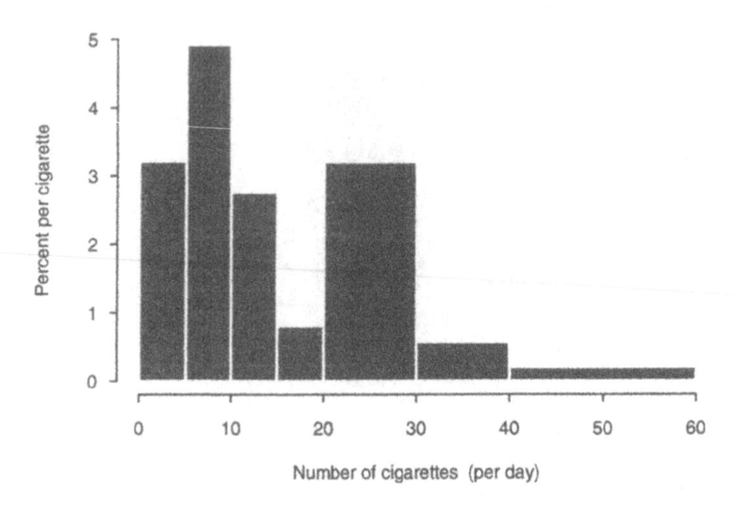

FIGURE 1.4. Histogram of the number of cigarettes smoked per day for the 484 mothers who smoked in the CHDS subset.

TABLE 1.3. Distribution of the number of cigarettes smoked per day for 484 mothers in the CHDS subset who smoked during their pregnancy, rounded to the nearest percent.

Number of cigarettes	Percent of smokers
0–5	16
5–10	25
10–15	14
15–20	4
20–30	32
30–40	5
40–60	4
Total	100

This bar is the same height as the bar above 20–30 cigarettes even though it has twice the number of mothers in it. This is because the 20–30 bar is twice as wide. Both bars have the same density of mothers per cigarette (i.e., 3.2% per cigarette).

Histograms can also be used to answer distributional questions such as: what proportion of the babies weigh under 100 ounces or what percentage of the babies weigh more than 138 ounces. From the histogram in Figure 1.5, we sum the areas of the bars to the left of 100 and find that 14% of the babies weigh under 100

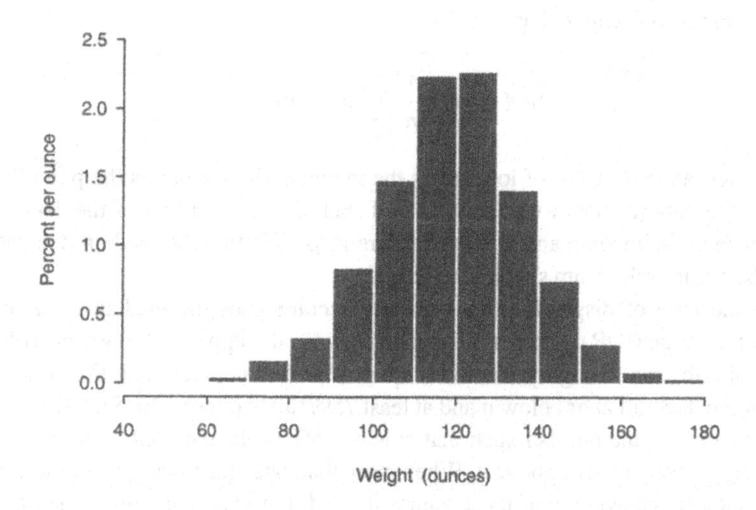

FIGURE 1.5. Histogram of infant birth weight for 1236 babies in the CHDS subset.

ounces. However, to answer the second question, we note that 138 does not fall at an interval endpoint of the histogram, so we need to approximate how many babies weigh between 138 and 140 ounces. To do this, split up the interval that runs from 130 to 140 into 10 one-ounce subintervals. The bar contains 14.2% of the babies, so we estimate that each one-ounce subinterval contains roughly 1.4% of the babies and 2.8% of the babies weigh 138–140 ounces. Because 12.5% of the babies weigh over 140 ounces, our estimate is that 15.3% of the babies weigh more than 138 ounces. In fact, 15.1% of the babies weighed more than this amount. The approximation was quite good.

Numerical Summaries

A measure of location is a statistic that represents the center of the data distribution. One such measure is the mean, which is the average of the data. The mean can be interpreted as the balance point of the histogram. That is, if the histogram were made up of bars sitting on a weightless balance beam, the mean would be the point at which the histogram would balance on the beam.

For a list of numbers $x_1, \ldots x_n$, the mean $\bar{x}$ is computed as follows:

$$\bar{x} = \frac{1}{n}\sum_{i=1}^{n} x_i.$$

A measure of location is typically accompanied by a measure of dispersion that gives an idea as to how far an individual value may vary from the center of the data. One such measure is the standard deviation (SD). The standard deviation is the root mean square (r.m.s.) of the deviations of the numbers on the list from the list average. It is computed as

$$\mathrm{SD}(x) = \sqrt{\frac{1}{n}\sum_{i=1}^{n}(x_i - \bar{x})^2}.$$

An alternative measure of location is the median. The median is the point that divides the data (or list of numbers) in half such that at least half of the data are smaller than the median and at least half are larger. To find the median, the data must be put in order from smallest to largest.

The measure of dispersion that typically accompanies the median is the interquartile range (IQR). It is the difference between the upper and lower quartiles of the distribution. Roughly, the lower quartile is that number such that at least 25% of the data fall at or below it and at least 75% fall at or above it. Similarly, the upper quartile is the number such that at least 75% of the data fall at or below it and at least 25% fall at or above it. When more than one value meets this criterion, then typically the average of these values is used. For example, with a list of 10 numbers, the median is often reported as the average of the 5th and 6th largest numbers, and the lower quartile is reported as the 3rd smallest number.

For infant birth weight, the mean is 120 ounces and the SD is 18 ounces. Also, the median is 120 ounces and the IQR is 22 ounces. The mean and median are very close due to the symmetry of the distribution. For heavily skewed distributions, they can be very far apart. The mean is easily affected by outliers or an asymmetrically long tail.

Five-Number Summary

The five-number summary provides a measure of location and spread plus some additional information. The five numbers are: the median, the upper and lower quartiles, and the extremes (the smallest and largest values). The five-number summary is presented in a box, such as in Table 1.4, which is a five-number summary for the weights of 1200 mothers in the CHDS.

From this five-number summary, it can be seen that the distribution of mother's weight seems to be asymmetric. That is, it appears to be either skewed to the right or to have some large outliers. We see this because the lower quartile is closer to the median than the upper quartile and because the largest observation is very far

TABLE 1.4. Five-number summary for the weights (in pounds) of 1200 mothers in the CHDS subset.

Median	125	
Quartiles	115	139
Extremes	87	250

from the upper quartile. Half of the mothers weigh between 115 and 139 pounds, but at least one weighs as much as 250 pounds.

Box-and-Whisker Plot

A box-and-whisker plot is another type of graphical representation of data. It contains more information than a five-number summary but not as much information as a histogram. It shows location, dispersion and outliers, and it may indicate skewness and tail size. However, from a box-and-whisper plot it is not possible to ascertain whether there are gaps or multiple modes in a distribution.

In a box-and-whisker plot, the bottom of the box coincides with the lower quartile and the the top with the upper quartile; the median is marked by a line through the box; the whiskers run from the quartiles out to the smallest (largest) number that falls within $1.5 \times$ IQR of the lower (upper) quartile; and smaller or larger numbers are marked with a special symbol such as a * or –.

Figure 1.6 contains a box-and-whisker plot of mother's weight. The right skewness of the distribution is much more apparent here than in the five-number summary. There are many variants on the box-and-whisker plot, including one that simply draws whiskers from the sides of the box to the extremes of the data.

The Normal Curve

The standard normal curve (Figure 1.7), known as the bell curve, sometimes provides a useful method for summarizing data.

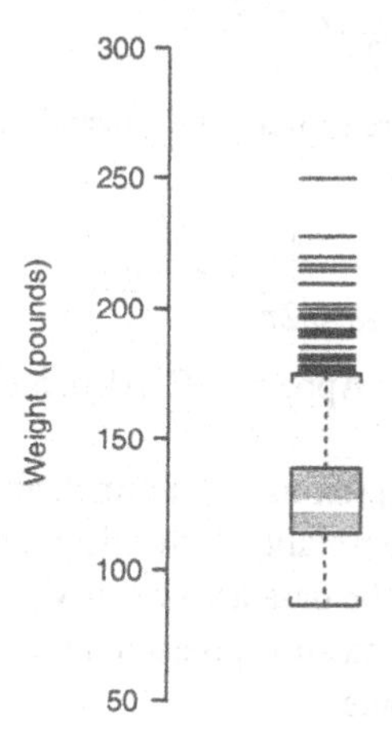

FIGURE 1.6. Box-and-whisker plot of mother's weight for 1200 mothers in the CHDS subset.

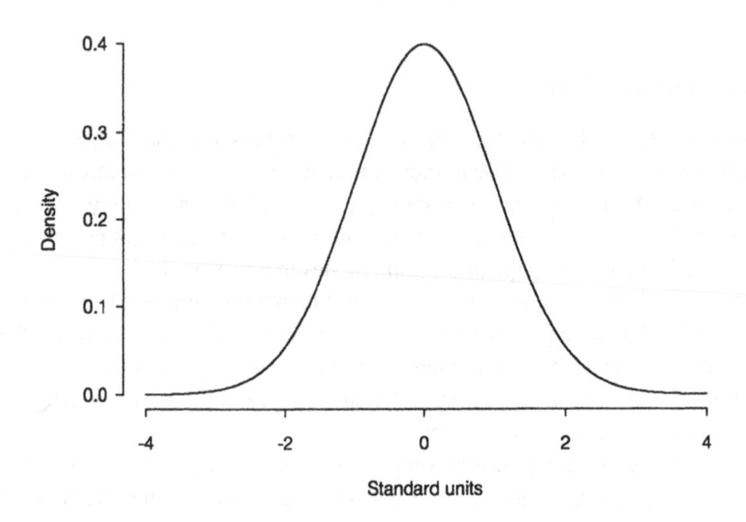

FIGURE 1.7. The standard normal curve.

The normal curve is unimodal and symmetric around 0. It also follows the 68-95-99.7 rule. The rule states that 68% of the area under the curve is within 1 unit of its center, 95% is within 2 units of the center, and 99.7% is within 3 units of its center. These areas and others are determined from the following analytic expression for the curve:

$$\frac{1}{\sqrt{2\pi}}e^{-x^2/2}.$$

Traditionally, $\Phi(z)$ represents the area under the normal curve to the left of z, namely,

$$\Phi(z) = \int_{-\infty}^{z} \frac{1}{\sqrt{2\pi}}e^{-x^2/2}dx.$$

A table of these areas can be found in Appendix C. Also, most statistical software provides these numbers.

Many distributions for data are approximately normal, and the 68-95-99.7 rule can be used as an informal check of normality. If the histogram looks normal, then this rule should roughly hold when the data are properly standardized. Note that to standardize the data, subtract the mean from each number and then divide by the standard deviation; that is, compute

$$\frac{x_i - \bar{x}}{\mathrm{SD}(x)}.$$

Notice that a value of $+1$ for the standard normal corresponds to an x-value that is 1 SD above $\bar{x}$. We saw in Figure 1.2 that standardizing the birth weights of babies led to a more informative comparison of mortality rates for smokers and nonsmokers.

For birth weight, we find that 69% of the babies have weights within 1 standard deviation of the average, 96% are within 2 SDs, and 99.4% are within 3 SDs. It looks pretty good. When the normal distribution fits well and we have summarized the data by its mean and SD, the normal distribution can be quite handy for answering such questions as what percentage of the babies weigh more than 138 ounces. The area under the normal curve can be used to approximate the area of the histogram. When standardized, 138 is 1 standard unit above average. The area under a normal curve to the right of 1 is 16%. This is close to the actual figure of 15%.

Checks for normality that are more formal than the 68-95-99.7 rule are based on the coefficients of skewness and kurtosis. In standard units, the coefficient of skewness is the average of the third power of the standardized data, and the coefficient of kurtosis averages the 4th power of the standardized list. That is,

$$\text{skewness} = \frac{1}{n}\sum_{i=1}^{n}\left(\frac{x_i - \bar{x}}{\mathrm{SD}(x)}\right)^3 \qquad \text{kurtosis} = \frac{1}{n}\sum_{i=1}^{n}\left(\frac{x_i - \bar{x}}{\mathrm{SD}(x)}\right)^4.$$

For a symmetric distribution, the skewness coefficient is 0. The kurtosis is a measure of how pronounced is the peak of the distribution. For the normal, the kurtosis should be 3. Departures from these values (0 for skewness and 3 for kurtosis) indicate departures from normality.

To decide whether a given departure is big or not, simulation studies can be used. A simulation study generates pseudo-random numbers from a known distribution, so we can check the similarity between the simulated observations and the actual data. This may show us that a particular distribution would be unlikely to give us the data we see. For example, the kurtosis of birth weight for the 484 babies born to smokers in the CHDS subset is 2.9. To see if 2.9 is a typical kurtosis value for a sample of 484 observations from a normal distribution, we could repeat the following a large number of times: generate 484 pseudo-random observations from a normal distribution and calculate the sample kurtosis. Figure 1.8 is a histogram of 1000 sample values of kurtosis computed for 1000 samples of size 484 from the standard normal curve. From this figure, we see that 2.9 is a very typical kurtosis value for a sample of 484 from a standard normal.

Quantile Plots

For a distribution such as the standard normal, the qth quantile is z_q, where

$$\Phi(z_q) = q, \quad 0 < q < 1.$$

The median, lower, and upper quartiles are examples of quantiles. They are, respectively, the 0.50, 0.25, and 0.75 quantiles.

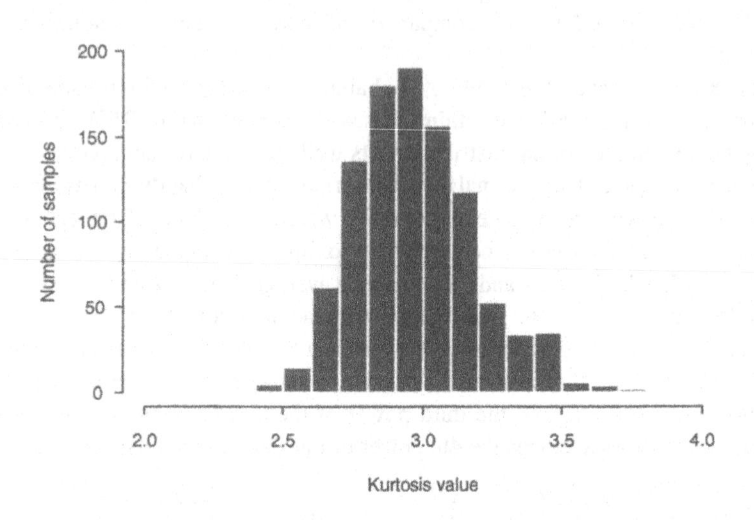

FIGURE 1.8. Histogram of kurtosis values for 1000 samples of size 484 from the standard normal.

For data $x_1, \ldots, x_n$, the sample quantiles are found by ordering the data from smallest to largest. We denote this ordering by $x_{(1)}, \ldots, x_{(n)}$. Then $x_{(k)}$ is considered the $k/(n+1)$th sample quantile. We divide by $n+1$ rather than n to keep q less than 1.

The normal-quantile plot, also known as the normal-probability plot, provides a graphical means of comparing the data distribution to the normal. It graphs the pairs $(z_{k/(n+1)}, x_{(k)})$. If the plotted points fall roughly on a line, then it indicates that the data have an approximate normal distribution. See the Exercises for a more formal treatment of quantiles. Figure 1.9 is a normal-quantile plot of the weights of mothers in the CHDS. The upward curve in the plot identifies a long right tail, in comparison to the normal, for the weight distribution.

Departures from normality are indicated by systematic departures from a straight line. Examples of different types of departures are provided in Figure 1.10. Generally speaking, if the histogram of the data does not decrease as quickly in the right tail as the normal, this is indicated by an upward curve on the right side of the normal-quantile plot. Similarly, a long left tail is indicated by a downward curve to the left (bottom right picture in Figure 1.10). On the other hand, if the tails decrease more quickly than the normal, then the curve will be as in the bottom left plot in Figure 1.10. Granularity in the recording of the data appears as stripes in the plot (top left plot in Figure 1.10). Bimodality is shown in the top right plot of Figure 1.10.

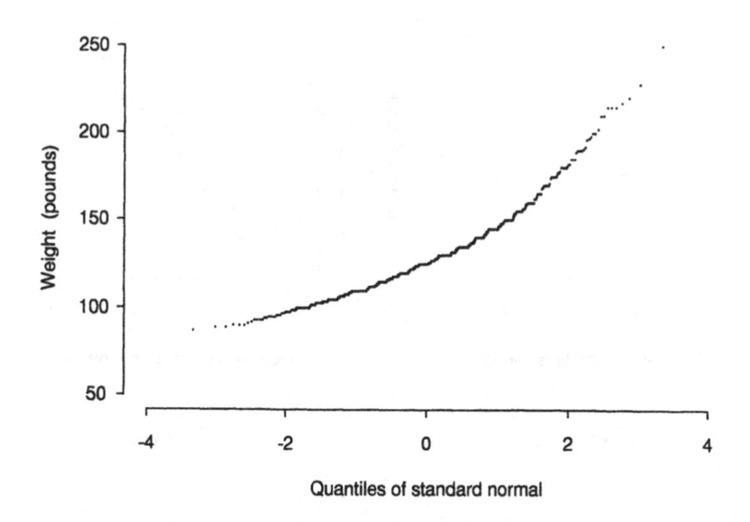

FIGURE 1.9. Normal quantile plot of mother's weight for 1200 mothers in the CHDS subset.

Quantile plots can be made for any distribution. For example, a uniform-quantile plot for mother's weight appears in Figure 1.11, where the sample quantiles of mother's weight are plotted against the quantiles of the uniform distribution. It is evident from the plot that both the left and right tails of the weight distribution are long in comparison to the uniform.

To compare two data distributions — such as the weights of smokers and nonsmokers — plots known as quantile-quantile plots can be made. They compare two sets of data to each other by pairing their respective sample quantiles. Again, a departure from a straight line indicates a difference in the shapes of the two distributions. When the two distributions are identical, the plot should be linear with slope 1 and intercept 0 (roughly speaking, of course). If the two distributions are the same shape but have different means or standard deviations, then the plot should also be roughly linear. However, the intercept and slope will not be 0 and 1, respectively. A nonzero intercept indicates a shift in the distributions, and a nonunit slope indicates a scale change. Figure 1.12 contains a quantile-quantile plot of mother's weight for smokers and nonsmokers compared with a line of slope 1 and intercept 0. Over most of the range there appears to be linearity in the plot, though lying just below the line: smokers tend to weigh slightly less than nonsmokers. Notice that the right tail of the distribution of weights is longer for the nonsmokers, indicating that the heaviest nonsmokers weigh quite a bit more than the heaviest smokers.

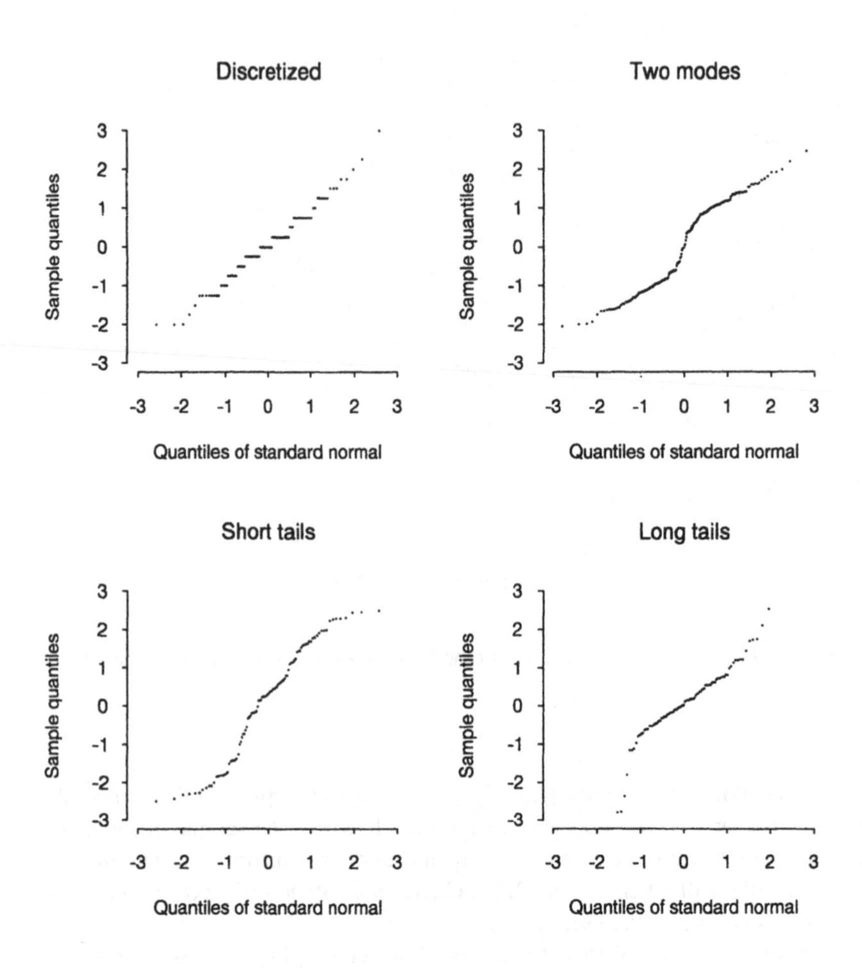

FIGURE 1.10. Examples of normal quantile plots.

Cross-tabulations

Distribution tables for subgroups of the data are called cross-tabulations. They allow for comparisons of distributions across more homogeneous subgroups. For example, the last row of Table 1.5 contains the distribution of body length for a sample of 663 babies from the CHDS. The rows of the table show the body-length distribution for smokers and nonsmokers separately. Notice that the babies of the smokers seem to be shorter than the babies of nonsmokers. It looks as though the distribution for the smokers is shifted to the left.

Bar Charts and Segmented Bar Charts

A bar chart is often used as a graphical representation of a cross-tabulation. It depicts the count (or percent) for each category of a second variable within each

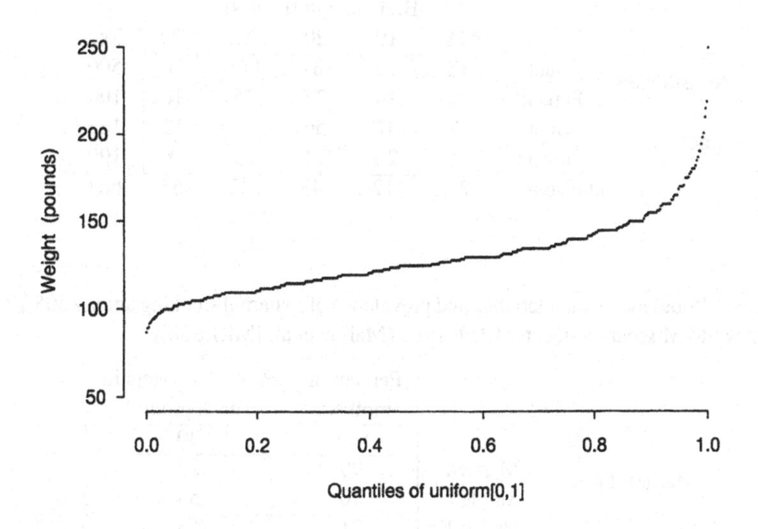

FIGURE 1.11. Uniform-quantile plot of mother's weight for 1200 mothers in the CHDS subset.

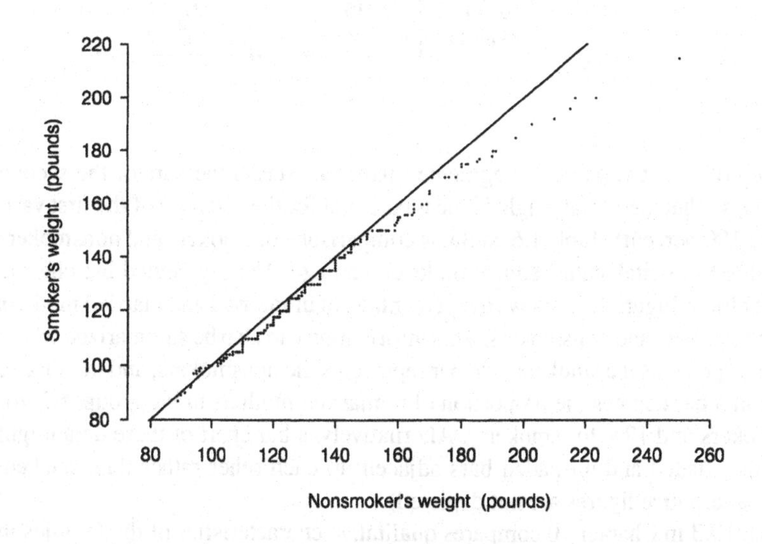

FIGURE 1.12. Quantile-quantile plot of mother's weight for smokers (484) and nonsmokers (752) in the CHDS subset; superimposed is a line of slope 1 and intercept 0.

TABLE 1.5. Cross-tabulation of infant body length (in inches) for smokers and nonsmokers for a sample of 663 babies from the CHDS.

		Body length (inches)					
		≤18	19	20	21	≥22	Total
Nonsmokers	Count	18	70	187	175	50	500
	Percent	4	14	37	35	10	100
Smokers	Count	5	42	56	47	13	163
	Percent	3	26	34	29	8	100
	Total count	23	112	243	222	63	663

TABLE 1.6. Population characteristics and prevalence of maternal smoking among 305,730 births to white Missouri residents, 1979–1983 (Malloy et al. [MKLS88]).

		Percent of mothers	Percent smokers in each group
All		100	30
Marital status	Married	90	27
	Single	10	55
Educational level (years)	Under 12	21	55
	12	46	29
	Over 12	33	15
Maternal age (years)	Under 18	5	43
	18–19	9	44
	20–24	35	34
	25–29	32	23
	30–34	15	21
	Over 34	4	26

category of a first variable. A segmented bar chart stacks the bars of the second variable, so that their total height is the total count for the category of the first variable (or 100 percent). Table 1.6 contains comparisons of smokers and nonsmokers according to marital status, education level, and age. The segmented bar chart in the left plot of Figure 1.13 shows the percentage of unmarried and married mothers who are smokers and nonsmokers. This information can also be summarized where one bar represents the smokers, one bar represents the nonsmokers, and the shaded region in a bar denotes the proportion of unmarried mothers in the group (6% for nonsmokers and 19% for smokers). Alternatively, a bar chart of these data might show the shaded and unshaded bars adjacent to each other rather than stacked. (These alternative figures are not depicted).

Table 10.3 in Chapter 10 compares qualitative characteristics of the families in the CHDS study according to whether the mother smokes or not. One of these characteristics, whether the mother uses contraceptives or not, is pictured in the segmented bar chart in the right plot of Figure 1.13.

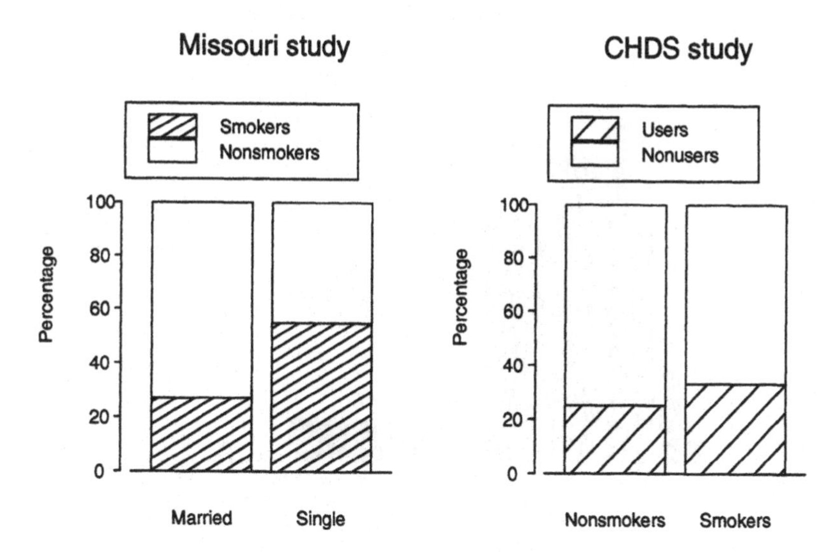

FIGURE 1.13. Bar charts of smoking prevalence by marital status (left) for mothers in the Missouri study (Malloy et al. [MKLS88]) and contraceptive use by smoking prevalence (right) for mothers in the CHDS study (Yerushalmy [Yer71]).

Exercises

1. Use Table 1.3 to find the approximate quartiles of the distribution of the number of cigarettes smoked per day for the mothers in the CHDS who smoked during their pregnancy.
2. Combine the last four categories in Table 1.3 of the distribution of the number of cigarettes smoked by the smoking mothers in the CHDS. Make a new histogram using the collapsed table. How has the shape changed from the histogram in Figure 1.4? Explain.
3. Consider the histogram of father's age for the fathers in the CHDS (Figure 1.14). The bar over the interval from 35 to 40 years is missing. Find its height.
4. Consider the normal quantile plots of father's height and weight for fathers in the CHDS (Figure 1.15). Describe the shapes of the distributions.
5. Following are the quantiles at 0.05, 0.10, ..., 0.95 for the gestational ages of the babies in the CHDS. Plot these quantiles against those of the uniform distribution on (0, 1). Describe the shape of the distribution of gestational age in comparison to the uniform.
252, 262, 267, 270, 272, 274, 276, 277, 278, 280, 281, 283, 284, 286, 288, 290, 292, 296, 302.

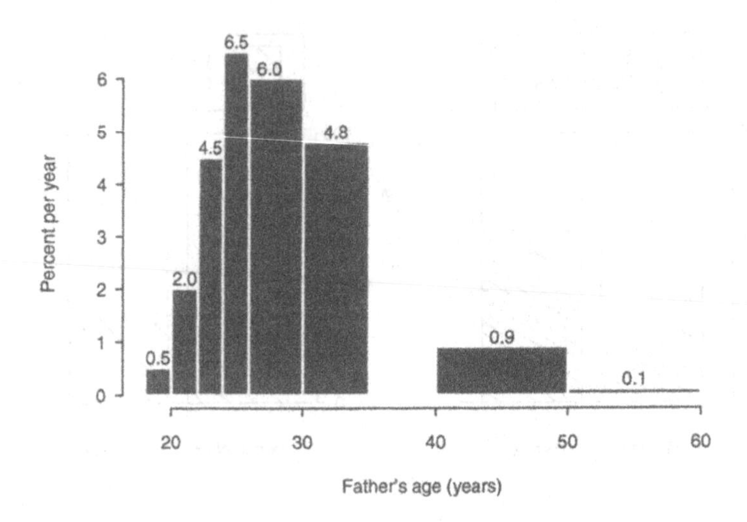

FIGURE 1.14. Histogram of father's age for fathers in the CHDS, indicating height of the bars. The bar over the interval from 35 to 40 years is missing.

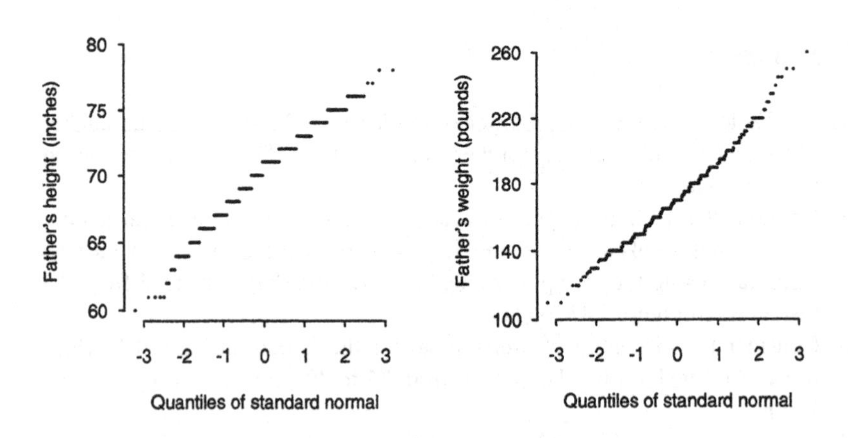

FIGURE 1.15. Normal quantile plots of father's height (left) and weight (right) for fathers in the CHDS.

6. Use the normal approximation to estimate the proportion of mothers in the CHDS between 62 and 64 inches tall to the nearest half inch (i.e., between 61.5 and 64.5 inches). The average height is 64 inches and the SD is 2.5 inches.

7. In the Missouri study, the average birth weight for babies born to smokers is 3180 grams and the SD 500 grams, and for nonsmokers the average is 3500 grams and the SD 500 grams. Consider a baby who is born to a smoker. If the baby's weight is 2 SDs below average weighs, then the baby weighs ________ grams. Suppose another baby weighs this same number of grams, but is born to a nonsmoker. This baby has a weight that falls ________ SDs below the average of its group. According to the normal approximation, approximately what percentage of babies born to nonsmokers are below this weight?
8. Suppose there are 100 observations from a standard normal distribution. What proportion of them would you expect to find outside the whiskers of a box-and-whisker plot?
9. Make a table for marital status that gives the percentage of smokers and nonsmokers in each marital category for the mothers in the Missouri study (Table 1.6).
10. Make a segmented bar graph showing the percentage at each education level for both smokers and nonsmokers for the mothers in the Missouri study (Table 1.6).
11. Make a bar graph of age and smoking status for the mothers in the Missouri study (Table 1.6). For each age group, the bar should denote the percentage of mothers in that group who smoke. How are age and smoking status related? Is age a potential confounding factor in the relationship between a mother's smoking status and her baby's birth weight?
12. In the Missouri study, the average birth weight for babies born to smokers is 3180 grams and the SD is 500 grams. What is the average and SD in ounces? There are 0.035 ounces in 1 gram.
13. Consider a list of numbers $x_1, \ldots, x_n$. Shift and rescale each x_i as follows:

$$y_i = a + bx_i.$$

Find the new average and SD of the list $y_1, \ldots y_n$ in terms of the average and SD of the original list $x_1, \ldots, x_n$.
14. Consider the data in Exercise 13. Express the median and IQR of $y_1, \ldots, y_n$ in terms of the median and IQR of $x_1, \ldots, x_n$. For simplicity, assume $y_1 < y_2 < \cdots < y_n$ and assume n is odd.
15. For a list of numbers $x_1, \ldots, x_n$ with $x_1 < x_2 \cdots < x_n$, show that by replacing x_n with another number, the average and SD of the list can be made arbitrarily large. Is the same true for the median and IQR? Explain.
16. Suppose there are n observations from a normal distribution. How could you use the IQR of the list to estimate σ?
17. Suppose the quantiles y_q of a $\mathcal{N}(\mu, \sigma^2)$ distribution are plotted against the quantiles z_q of a $\mathcal{N}(0, 1)$ distribution. Show that the slope and intercept of the line of points are σ and μ, respectively.
18. Suppose $X_1, \ldots, X_n$ form a sample from the standard normal. Show each of the following:

a. $\Phi(X_1), \ldots \Phi(X_n)$ is equivalent to a sample from a uniform distribution on (0, 1). That is, show that for X a random variable with a standard normal distribution,

$$\mathbb{P}(\Phi(X) \leq q) = q.$$

b. Let $U_1, \ldots, U_n$ be a sample from a uniform distribution on (0, 1). Explain why

$$\mathbb{E}(U_{(k)}) = \frac{k}{n+1},$$

where $U_{(1)} \leq \ldots \leq U_{(n)}$ are the ordered sample.

c. Use (a) and (b) to explain why $X_{(k)} \approx z_{k/n+1}$.

19. Prove that $\bar{x}$ is the constant that minimizes the following squared error with respect to c:

$$\sum_{i=1}^{n}(x_i - c)^2.$$

20. Prove that the median $\tilde{x}$ of $x_1, \ldots, x_n$ is the constant that minimizes the following absolute error with respect to c:

$$\sum_{i=1}^{n}|x_i - c|.$$

You may assume that there are an odd number of distinct observations. *Hint*: Show that if $c < c_o$, then

$$\sum_{i=1}^{n}|x_i - c_o| = \sum_{i=1}^{n}|x_i - c| + (c - c_o)(r - s) + 2\sum_{x \in (c, c_o)}(c - x_i),$$

where r = number of $x_i \geq c_o$, and $s = n - r$.

Notes

Yerushalmy's original analysis of the CHDS data ([Yer64], [Yer71]) and Hodges et al. ([HKC75]) provide the general framework for the analysis found in this lab and its second part in Chapter 10.

The data for the lab are publicly available from the School of Public Health at the University of California at Berkeley. Brenda Eskanazi and David Lein of the School of Public Health provided valuable assistance in the extraction of the data used in this lab.

The information on fetal development is adapted from Samuels and Samuels ([SS86]).

References

[Gre41] N.M. Gregg. Congenital cataract following German measles in the mother. *Trans. Opthalmol. Soc. Aust.*, **3**:35–46, 1941.

[HKC75] J.L. Hodges, D. Krech, and R.S. Crutchfield. *Instructor's Handbook to Accompany StatLab*. McGraw–Hill Book Company, New York, 1975.

[Lon76] L. Longo. Carbon monoxide: Effects on oxygenation of the fetus in utero. *Science*, **194**: 523–525, 1976.

[MKLS88] M. Malloy, J. Kleinman, G. Land, and W. Schram. The association of maternal smoking with age and cause of infant death. *Am. J. Epidemiol.*, **128**:46–55, 1988.

[MT77] M.B. Meyer and J.A. Tonascia. Maternal smoking, pregnancy complications, and perinatal mortality. *Am. J. Obstet. Gynecol.*, **128**: 494–502, 1977.

[MT90] I. Merkatz and J. Thompson. *New Perspectives on Prenatal Care*. Elsevier, New York, 1990.

[SS86] M. Samuels and N. Samuels. *The Well Pregnancy Book*. Summit Books, New York, 1986.

[Wil93] A.J. Wilcox. Birthweight and perinatal mortality: The effect of maternal smoking. *Am. J. Epidemiol.*, **137**:1098–1104, 1993.

[WR86] A.J. Wilcox and I.T. Russell. Birthweight and perinatal mortality, III: Towards a new method of analysis. *Int. J. Epidemiol.*, **15**:188–196, 1986.

[Yer64] J. Yerushalmy. Mother's cigarette smoking and survival of infant. *Am. J. Obstet. Gynecol.*, **88**:505–518, 1964.

[Yer71] J. Yerushalmy. The relationship of parents' cigarette smoking to outcome of pregnancy—implications as to the problem of inferring causation from observed associations. *Am. J. Epidemiol.*, **93**:443–456, 1971.

2

Who Plays Video Games?

TUESDAY, MARCH 3, 1998 San Francisco Chronicle[1]

Computing Computers' Effect

Critics ask if there's too much technology in the classroom

By John Wildermuth

When Bill Gates, founder of Microsoft, cruised through the Bay Area in January, he took time out to visit some of his company's youngest customers.

The billionaire software developer spent part of the morning at Cesar Chavez Academy in East Palo Alto, watching a giggling bunch of third-graders work on computer art programs and search for information on the Internet.

"I need to see what you're doing with the computers and how we can make it better," Gates told the twenty 8- and 9-year-olds. But some people are asking the question, better for whom? Better for the students whose parents desperately want them to get the education that best prepares them for the future, or better for the school districts, politicians and technology companies carrying on a wild love affair with a high-tech vision of education's future?

President Clinton says in a speech that he wants to see "the day when computers are as much a part of a classroom as blackboards." California legislators plan to commit about $460 million over the next four years to get technology into every one of the state's 1,400 high schools. Schools throughout the state are spending millions of dollars to wire classrooms, buy computers and show children from kindergarten on up how to surf the Net....

Making computers part of the daily classroom activity is one of the toughest parts of the new technology, said Rhonda Neagle, technology coordinator for the New Haven Unified School District in Union City....

"What we're asking teachers to do is change the way they teach," she said. "With only six (student) computers in a room, teachers need to provide more group work and maybe use a different setup for their course."

[1]Reprinted by permission.

Introduction

Every year, three to four thousand students enroll in statistics courses at the University of California, Berkeley. About half of these students are taking an introductory statistics course in order to satisfy the university's quantitative reasoning requirement. To aid the instruction of these students, a committee of faculty and graduate students of the Statistics Department have designed a series of computer labs.

The labs are meant to extend the traditional syllabus for a course by providing an interactive learning environment that offers students an alternative method for learning the concepts of statistics and probability. Some have likened the labs to educational video games. To help the committee design the labs, a survey of undergraduate students who were enrolled in a lower-division statistics course was conducted. The survey's aim was to determine the extent to which the students play video games and which aspects of video games they find most and least fun.

Students who were enrolled in an advanced statistics course conducted the study. They developed the questionnaire, selected the students to be sampled, and collected the data. Their work is outlined in the Background section of this chapter.

In this lab, you will have the opportunity to analyze the results from the sample survey to offer advice to the design committee.

The Data

Ninety-five of the 314 students in Statistics 2, Section 1, during Fall 1994 were selected at random to participate in the survey. Completed questionnaires were obtained from 91 of the 95 students. The data available here are the students' responses to the questionnaire.

The survey asks students to identify how often they play video games and what they like and dislike about the games. The answers to these questions were coded numerically as described in Table 2.1. Also provided in Table 2.1 are a few sample observations. If a question was not answered or improperly answered, then it was coded as a 99. All questions with yes/no answers recorded a 1 for a "Yes" and a 0 for a "No." For the exact wording of the questions, see the questionnaire at the end of this chapter. Those respondents who had never played a video game or who did not at all like playing video games were asked to skip many of the questions.

The survey can be roughly divided into three parts. Two of them pertain to the students' use of video games. One of these parts ascertains the frequency of play. This information is requested via two questions. One question asks how much time a student actually spent playing video games in the week prior to the survey, and the other asks how often a student usually plays (daily, weekly, monthly, semesterly).

The second part of the survey covers whether the student likes or dislikes playing games, and why. A summary of responses to three of these questions appears in Tables 2.2, 2.3, and 2.4. These questions are different from the others in that more

TABLE 2.1. Sample observations and data description for the survey of 91 undergraduate statistics students.

Time	2	0	0	.5	0	0	0	0	2
Like to play	3	3	3	3	3	3	4	3	3
Where play	3	3	1	3	3	2	3	3	2
How often	2	3	3	3	4	4	4	4	1
Play if busy	0	0	0	0	0	0	0	0	1
Playing educational	1	0	0	1	1	0	0	0	1
Sex	0	0	1	0	0	1	1	0	1
Age	19	18	19	19	19	19	20	19	19
Computer at home	1	1	1	1	1	0	1	1	0
Hate math	0	1	0	0	1	0	1	0	0
Work	10	0	0	0	0	12	10	13	0
Own PC	1	1	1	1	0	0	1	0	0
PC has CD-Rom	0	1	0	0	0	0	0	0	0
Have email	1	1	1	1	1	0	1	1	0
Grade expected	4	2	3	3	3	3	3	3	4

Variable	Description
Time	Number of hours played in the week prior to survey.
Like to play	1=never played; 2=very much; 3=somewhat; 4=not really; 5=not at all.
Where play	1=arcade; 2=home system; 3=home computer; 4=arcade and either home computer or system; 5=home computer and system; 6= all three.
How often	1=daily; 2=weekly; 3=monthly; 4=semesterly.
Play if busy	1=yes; 0=no.
Playing educational	1=yes; 0=no.
Sex	1=male; 0=female.
Age	Student's age in years.
Computer at home	1=yes; 0=no.
Hate math	1=yes; 0=no.
Work	Number of hours worked the week prior to the survey.
Own PC	1=yes; 0=no.
PC has CD-Rom	1=yes; 0=no.
Have email	1=yes; 0=no.
Grade expected	4=A; 3=B; 2=C; 1=D; 0=F.

than one response may be given. Table 2.2 summarizes the types of games played. The student is asked to check all types that he or she plays. So, for example, 50% of the students responding to this question said that they play action games. Not all students responded to this question, in part because those who said that they have never played a video game or do not at all like to play video games were

TABLE 2.2. What types of games do you play?

Type	Percent
Action	50
Adventure	28
Simulation	17
Sports	39
Strategy	63

(Each respondent was asked to check all that applied. Percentages are based on the 75 respondents to this question.)

TABLE 2.3. Why do you play the games you checked above?

Why?	Percent
Graphics/realism	26
Relaxation	66
Eye/hand coordination	5
Mental challenge	24
Feeling of mastery	28
Bored	27

(Each respondent was asked to check at most three responses. Percentages are based on the 72 respondents to this question.)

TABLE 2.4. What don't you like about video game playing?

Dislikes	Percent
Too much time	48
Frustrating	26
Lonely	6
Too many rules	19
Costs too much	40
Boring	17
Friends don't play	2
It's pointless	33

(Each respondent was asked to check at most three responses. Percentages are based on the 83 respondents to this question.)

instructed to skip this question. Students who did answer this question were also asked to provide reasons why they play the games they do. They were asked to select up to three such reasons. Their responses are presented in Table 2.3. Finally, Table 2.4 contains a summary of what the students do not like about video games. All students were asked to answer this question, and again they were asked to select up to three reasons for not liking video games.

The third part of the questionnaire collects general information about the student, such as their age and sex. The students were also asked whether or not they had an e-mail account, what they thought of math, and what grade they expected in the course.

Background

The Survey Methodology

All of the population studied were undergraduates enrolled in *Introductory Probability and Statistics*, Section 1, during Fall 1994. The course is a lower-division prerequisite for students intending to major in business. During the Fall semester the class met Monday, Wednesday, and Friday from 1 to 2 pm in a large lecture hall that seats four hundred. In addition to three hours of lecture, students attended a small, one-hour discussion section that met on Tuesday and Thursday. There were ten discussion sections for the class, each with approximately 30 students.

The list of all students who had taken the second exam of the semester was used to select the students to be surveyed. The exam was given the week prior to the survey. A total of 314 students took the exam. To choose 95 students for the study, each student was assigned a number from 1 to 314. A pseudo-random number generator selected 95 numbers between 1 and 314. The corresponding students were entered into the study.

To encourage honest responses, the students' anonymity was preserved. No names were placed on the surveys, and completed questionnaires were turned in for data entry without any personal identification on them.

To limit the number of nonrespondents, a three-stage system of data collection was employed. Data collectors visited both the Tuesday and Thursday meetings of the discussion sections in the week the survey was conducted. The students had taken an exam the week before the survey, and the graded exam papers were returned to them during the discussion section in the week of the survey. On Friday, those students who had not been reached during the discussion section were located during the lecture. A total of 91 students completed the survey.

Finally, to encourage accuracy in reporting, the data collectors were asked to briefly inform the students of the purpose of the survey and of the guarantee of anonymity.

Video Games

Video games can be classified according to the device on which they are played and according to the kinds of skills needed to play the game. With regard to the device, there are three basic types of games: arcade, console, and personal computer (PC). An arcade is a commercial establishment where video games are played outside the home. In an arcade, players pay each time they play the game. Console games are played on a television that has a port connection (e.g., Nintendo and Sega). Games for the console can be purchased and then played repeatedly at no cost. The PC games typically require the computer to have a large memory, fast processor, and

TABLE 2.5. Classification of five main types of video games.

	Eye/hand	Puzzle	Plot	Strategy	Rules
Action	×				
Adventure		×	×		
Simulation				×	×
Strategy				×	×
Role-Play		×	×		×

CD-Rom. As with console games, once the basic equipment is available, software can be purchased and games are then played at no additional cost.

Video games are divided into five categories: action, adventure, simulation, strategy, and role-play. Each of these can be described in terms of a few attributes: eye/hand coordination, puzzle solving, intricate plot line, strategy, and rule learning. Table 2.5 summarizes the attributes typically found in each category.

Most arcade games are fast action games that emphasize eye/hand coordination and have a short learning curve. Console games are usually either action, adventure, or strategy games. Simulation and role-playing games are found almost exclusively on the PC. They are unsuitable for the arcade and console because they often take many hours to play. All five types of games are made for the PC.

Investigations

The objective of this lab is to investigate the responses of the participants in the study with the intention of providing useful information about the students to the designers of the new computer labs.

- Begin by providing an estimate for the fraction of students who played a video game in the week prior to the survey. Provide an interval estimate as well as a point estimate for this proportion.
- Check to see how the amount of time spent playing video games in the week prior to the survey compares to the reported frequency of play (i.e., daily, weekly, etc.). How might the fact that there was an exam in the week prior to the survey affect your previous estimates and this comparison?
- Consider making an interval estimate for the average amount of time spent playing video games in the week prior to the survey. Keep in mind the overall shape of the sample distribution. A simulation study may help determine the appropriateness of an interval estimate.
- Next consider the "attitude" questions. In general, do you think the students enjoy playing video games? If you had to make a short list of the most important reasons why students like (or dislike) video games, what would you put on the list? Don't forget that those students who say that they have never played a video game or do not at all like to play video games are asked to skip over some of these questions. So, there may be many nonresponses to the questions as to

whether they think video games are educational, where they play video games, etc.

- Look for differences between those who like to play video games and those who do not. To do this, use the questions in the last part of the survey, and make comparisons between male and female students, those who work for pay and those who do not, those who own a computer and those who do not, or those who expect A's in the class and those who do not. Graphical displays and cross-tabulations are particularly helpful in making these kinds of comparisons. Also, you may want to collapse the range of responses to a question down to two or three possibilities before making these comparisons.
- Just for fun, further investigate the grade that students expect in the course. How well does it match the target distribution used in grade assignment of 20% As, 30% Bs, 40% Cs, and 10% D or lower? If the nonrespondents were failing students who no longer bothered to come to the discussion section, would this change the picture?

Summarize your findings in a memo to the committee in charge of designing the new computer labs.

Theory

In this section we will use as our primary example the problem of estimating the average amount of time students in the class spent playing video games in the week prior to the survey. To determine the exact amount of time for the entire class, we would need to interview all of the students (there are over three hundred of them). Alternatively, a subset of them could be interviewed, and the information collected from this subset could provide an approximation to the full group.

In this section, we discuss one rule for selecting a subset of students to be surveyed, the *simple random sample.* The simple random sample is a probability method for selecting the students. Probability methods are important because through chance we can make useful statements about the relation between the sample and the entire group. With a probability method, we know the chance of each possible sample.

We begin by introducing some terminology for describing the basic elements of a sampling problem. For the population:

- *Population units* make up the group that we want to know more about. In this lab, the units are the students enrolled in the 1994 Fall semester class of *Introductory Probability and Statistics.*
- *Population size*, usually denoted by N, is the total number of units in the population. For very large populations, often the exact size of the population is not known. Here we have 314 students in the class.
- *Unit characteristic* is a particular piece of information about each member of the population. The characteristic that interests us in our example is the amount of time the student played video games in the week prior to the survey.

- *Population parameter* is a summary of the characteristic for all units in the population, such as the average value of the characteristic. The population parameter of interest to us here is the average amount of time students in the class spent playing video games in the week prior to the survey.

In parallel, for the sample, we have the following:

- *Sample units* are those members of the population selected for the sample.
- *Sample size*, usually denoted by n, is the number of units chosen for the sample. We will use 91 for our sample size, and ignore the four who did not respond.
- *Sample statistic* is a numerical summary of the characteristic of the units sampled. The statistic estimates the population parameter. Since the population parameter in our example is the average time spent playing video games by all students in the class in the week prior to the survey, a reasonable sample statistic is the average time spent playing video games by all students in the sample.

Finally, there is the *selection rule*, the method for choosing members of the population to be surveyed. In this case, each student was assigned a number from 1 to 314, and the computer was used to choose numbers (students) between 1 and 314 one at a time. Once a number was chosen, it was eliminated from the list for future selections, thus ensuring that a student is chosen at most once for the sample. Also, at each stage in the selection process, all numbers remaining on the list were equally likely to be chosen. This method for selecting the sample is equivalent to the *simple random sample*, which is described in more detail below.

The Probability Model

The simple random sample is a very simple probability model for assigning probabilities to all samples of size n from a population of size N.

In our case, n is 91 and N is 314. There are very many different sets of 91 students that could be chosen for the sample. Any one of the 314 students could be the first to be selected. Once the first person is chosen and removed from the selection pool, there are 313 students left for the second selection, and after that student is chosen there are 312 students remaining for the third selection, etc. Altogether there are $314 \times 313 \times \cdots \times 224$ different ways to choose 91 students from 314, if we keep track of the order in which they are chosen (i.e., who was chosen first, second, etc). But since we only care which students are selected, not the order in which they are chosen, there are

$$\frac{314 \times 313 \times \cdots \times 224}{91 \times 90 \times \cdots \times 1}$$

different subsets of 91 from 314. A shorter way to write this is

$$\binom{314}{91} = \frac{314!}{223!\,91!},$$

where we say the number of subsets is "314 choose 91."

In general, with N population units and a sample of size n, there are N choose n possible samples. The probability rule that defines the simple random sample is that each one of the $\binom{N}{n}$ samples is equally likely to be selected. That is, each unique sample of n units has the same chance, $1/\binom{N}{n}$, of being selected. From this probability, we can make statements about the variations we would expect to see across repeated samples.

One way to conceptualize the simple random sample is to assign every unit a number from 1 to N. Then, write each number on a ticket, put all of the tickets in a box, mix them up, and draw n tickets one at a time from the box without replacement. The chance that unit #1 is the first to be selected for the sample is $1/N$. Likewise, unit #1 has chance $1/N$ of being the second unit chosen for the sample, and all together unit #1 has chance n/N of being in the sample. By symmetry, the chance that unit #2 is chosen first is $1/N$, and unit #2 has chance n/N of appearing in the sample. That is, each unit has the same chance ($1/N$) of being the first chosen for the sample and the same chance (n/N) of being selected for the sample. However, there is dependence between the selections. We see this when we compute the chance that unit #1 is chosen first and unit #2 is chosen second. It is $1/N(N-1)$. This chance is the same for any two units in the population, and the chance that #1 and #2 are both in the sample is

$$\frac{n(n-1)}{N(N-1)}.$$

In our example,

$$\mathbb{P}(\text{unit \#1 in the sample}) = \frac{91}{314},$$

$$\mathbb{P}(\text{units \#1 and \#2 are in the sample}) = \frac{91 \times 90}{314 \times 313}.$$

The probability distribution for the units chosen at random from the population can be concisely described as follows. Let $I(1)$ represent the first number drawn from the list $1, 2, \ldots, N$, $I(2)$ the second number drawn, ... and $I(n)$ the last number drawn. Each $I(\cdot)$ is called a *random index*. Then

$$\mathbb{P}(I(1) = 1) = \frac{1}{N},$$

$$\mathbb{P}(I(1) = 1 \text{ and } I(2) = N) = \frac{1}{N \times (N-1)},$$

and, in general, for $1 \leq j_1 \neq j_2 \ldots \neq j_n \leq N$

$$\mathbb{P}(I(1) = j_1, I(2) = j_2, \ldots, I(n) = j_n) = \frac{1}{N \times (N-1) \times \cdots (N-n+1)}.$$

The simple random sample method puts a probability structure on the sample. Different samples have different characteristics and different sample statistics. This means that the sample statistic has a probability distribution related to the sampling procedure. We can find the expected value and variance for the sample statistic under the random sampling method used.

Sample Statistics

Suppose we let x_1 be the value of the characteristic for unit #1, x_2 the value for unit #2, ..., and x_N the value for unit #N. In our example, x_i is time spent playing video games by student #i, $i = 1, \ldots, 314$. Take the population average,

$$\mu = \frac{1}{N}\sum_{i=1}^{N} x_i,$$

as our population parameter.

To specify the values of the characteristic for the units sampled, we use the random indices, $I(1), \ldots, I(n)$. That is, $x_{I(1)}$ represents the value of the characteristic for the first unit sampled. The value $x_{I(1)}$ is random. If $I(1) = 1$ then the value will be x_1, if $I(1) = 2$ then it will be x_2, and so on. In our example, $x_{I(j)}$ represents the time spent playing video games by the jth unit sampled, $j = 1, \ldots, 91$. We find the expectation of $x_{I(1)}$ as follows:

$$\begin{aligned} \mathbb{E}(x_{I(1)}) &= \sum_{i=1}^{N} x_i \mathbb{P}(I(1) = i) \\ &= \sum_{i=1}^{N} x_i \frac{1}{N} \\ &= \mu. \end{aligned}$$

Similarly, since each unit is equally likely to be the jth unit chosen, $j = 1, \ldots, n$,

$$\mathbb{E}(x_{I(j)}) = \mu.$$

The sample average,

$$\bar{x} = \frac{1}{n}\sum_{j=1}^{n} x_{I(j)},$$

is the sample statistic that estimates the population parameter. It is random too, and from the computations above we find its expected value to be

$$\begin{aligned} \mathbb{E}(\bar{x}) &= \frac{1}{n}\mathbb{E}(x_{I(1)} + \cdots + x_{I(n)}) \\ &= \mathbb{E}(x_{I(1)}) \\ &= \mu. \end{aligned}$$

We have shown that the expected value of the sample average is the population parameter; that is, the sample average is an *unbiased* estimator of the population parameter.

Next we find the standard deviation of $\bar{x}$. To do this, we first find the variance of $x_{I(1)}$,

$$\begin{aligned}\text{Var}(x_{I(1)}) &= \mathbb{E}(x_{I(1)} - \mu)^2 \\ &= \frac{1}{N}\sum_{i=1}^{N}(x_i - \mu)^2 \\ &= \sigma^2,\end{aligned}$$

where we use σ^2 to represent the *population variance*. Then we compute the variance of the sample average $\bar{x}$ as follows:

$$\begin{aligned}\text{Var}(\bar{x}) &= \frac{1}{n^2}\text{Var}\left(\sum_{j=1}^{n} x_{I(j)}\right) \\ &= \frac{1}{n^2}\sum_{j=1}^{n}\text{Var}(x_{I(j)}) + \frac{1}{n^2}\sum_{j\neq k}^{n}\text{Cov}(x_{I(j)}, x_{I(k)}) \\ &= \frac{1}{n}\sigma^2 + \frac{n-1}{n}\text{Cov}(x_{I(1)}, x_{I(2)}).\end{aligned}$$

The last equality follows from noting that all pairs $(x_{I(j)}, x_{I(k)})$ are identically distributed. The covariance between any two sampled units $x_{I(j)}$ and $x_{I(k)}$ is not 0 because the sampling procedure makes them dependent. We leave it as an exercise to show that this covariance is $-\sigma^2/(N-1)$, and

$$\text{Var}(\bar{x}) = \frac{1}{n}\sigma^2\frac{N-n}{N-1},$$

$$\text{SD}(\bar{x}) = \frac{1}{\sqrt{n}}\sigma\frac{\sqrt{N-n}}{\sqrt{N-1}}.$$

The factor $(N-n)/(N-1)$ in the variance and SD is called the *finite population correction factor*. It can also be expressed as

$$1 - \frac{n-1}{N-1},$$

which is roughly $1 - n/N$. The ratio n/N is called the *sampling fraction*. It is very small when the sample size is small relative to the population size. This is frequently the case in sampling, and when this happens $\text{Var}(\bar{x}) \approx \sigma^2/n$, and the finite population correction factor is often ignored. In our example, the factor cannot be ignored;

$$\frac{\sqrt{314-91}}{\sqrt{314-1}} = 0.84.$$

Notice that without the correction factor we have the same variance as when the draws are made with replacement. If the draws from the box are made with replacement, the $x_{I(j)}$ are independent and the $\text{Var}(\bar{x}) = \sigma^2/n$. Sampling with replacement is like sampling from an infinite population. To see this, let N tend to ∞ in the correction factor, and keep n fixed. The factor tends to 1. This is why $(N-n)/(N-1)$ is called the finite population correction factor.

With a simple random sample, the standard deviation for the estimator can be computed in advance, dependent on the population variance σ^2. If σ^2 is known approximately, then the sample size can be chosen to give an acceptable level of accuracy for the estimator. Often a pilot study, results from a related study, or a worst-case estimate of σ^2 is used in planning the sample size for the survey.

Estimators for Standard Errors

Standard deviations for estimators are typically called *standard errors* (SEs). They indicate the size of the deviation of the estimator from its expectation.

When σ^2 is unknown, a common estimator for it is

$$s^2 = \frac{1}{n-1}\sum_{j=1}^{n}(x_{I(j)} - \bar{x})^2.$$

To estimate $\text{Var}(\bar{x})$, we can then use

$$\frac{s^2}{n}\frac{N-n}{N-1}.$$

The reason for using s^2 is that the sample, when chosen by the simple random sample method, should look roughly like a small-scale version of the population, so we plug in s^2 for σ^2 in the variance of $\bar{x}$.

In fact, we can make a slightly better estimate for $\text{Var}(\bar{x})$. When we take the expectation of s^2 we find that it is not exactly σ^2,

$$\mathbb{E}(s^2) = \frac{N}{N-1}\sigma^2.$$

An unbiased estimator of σ^2 is then

$$s^2\frac{N-1}{N},$$

and an unbiased estimator of $\text{Var}(\bar{x})$ is

$$\frac{s^2}{n}\frac{N-n}{N}.$$

Notice that there is essentially no difference between these two estimators of $\text{Var}(\bar{x})$ for any reasonably sized population.

Population Totals and Percentages

Sometimes the population parameter is a proportion or percentage, such as the proportion of students who played a video game in the week prior to the survey or the percentage of students who own PCs.

When the parameter is a proportion, it makes sense for the characteristic value x_i to be 1 or 0 to denote the presence or absence of the characteristic, respectively. For example, for $i = 1, \ldots, 314$,

$$x_i = \begin{cases} 1 & \text{if the } i\text{th student in the population owns a PC,} \\ 0 & \text{if not.} \end{cases}$$

Then $\tau = \sum x_i$ counts all of the students who own PCs in the population, and $\pi = \sum x_i / N$ is the proportion of students in the population who own PCs.

In this case $\bar{x}$ remains an unbiased estimator for π, the population average, and $N\bar{x}$ estimates τ. A simpler form for the population variance and the unbiased estimator of $\mathrm{Var}(\bar{x})$ can be obtained because

$$\begin{aligned} \sigma^2 &= \frac{1}{N} \sum_{i=1}^{N} (x_i - \pi)^2 \\ &= \pi(1 - \pi). \end{aligned}$$

Then an estimator for the standard error is

$$\widehat{\mathrm{SE}}(\bar{x}) = \frac{\sqrt{\bar{x}(1-\bar{x})}}{\sqrt{n-1}} \frac{\sqrt{N-n}}{\sqrt{N}}.$$

See the Exercises for a derivation.

Often the symbols $\hat{\mu}$, $\hat{\pi}$, and $\hat{\tau}$ are used in place of $\bar{x}$ and $N\bar{x}$ to denote sample estimates of the parameters μ, π, and τ. Table 2.6 contains the expectations and standard errors for estimators of a population average, proportion, and total.

The Normal Approximation and Confidence Intervals

If the sample size is large, then the probability distribution of the sample average is often well approximated by the normal curve. This follows from the Central Limit Theorem.

CENTRAL LIMIT THEOREM: *If $X_1, \ldots, X_n$ are independent, identically distributed with mean μ and variance σ^2 then, for n large, the probability*

TABLE 2.6. Properties of sample statistics.

	Average	Proportion	Total
Parameter	μ	π	τ
Estimator	$\bar{x}$	$\bar{x}$	$N\bar{x}$
Expectation	μ	π	τ
Standard error	$\frac{\sigma}{\sqrt{n}}\sqrt{\frac{N-n}{N-1}}$	$\frac{\sqrt{\pi(1-\pi)}}{\sqrt{n}}\sqrt{\frac{N-n}{N-1}}$	$N\frac{\sigma}{\sqrt{n}}\sqrt{\frac{N-n}{N-1}}$
Estimator of SE	$\frac{s}{\sqrt{n}}\sqrt{\frac{N-n}{N}}$	$\frac{\sqrt{\bar{x}(1-\bar{x})}}{\sqrt{n-1}}\sqrt{\frac{N-n}{N}}$	$N\frac{s}{\sqrt{n}}\sqrt{\frac{N-n}{N}}$

distribution of

$$Z = \frac{(\bar{X} - \mu)}{\sigma/\sqrt{n}}$$

is approximately standard normal.

In simple random sampling, the $x_{I(j)}$ are identically distributed but not independent. However, the normal approximation can still hold if, in addition to the sample size being large, it is not too large relative to the population size. That is, if the sampling fraction n/N is small, then the $x_{I(j)}$ are nearly independent. There are no hard and fast rules for how large n must be or how small n/N must be before we can use the normal approximation. You may want to test a few cases by simulation.

The Central Limit Theorem is a very powerful result. It implies that for any population distribution, under simple random sampling (for appropriate n and n/N), the sample average has an approximate normal distribution.

The normal distribution can be used to provide interval estimates for the population parameter. One interval estimate for μ is

$$(\bar{x} - \sigma/\sqrt{n}, \bar{x} + \sigma/\sqrt{n}).$$

This interval is called a 68% *confidence interval* for the population parameter. A 95% confidence interval for μ is $(\bar{x} - 2\sigma/\sqrt{n}, \bar{x} + 2\sigma/\sqrt{n})$. These interval estimates derive their names from the fact that, by the Central Limit Theorem, the chance that $\bar{x}$ is within one (or two) standard error(s) of μ is approximately 68% (or 95%). That is, ignoring the finite population correction factor,

$$\begin{aligned}\mathbb{P}\left(\bar{x} - 2\frac{\sigma}{\sqrt{n}} \leq \mu \leq \bar{x} + 2\frac{\sigma}{\sqrt{n}}\right) &= \mathbb{P}\left(\mu - 2\frac{\sigma}{\sqrt{n}} \leq \bar{x} \leq \mu + 2\frac{\sigma}{\sqrt{n}}\right) \\ &= \mathbb{P}\left(-2 \leq \frac{\bar{x} - \mu}{\sigma/\sqrt{n}} \leq 2\right) \\ &\approx 0.95\,.\end{aligned}$$

In practice, the population variance is rarely known, and we substitute s for σ in order to make the confidence interval. With this substitution, we sometimes refer to the interval as an approximate confidence interval.

The sample statistic $\bar{x}$ is random, so we can think of the interval estimates as random intervals. Just as different samples lead to different sample statistics, they also lead to different confidence intervals. If we were to take many simple random samples over and over, where for each sample we compute the sample average and make a confidence interval, then we expect about 95% of the 95% confidence intervals to contain μ.

An Example

To summarize the ideas introduced in this section, consider the problem of estimating the proportion of females in the class. In this case, we happen to know that there are 131 females in the class. Therefore we have all of the population information:

$$\pi = 131/314 = 0.4172;$$
$$\sigma^2 = 0.2431;$$
$$N = 314.$$

Because we have a simple random sample, the probability distribution for $x_{I(1)}$ matches the population distribution; that is, $\mathbb{P}(x_{I(1)} = 1) = 0.4172$ and $\mathbb{P}(x_{I(1)} = 0) = 0.5818$. This means that

$$\begin{aligned}\mathbb{E}(x_{I(1)}) &= 0.4172,\\ \mathrm{Var}(x_{I(1)}) &= 0.2431,\\ \mathbb{E}(\bar{x}) &= 0.4172,\\ \mathrm{SE}(\bar{x}) &= \sqrt{\frac{0.2431}{91} \times \frac{223}{313}} = 0.044.\end{aligned}$$

Also, the exact probability distribution of $\bar{x}$ can be found:

$$\begin{aligned}\mathbb{P}(\bar{x} = m/91) &= \mathbb{P}(\text{the sample has } m \text{ females})\\ &= \frac{\binom{131}{m}\binom{183}{91-m}}{\binom{314}{91}}.\end{aligned}$$

This is known as the hypergeometric distribution.

In a real sampling problem, the exact distribution of $\bar{x}$ is not known. However, it is known that the probability distribution of the $x_{I(j)}$s matches the population distribution, that $\bar{x}$ is an unbiased estimator of the population parameter, and that, provided n is large and n/N is small, the probability distribution of $\bar{x}$ is roughly normal. This is enough for us to make confidence intervals for the population parameter. In this example, $n = 91$ is large, but $n/N = 91/314$ is also relatively large, so we should check that we can use a normal approximation before making confidence intervals. The next section, on the bootstrap, discusses how this can be done, but here the approximation is satisfactory.

In our sample, 38 of the 91 students responding were female. Our estimate for the population parameter is $\bar{x} = 38/91 = 0.42$, our estimate of the standard error for $\bar{x}$ is

$$\frac{\sqrt{\frac{38}{91}(1 - \frac{38}{91})}}{\sqrt{91 - 1}}\sqrt{\frac{314 - 91}{314}} = 0.044,$$

and our approximate 95% confidence interval for the population parameter runs from 0.33 to 0.51. This includes or "covers" the known proportion $\pi = 0.4172$ of females in the class. In general, we will not know whether the confidence interval covers the population parameter when π is unknown.

Finally, in this example the actual proportion of women in the sample was very close to the expected proportion of women, $\mathbb{E}(\bar{x})$. We might want to calculate the probability that we would get as close or closer to the expected proportion. Sometimes samples may seem to be "too close" to what we expect: if the chance of getting data as close or closer to the expected value is small, say $\frac{1}{100}$, then we might suspect the sampling procedure. Here the expected number of women in the sample is $91 \times \pi = 37.97$, so we may want to know the chance that exactly 38 of the 91 students were women: $\binom{131}{38}\binom{183}{91-38}/\binom{314}{91} = 0.10$. Therefore about 1 time in 10 we would expect to get this close to the expected number of women. In practice, we could approximate this chance using the normal curve; this is left as an exercise.

The Bootstrap

From the histogram (Figure 2.1) of the time spent playing video games by the students in the sample, we see that the sample distribution is extremely skewed. This observation raises the question of whether the probability distribution of the sample average follows a normal curve. Without knowledge of the population, we cannot answer this question completely. However, the bootstrap can help us.

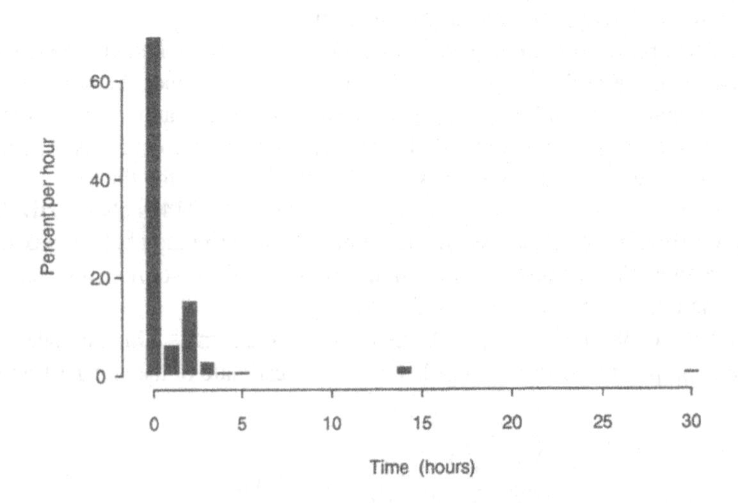

FIGURE 2.1. Histogram of the number of hours spent playing video games for the 91 statistics students in the sample.

TABLE 2.7. Distribution of time, in hours, spent playing video games in the week prior to the survey for the 91 statistics students in the sample.

Time	Count	Bootstrap population
0	57	197
0.1	1	3
0.5	5	17
1	5	17
1.5	1	4
2	14	48
3	3	11
4	1	3
5	1	4
14	2	7
30	1	3
Total	91	314

According to the simple random sample probability model, the distribution of the sample should look roughly similar to that of the population. We could create a new population of 314 based on the sample and use this population, which we call the bootstrap population, to find the probability distribution of the sample average. Table 2.7 shows how to do this. For every unit in the sample, we make $314/91 = 3.45$ units in the bootstrap population with the same time value and round off to the nearest integer.

Next, to determine the probability distribution of the sample average when the sample is taken from the bootstrap population, we use a computer. With a computer, we can select a simple random sample of 91 from the bootstrap population, called a bootstrap sample, and take its average. Then we take a second sample of 91 and take its average, and so on. A histogram of bootstrap sample averages, each constructed from a simple random sample of 91 from the bootstrap population, appears in Figure 2.2. We took 400 bootstrap samples from the bootstrap population in order to make a reasonable simulation of the probability distribution of the bootstrap sample average. From the normal-quantile plot in Figure 2.3, we see that it closely follows the normal curve. Of course, to further validate this claim of approximate normality, we could compare the skewness (0.19) and kurtosis (2.67) for the 400 bootstrap sample averages to a simulated distribution of skewness and kurtosis for samples of size 400 from a normal distribution. See Figure 2.4 for these simulated distributions. (Chapter 1 contains more details on the use of simulations for approximating complex statistical distributions.) Our sample values for skewness and kurtosis are in the tails of the simulated distributions but seem reasonable.

The method described here is one version of the bootstrap. The bootstrap technique derives its name from the expression, "to pull yourself up by your own bootstraps" (Diaconis and Efron [DE83]). In the sampling context, we study the

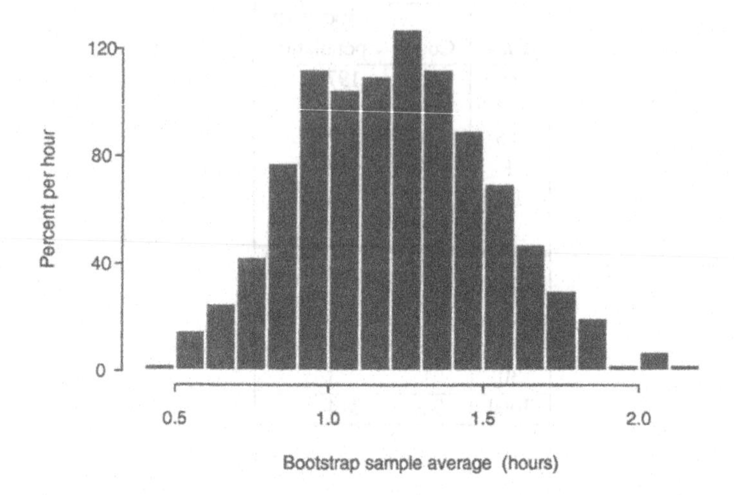

FIGURE 2.2. Histogram of the bootstrap sample averages, from 400 bootstrap samples of size 91 from the bootstrap population (Table 2.7), for the 91 statistics students in the sample.

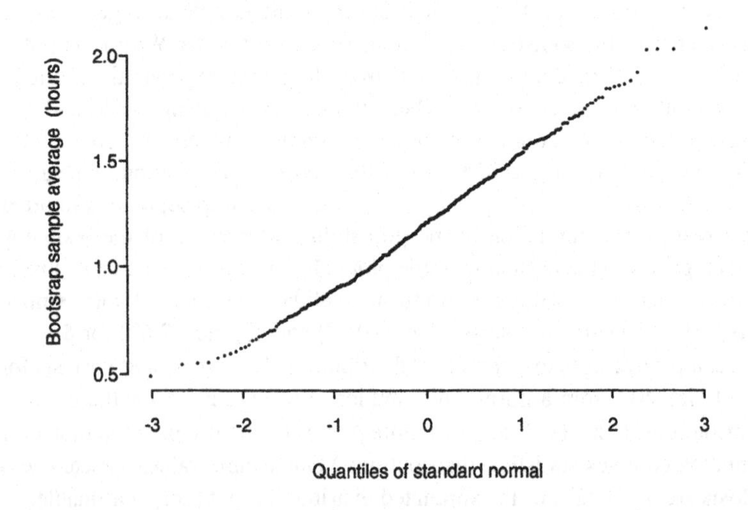

FIGURE 2.3. Normal-quantile plot of the bootstrap sample averages, from 400 bootstrap samples of size 91 from the bootstrap population (Table 2.7), for the 91 statistics students in the sample.

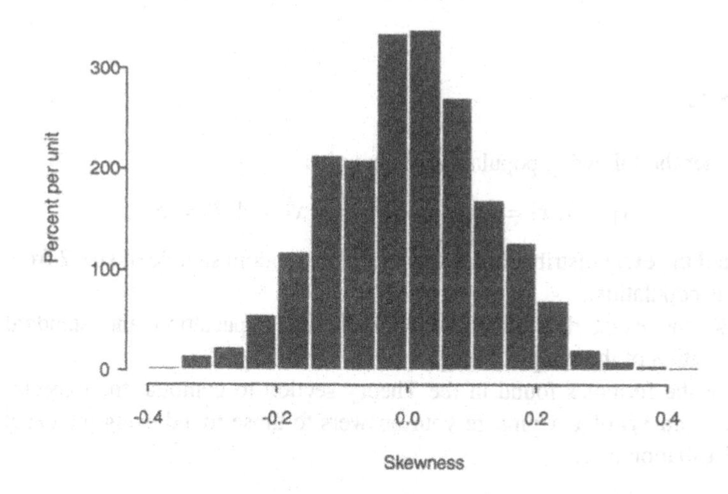

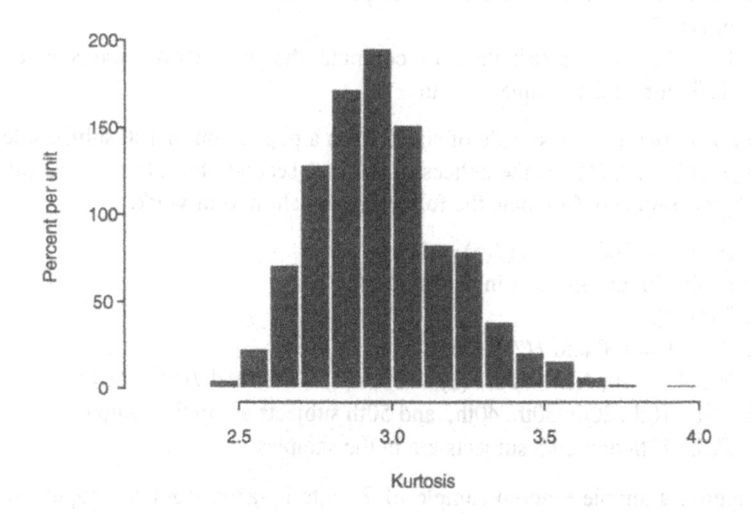

FIGURE 2.4. Approximate distribution of skewness (top plot) and kurtosis (bottom plot) for samples of size 400 from the normal. (Based on 1000 samples of size 400.)

relation between bootstrap samples and the bootstrap population, where both the samples and the population are known, in order to learn about the relationship between our actual sample and the population, where the latter is unknown.

Exercises

1. Consider the following population of six units:

$$x_1 = 1, x_2 = 2, x_3 = 2, x_4 = 4, x_5 = 4, x_6 = 5.$$

a. Find the exact distribution of $\bar{x}$ for a simple random sample of size 2 from this population.
b. Use the exact distribution to compute the expectation and standard deviation of the sample average.
c. Use the formulas found in the Theory section to compute the expectation and SD of $\bar{x}$. Compare your answers to those found using the exact distribution of $\bar{x}$.

2. Consider the following population of five units:

$$x_1 = 1, x_2 = 2, x_3 = 2, x_4 = 4, x_5 = 4.$$

a. Find the exact distribution of the sample median of a simple random sample of size 3.
b. Use the exact distribution to compute the expectation and standard deviation of the sample median.

3. For a simple random sample of size 5 from a population of 100 subjects, let $I(1), I(2), \ldots, I(5)$ be the indices of the first, second, third, fourth, and fifth subjects sampled. Compute the following and show your work.

a. $\mathbb{P}(I(1) = 100), \ldots, \mathbb{P}(I(5) = 100)$.
b. $\mathbb{P}$ (the 100th subject is in the sample).
c. $\mathbb{E}[I(1)]$.
d. $\mathbb{P}(I(1) = 100$ and $I(2) = 2)$.
e. $\mathbb{P}(I(1) = 10, I(2) = 20, I(3) = 30, I(4) = 40,$ and $I(5) = 50)$.
f. $\mathbb{P}$(the 10th, 20th, 30th, 40th, and 50th subjects are in the sample).
g. $\mathbb{P}$(the 10th and 20th subjects are in the sample).

4. Suppose a simple random sample of 2 units is taken from the population described in Exercise 1, find

a. $\mathbb{P}(x_{I(2)} = 5)$.
b. $\mathbb{E}[x_{I(1)}]$.
c. $\mathbb{P}(x_{I(1)} = 2$ and $x_{I(2)} = 2)$.

5. Consider the following list of quantities:

$$x_1,\ x_{I(1)},\ \bar{x},\ N,\ \mu,\ I(1),\ n.$$

Determine which elements on the list are random and which are not. Explain.

6. For $I(1), I(2), \ldots, I(5)$ defined in Exercise 3, suppose $x_i = 0$ or 1 and the proportion of 1's in the population is π. Find

$$\mathbb{E}[x_{I(1)}x_{I(2)}] .$$

Use your computation to find the covariance of $x_{I(1)}$ and $x_{I(2)}$ in this special case when they have values 0 or 1.

7. In the survey of the statistics class, 67 of 91 respondents said they owned a PC. Construct a 95% confidence interval for the proportion of students in the statistics class who own a PC.
8. The average age of the respondents in the statistics class survey was 19.5 years, and the sample standard deviation was 1.85 years. Find a 95% confidence interval for the average age of the students in the class.
9. Suppose that an unknown fraction π of the population of students at a particular university own a PC. What value of π gives the largest population variance? Explain.
10. Suppose a survey of the 32,000 students at a university is planned. The goal of the survey is to estimate the percentage of the students who own PCs. Find the minimum sample size required to make a 95% confidence interval for the population percentage that is at most 4 percentage points wide. The variance of the population is unknown. Ignore the finite population correction factor.

   a. In your calculation of the sample size use the percentage of students who own PCs in the survey of the statistics class to estimate the population variance for the university students.
   b. Instead of using the percentage from the statistics class sample in your calculation, assume the worst-case scenario — that is, the largest variance possible for the parameter.

11. Continue with Exercise 10, and suppose that the survey is to estimate two characteristics of the university students. One characteristic is thought to have a prevalence of roughly 50% of the population and the other only 10%. For each of the following conditions, find the sample size required.

   a. Both estimates are to be accurate to about 1% (i.e., the standard error of each estimator should be at most 1%).
   b. Each statistic is to be accurate to about 1/10 of its population parameter.

12. In a survey of students at a large university, graduate students and undergraduates are to be surveyed separately. A simple random sample of 100 of the 4000 graduate students is planned. Because there are 8 times as many undergraduates, does the sample size for the undergraduate survey need to be 800 to get the same accuracy as in the graduate student survey? You may assume that the SDs for the two groups of students are the same.

13. In this exercise, we estimate the fraction π of women that played video games in the week prior to the survey. Our estimate is

$$\hat{\pi} = \frac{\bar{v}}{131/314},$$

where 131/314 is the known fraction of women in the course, and $\bar{v}$ is the fraction of the respondents who were both female and played a video game in the week prior to the survey.

a. Prove that $\hat{\pi}$ is an unbiased estimator.
b. What is the standard error of $\hat{\pi}$?

14. Of the 314 students in the STAT 21 class surveyed, 131 are female. Simulate many different samples of size 91 from the class, and use each sample to construct a 95% confidence interval for the percentage of females in the class. Is the distribution of the left endpoint of the 95% confidence interval normal? Make a normal-quantile plot of the left endpoints of the intervals. Explain why this would or would not be the case.

15. Use a normal-curve approximation to estimate the chance that exactly 38 of the 91 students in the sample would be female. You will need to use a continuity correction: the probability histogram for the hypergeometric distribution is discrete, but the normal curve is continuous. Is your answer similar to the exact calculation in the text? Discuss whether you would expect them to be similar.

16. Use the bootstrap to find a 95% confidence interval for the proportion of students who own PCs. Compare your bootstrapped confidence interval to the interval obtained assuming the sample proportion is approximately normally distributed.

17. In the video survey, 31 of the 91 (34%) students expect an A for their grade in the course, and

$$\sqrt{\frac{.34 \times .66}{90}}\sqrt{\frac{314-91}{314}} = 0.04\,.$$

Consider the following statements about confidence intervals. Determine which are true and which are false. Explain your reasoning carefully.

a. There is a 95% chance that the percentage of students in the STAT 21 class who expect an A for the course is between 26% and 42%.
b. There is a 95% chance that the sample percentage is in the interval (26%,42%).
c. About 95% of the STAT 21 students will be contained in a 95% confidence interval.
d. Out of one hundred 95% confidence intervals for a population parameter, we expect 95 to contain the population parameter.

18. For a random sample of n units with replacement from a population, consider the following estimates for the population average:

a. $\hat{x} = x_{I(1)}$.
b. $\tilde{x} = 2x_{I(1)} - x_{I(2)}$.
c. $x^* = 2\bar{x}$.

For each estimate, find its expectation and variance. Comment on these estimates.

19. For a simple random sample of size n from a population of size N, consider the following estimate of the population average:

$$\bar{x}_w = \sum_{i=1}^{n} w_i x_{I(i)},$$

where the w_i are fixed weights. Show that all of the estimators in Exercise 18 are special cases of $\bar{x}_w$. Prove that for the estimator to be unbiased, the weights must sum to 1.

20. With simple random sampling, is $\bar{x}^2$ an unbiased estimate of μ^2? If not, what is the bias? Avoid a lot of calculations by using facts you already know about $\bar{x}$.

21. A clever way to find the covariance between $x_{I(1)}$ and $x_{I(2)}$ is to note that if we sampled all the units from the population then the sample average would always be the population average (i.e., when $n = N$ then $\bar{x} = \mu$ always). In this case, Var($\bar{x}$)=0. Use this fact and the fact that

$$\text{Var}(\bar{x}) = \frac{1}{n}\sigma^2 + \frac{n-1}{n}\text{Cov}(x_{I(1)}, x_{I(2)})$$

to find the covariance of $x_{I(1)}$ and $x_{I(2)}$. Then show that

$$\text{Var}(\bar{x}) = \frac{1}{n}\sigma^2 \frac{N-n}{N-1}.$$

22. Show that when the x_i's in a population are 0s and 1s, and the proportion of 1's in the population is π, then

$$\sigma^2 = \pi(1-\pi).$$

23. Suppose the x_i values are a or b only and p is the proportion of x_i's taking the value a. Show that this implies

$$\sigma^2 = (b-a)^2 p(1-p).$$

24. A bootstrap method (Bickel and Freedman [BF84]) better than the one described in the Theory section would proceed as follows. For a sample of n from N, when N/n is not an integer, two bootstrap populations are created. The first takes k copies of each unit in the sample, and the second takes $k+1$ copies, where

$$N = k \times n + r \quad 0 < r < n.$$

Then to sample from the bootstrap populations, first one of the two populations is chosen. The first population is selected with probability

$$p = \left(1 - \frac{r}{n}\right)\left(1 - \frac{n}{N-1}\right),$$

and the second population is chosen with chance $1-p$. Then a simple random sample of n units is taken from the chosen bootstrap population to form a bootstrap sample, and the sample average is computed. This procedure is repeated many times, each time selecting a population at random and then a sample from the population, to simulate the distribution of the sample average.

a. Construct the two bootstrap populations for the sample in Table 2.7.

b. Take 400 bootstrap samples from each population and find the bootstrap sample averages. Compare the two bootstrap distributions.

Notes

The students in the Fall 1994 Statistics 131a, Section 1 class provided the leg work for conducting this survey. Cherilyn Chin, an avid gamer, was instrumental in designing the questionnaire. Roger Purves was also very helpful in designing the questionnaire and kindly let the Statistics 131 students survey his Statistics 21 students. Wesley Wang helped with the background material on video games.

References

[DE83] P. Diaconis and B. Efron. Computer-intensive methods in statistics. *Sci. Am.*, **248** (5):116–129, May, 1983.

[BF84] P. J. Bickel and D. A. Freedman. Asymptotic normality and the bootstrap in stratified sampling. *Ann. Statist.*, **12**: 470–482, 1984.

The Questionnaire

1. How much time did you spend last week playing video and/or computer games? (C'mon, be honest, this is confidential)

☐ No time   ______ Hours

2. Do you like to play video and/or computer games?
☐ Never played→ Question 9
☐ Very Much ☐ Somewhat ☐ Not Really ☐ Not at all→ Question 9

3. What types of games do you play? (Check all that apply)
☐ Action (Doom, Street Fighter)
☐ Adventure (King's Quest, Myst, Return to Zork, Ultima)
☐ Simulation (Flight Simulator, Rebel Assault)
☐ Sports (NBA Jam, Ken Griffey's MLB, NHL '94)
☐ Strategy/Puzzle (Sim City, Tetris)

4. Why do you play the games you checked above? (Choose at most 3)
☐ I like the graphics/realism
☐ relaxation/recreation/escapism
☐ It improves my hand-eye coordination
☐ It challenges my mind
☐ It's such a great feeling to master or finish a game
☐ I'll play anything when I'm bored
☐ Other (Please specify) ________________________________

5. Where do you usually play video/computer games?
☐ Arcade
☐ Home ☐ on a system (Sega, Nintendo, etc.)
☐ on a computer (IBM, MAC, etc.)

6. How often do you play?
☐ Daily ☐ Weekly ☐ Monthly ☐ Semesterly

7. Do you still find time to play when you're busy (i.e., during midterms)?
☐ Yes (can't stay away) ☐ No (school comes first!)

8. Do you think video games are educational?
☐ Yes (or else all my years of playing have gone to waste)
☐ No (I should have been reading books all those years)

9. What don't you like about video game playing?
Choose at most 3

□ It takes up too much time	□ It costs too much
□ It's frustrating	□ It's boring
□ It's lonely	□ My friends don't play
□ Too many rules to learn	□ It's pointless
□ Other (Please specify)	______________

10. Sex: □ Male □ Female

11. Age: _____

12. When you were in high school was there a computer in your home?
□ Yes □ No

13. What do you think of math? □ Hate it □ Don't hate it

14. How many hours a week do you work for pay? ________

15. Do you own a PC? □ Yes □ No
Does it have a CD-Rom? □ Yes □ No

16. Do you have an e-mail account? □ Yes □ No

17. What grade do you expect in this class? □A □B □C □D □F

3

Minnesota Radon Levels

[1]

WEDNESDAY, JULY 5, 1995 ★★★★★· San Francisco Chronicle

Does Your House Make You Sick?

How to detect and manage radon, lead, asbestos and carbon monoxide

Tara Aronson

Every day there's another frightening headline about dangerous substances in our homes.

Cancer-causing radon seeping up through the floorboards. Deadly asbestos sprayed on the ceiling. Toxic lead in our drinking water or painted walls. Poisonous carbon monoxide gas spewing from gas stoves.

How much of a threat are these hazards? And what should we do about them?

Many people simply bury their heads. And why not? If you know you have a problem, you either have to spend big bucks to fix it or disclose it when you sell your house.

And, if you aren't sick, why test your home? Ignorance can be bliss.

Or at least it used to be....

RADON - WE'RE NOT EXEMPT

Unlike asbestos, lead and carbon monoxide, radon is a naturally occurring toxin, which leads many people to mistakenly believe it is less dangerous than its man-made counterparts, says Groth of Consumer Reports.

"As one expert put it, it's 'God's radon', " Groth says. "It's a problem people tend not to worry about because it's natural and it's there. But radon is simply a very big risk." ...

Radon is produced by the normal decay of uranium, ... It is colorless, odorless and tasteless - and impossible to detect without a special test. ...

Radon is the second-leading cause of lung cancer, after smoking, according to the U.S. surgeon general. According to the EPA, an estimated 14,000 people nationwide die each year from radon-caused lung cancer ...

---

[1] Reprinted by permission.

Introduction

Radon is a radioactive gas with a very short half-life, yet it is considered a serious health risk to the general public. Because radon is a gas, it moves easily from soil into the air inside a house, where it decays. The decay products of radon are also radioactive, but they are not gases. They stick to dust particles and cigarette smoke and can lodge in lung tissue, where they irradiate the respiratory tract.

Radon has long been known to cause lung cancer. In the sixteenth century, a high incidence of fatal respiratory disease was reported among eastern European miners; this disease was identified as lung cancer in the nineteenth century. More recently, the U.S. National Cancer Institute conducted a large epidemiological study of 68,000 miners exposed to radon. It was found that the miners are dying of lung cancer at 5 times the rate of the general population. According to the U.S. Environmental Protection Agency (EPA), the level of radon exposure received in the mines by some of these miners is comparable to the exposure millions of people in the U.S. receive over their lifetime in their homes.

In the 1980s, the EPA began funding efforts to study indoor radon concentrations with the goal of finding homes with high levels of radon. As part of their search for these homes, the EPA constructed a map for each state in the U.S. indicating county radon levels. The purpose of the map is to assist national, state, and local organizations in targeting their resources for public awareness campaigns and in setting radon-resistant building codes.

In this lab, you will examine indoor radon concentrations for a sample of homes in Minnesota. The survey results are from a 1987 study of radon conducted by the EPA and the Minnesota Department of Health. You will have the opportunity to create a county radon map for the state of Minnesota based on these survey results and information about the state's geology.

The Data

The data were collected by the EPA and the Minnesota Department of Health in November, 1987, for 1003 households in Minnesota (Tate [Tat88]). Radon concentrations were monitored in each house for two days using a charcoal canister. To maintain anonymity, houses were identified by county only.

For each house in the survey, the county identification number and the two-day charcoal canister measurement of the radon level are available (Table 3.1).

Table 3.2 contains a key to match the county identification number with the county name, which can be used to locate counties on the map in Figure 3.1.

Survey Design

The houses that were included on the list to be surveyed were all those with permanent foundations, at least one floor at or below the ground level, owner

TABLE 3.1. Sample observations and data description for the EPA radon survey of 1003 Minnesota homes (Tate [Tat88]).

County ID	1	1	1	1	2	2	2	2	2	2
Radon	1.0	2.2	2.2	2.9	2.4	0.5	4.2	1.8	2.5	5.4

Variable	Description
County ID	Identifier for the county in which the house is located.
Radon	Radon measurement, in picoCuries per liter (pCi/l).

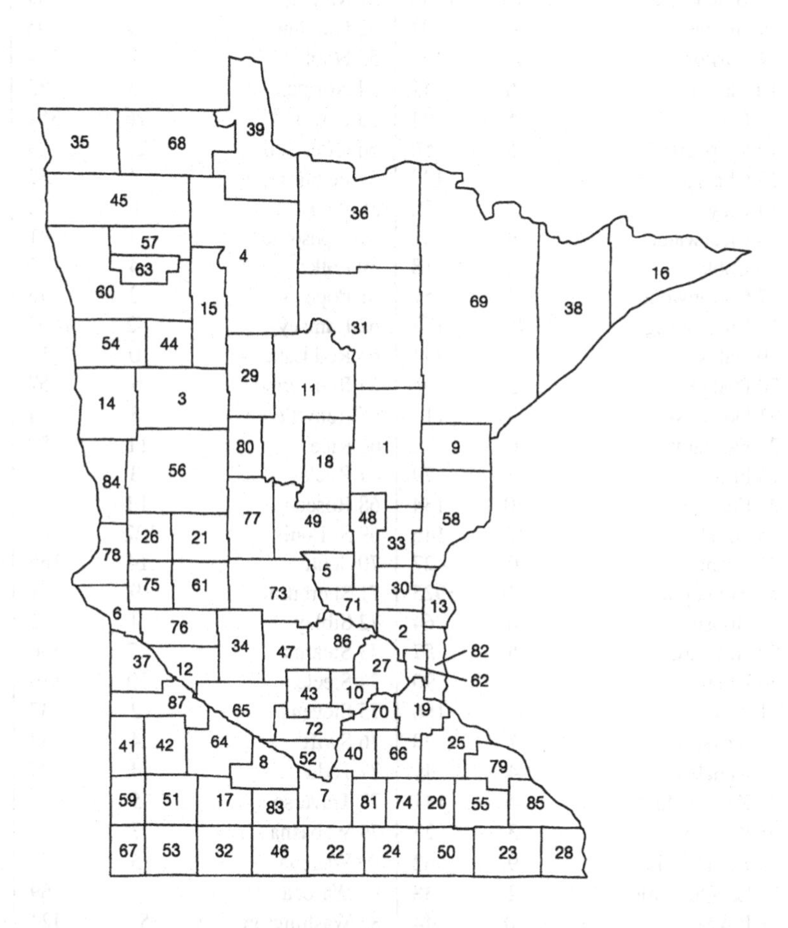

FIGURE 3.1. County map of Minnesota. See Table 3.2 to identify counties.

occupied, and with a listed phone number. The real population of interest is all occupied residences. These restrictions resulted from the difficulties in gaining permission to conduct the survey in rental units and finding houses without listed

TABLE 3.2. County populations and sample sizes for the EPA radon survey in Minnesota (Tate [Tat88]).

County ID Name	Houses sampled	Total (100s)	County ID Name	Houses sampled	Total (100s)
1 Aitkin	4	54	45 Martin	8	97
2 Anoka	57	719	46 McLeod	13	111
3 Becker	4	110	47 Meeker	5	77
4 Beltrami	7	115	48 Mille Lacs	3	72
5 Benton	4	95	49 Morrison	10	103
6 Big Stone	3	29	50 Mower	14	149
7 Blue Earth	14	186	51 Murray	1	39
8 Brown	4	102	52 Nicollet	4	95
9 Carlton	11	105	53 Nobles	3	78
10 Carver	6	141	54 Norman	3	33
11 Cass	5	84	55 Olmsted	26	361
12 Chippewa	5	56	56 Otter Tail	11	199
13 Chisago	6	100	57 Pennington	4	57
14 Clay	15	172	58 Pine	6	75
15 Clearwater	4	31	59 Pipestone	4	41
16 Cook	2	18	60 Polk	4	125
17 Cottonwood	4	52	61 Pope	2	46
18 Crow Wing	12	172	62 Ramsey	42	1809
19 Dakota	69	794	63 Red Lake	0	18
20 Dodge	3	55	64 Red Wood	5	67
21 Douglass	11	114	65 Renville	3	74
22 Faribault	6	73	66 Rice	11	159
23 Fillmore	2	79	67 Rock	3	38
24 Freeborn	10	134	68 Roseau	14	47
25 Goodhue	15	146	69 St Louis	122	81
26 Grant	0	27	70 Scott	14	165
27 Hennepin	119	3925	71 Sherburne	9	111
28 Houston	6	64	72 Sibley	4	55
29 Hubbard	5	57	73 Stearns	27	360
30 Istani	4	90	74 Steele	10	110
31 Itasca	12	166	75 Stevens	2	37
32 Jackson	7	48	76 Swift	4	47
33 Kanabec	4	49	77 Todd	4	94
34 Kandiyohi	4	141	78 Traverse	5	17
35 Kittson	3	23	79 Wabasha	7	74
36 Koochiching	9	62	80 Wadena	5	50
37 LacQui Parle	2	38	81 Waseca	4	69
38 Lake	10	44	82 Washington	50	424
39 Lake Of The Woods	5	15	83 Watonwan	3	47
40 LeSueur	6	87	84 Wilkin	1	28
41 Lincoln	4	29	85 Winona	13	161
42 Lyons	10	90	86 Wright	14	216
43 Mahnomen	1	16	87 Yellow Medicine	3	46
44 Marshall	9	45	Total	1003	14913

phone numbers, and from the fact that houses entirely above ground tend to have very low radon concentrations.

Houses were selected county by county for the sample. Within a county, each house had an equal chance of being included in the survey. The county population and a radon index were used to determine how many houses to choose from each county. Table 3.2 contains, for each county, the total number of households in the county and the number of houses sampled from the county.

To select the houses to be surveyed, telephone numbers were randomly chosen from a directory of listed telephone numbers. For each county, the list of randomly selected phone numbers was 5 times the desired number of households to be contacted in the county. The phone numbers were arranged in lists of 50, and the contacts for each county were made in waves of 50, until the desired number of participants was obtained.

The phone caller determined whether the candidate was eligible and willing to participate. If this was the case, the candidate was mailed a packet of materials containing the charcoal canister, an instruction sheet, a questionnaire, literature on radon, and a postage-paid return envelope. Eligible candidates who were unwilling to participate were mailed information about radon, and a second phone contact was made later to see if they had changed their minds about participating in the study.

The original survey design was to use sampling rates proportional to county populations, with the proportion determined by one of three factors according to whether the county was considered a potentially high-radon, medium-radon, or low-radon area. In reality, the sampling rates were far more varied. Nonetheless, for each county, a simple random sample of willing households was obtained.

Background

Radon

Radon is a radioactive gas that comes from soil, rock, and water. It is a by-product of the decay of uranium (U^{238}), which is naturally present in all rocks and soils. Because radon is a gas, it escapes into the air through cracks in rocks and pore spaces between soil grains. Outdoors, radon is diluted by other atmospheric gases and is harmless. Indoors, radon may accumulate to unsafe levels.

Uranium is the first element in a long series of radioactive decay that produces radium and then radon. Radon is called a daughter of radium because it is the element produced when radium decays. Radon is radioactive because it decays to form polonium. Polonium is also radioactive, but it is not a gas. Polonium easily sticks to dust particles in the air and to lung tissue, and for this reason radon is considered a health hazard.

Radioactive decay is a natural spontaneous process in which an atom of one element breaks down to form another element by losing protons, neutrons, or electrons. In this breakdown, radioactive elements emit one of three kinds of rays:

alpha, beta, or gamma. An alpha particle consists of two protons and two neutrons, beta rays are electrons, and gamma rays are photons. When an alpha particle is emitted by an atomic nucleus, the element changes into a new element two columns to the left in the periodic table. When a beta particle is emitted, the element changes into an element one column to the right in the periodic table. The emission of gamma rays does not change the element. Each of these types of rays can sensitize photographic plates. When radium and radon decay, an alpha particle is given off.

Measuring Radon

Radioactive elements are measured in terms of their half-life. A half-life is the time it takes a large quantity of a radioactive element to decay to half its original amount. Radon decays very quickly. Its half-life is about four days. In contrast, the half-life of uranium is about four billion years, and radium's half-life is about 1600 years.

Radon concentrations are measured in Becquerels per cubic meter or in picoCuries per liter. One Becquerel is one radioactive decay per second. A picoCurie per liter (pCi/l) equals 37 Becquerels per cubic meter (Bq/m^3). In the U.S., concentrations of radon in single-family houses vary from 1 Becquerel per cubic meter to 10,000 Becquerels per cubic meter.

These units of measurement are named after the pioneers in the discovery of radioactivity (Pauling [Pau70]). In 1896, Becquerel showed that uranium could blacken a photographic plate that was wrapped in paper. Later that same year, Marie and Pierre Curie discovered radium. One gram of radium is approximately one Curie.

Table 3.3 contains the U.S. EPA guidelines for safe radon levels.

Radon in the Home

Indoor radon concentrations are related to the radium content of the soil and rock below the building. All rock contains uranium. Most contain a very small amount, say 1–3 parts per million (ppm). Some rocks have high uranium content, such as 100 ppm. Examples of these rocks include light-colored volcanic rock, granite, shale, and sedimentary rocks that contain phosphate.

When radium decays and shoots off an alpha particle in one direction, it also shoots off a radon atom in the opposite direction. If the radium atom is close to the surface of the rock grain, then the radon atom can more easily leave the mineral grain and enter the pore space between the grains. Once it has escaped the rock or soil grain, it can travel a long way before it decays. Radon moves more quickly through coarse sand and gravel in comparison to clay and water. The interconnectedness of the pore spaces determines the ability of the soil to transmit a gas such as radon — we call this property permeability.

The permeability and radioactive levels of the soil and rock beneath a house affect the quantity of radon that enters the house. For example, some granite contains high concentrations of radium; however, because it is rock, a large fraction of the

TABLE 3.3. EPA Action Guidelines (Otton [Ott92].)

Radon concentration (pCi/l)	Recommended urgency of reduction efforts
200 or above	Action to reduce levels as far below 200 pCi/l as possible is recommended within several weeks after measuring these levels.
20 to 200	Action to reduce levels as far below 20 pCi/l as possible is recommended within several months.
4 to 20	Action to reduce levels to under 4 pCi/l is recommended within a few years, and sooner if levels are at the upper end of this range.
Less than 4	While these levels are at or below the EPA guideline, some homeowners might wish to attempt further reductions.

radon generated in the mineral grain is absorbed in neighboring grains, not reaching the mineral pores. In contrast, buildings on drier, highly permeable soils such as hillsides, glacial deposits, and fractured bedrock may have high radon levels even if the radon content of the soil pores is in the normal range. This is because the permeability of the soils permits the radon to travel further before decaying.

Radon enters a building via the air drawn from the soil beneath the structure. This air movement is caused by small differences between indoor and outdoor air pressures due to wind and air temperature. The air pressure in the ground around a house is often greater than the air pressure inside the house so air tends to move from the ground into the house. A picture of how radon may enter a house is shown in Figure 3.2. Most houses draw less than 1% of their air from the soil. However, houses with low indoor air pressures and poorly sealed foundations may draw up to 20% of their air from the soil.

Indoor radon concentrations are also affected by ventilation and by reactions with other airborne particles in the house, such as cigarette smoke.

The Geology and Geography of Minnesota

Minnesota's landscape is diverse and influenced in most of the state by the action of glaciers. The map in Figure 3.3 (Schumann and Schmidt [SS88]) shows several distinct areas characterized by different landscape features.

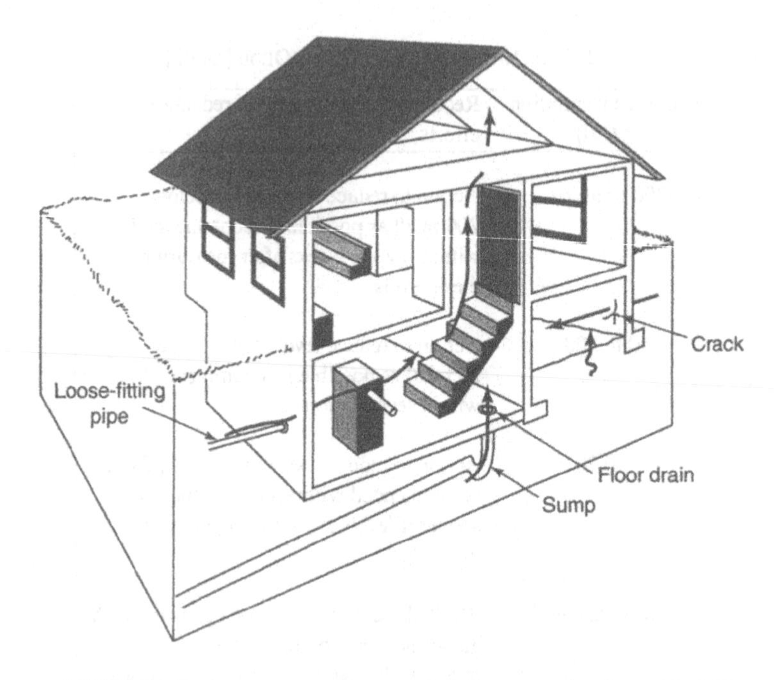

FIGURE 3.2. Radon can enter a house through many paths (Otton [Ott92]).

The Superior Upland is primarily an area of glacial erosion, where exposed bedrock has been smoothed and grooved by overriding glaciers, producing linear patterns of lakes and ridges. In the southern and western parts of the Superior Upland are the Giants range, a ridge of granite flanking the Mesabi Iron Range. The Mesabi Range has many large open-pit iron mines. Some of these mine pits are now lakes.

The Western Lake Section of the Central Lowlands province is characterized by glacial landscape features, including various types of glacial hills and ridges and depressions, most of which are filled with lakes and wetlands.

The Till Prairie Section of the Central Lowlands occupies most of the southern half of the state. This area is relatively flat and featureless, except where it is dissected by rivers and streams, the largest of which is the Minnesota River. Chains of lakes are common features and may be due to buried preglacial valleys. Along the southern border of the state are uplands. To the east, these areas are partially covered by windblown silt. In the southeast corner of the state, the windblown silt rests directly on bedrock and is called the "driftless area." In the southwest, the Prairie Couteau is an upland area that lies between the Minnesota River lowland and the James River basin in South Dakota. It is a bedrock highland that existed in preglacial times; part of it is covered by windblown silt.

The Red River Lowland, in the northwestern part of the state, is a relatively flat-lying lowland, which is the former bed of Lake Agassiz. The lake was one of the largest Wisconsin glacial lakes. The flatness and clay deposits make this area very poorly drained, and much of it is occupied by wetlands.

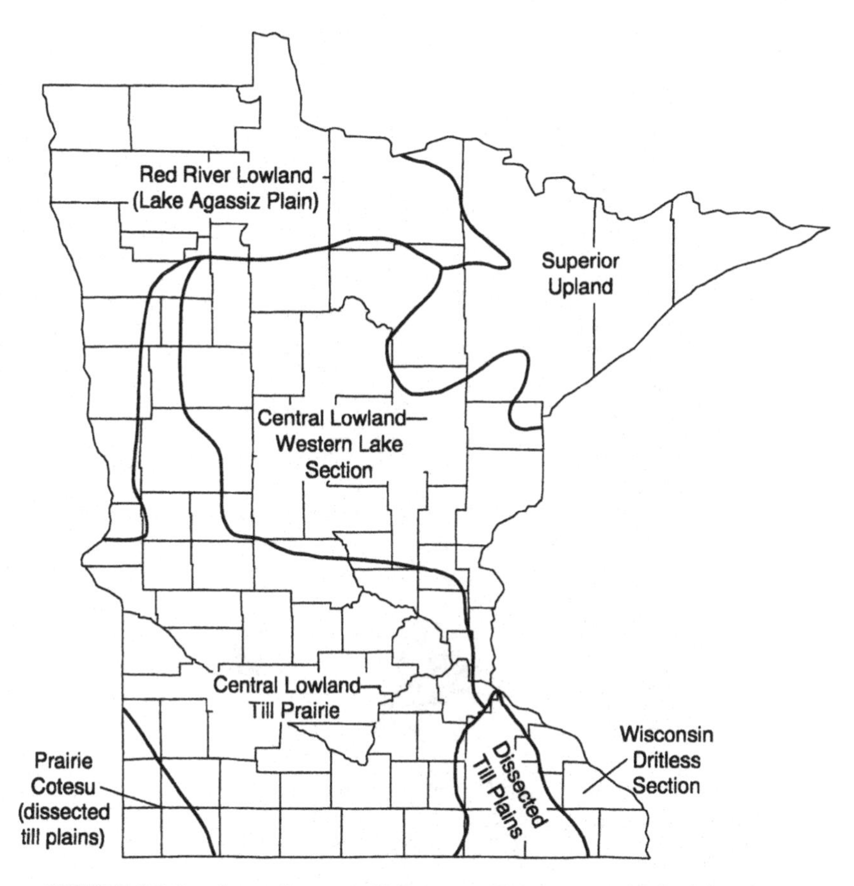

FIGURE 3.3. Landscape features of Minnesota (Schumann and Schmidt [SS88]).

A significant portion of Minnesota's population is clustered around urban centers such as Minneapolis, St. Paul, and Duluth (Figure 3.4). (Minneapolis is in Hennepin County, St. Paul is in Ramsey County, and Duluth is in St. Louis County.) Major land use in the state includes agriculture and manufacturing in the southern part of the state and logging, mining, and tourism in the north.

Many areas of Minnesota are underlaid by rocks that contain sufficient quantities of uranium to generate radon at levels of concern if they are at or near the surface (Figure 3.5). Most rocks of the greenstone-granite terrain in northern Minnesota have low uranium contents. Rocks in northeastern Minnesota are generally low in uranium. The rocks in Carlton and Pine Counties can contain significant amounts of uranium. Rocks of central Minnesota and the Minnesota River valley (e.g., Redwood, Renville, and Yellow Medicine Counties) have high concentrations of uranium. In general, soils from Cretaceous shales contain more uranium than those from metamorphic and other crystalline rocks. Sandstones in southeastern Minnesota are generally low in uranium, but these rocks contain small amounts of uranium-bearing heavy minerals that may be locally concentrated.

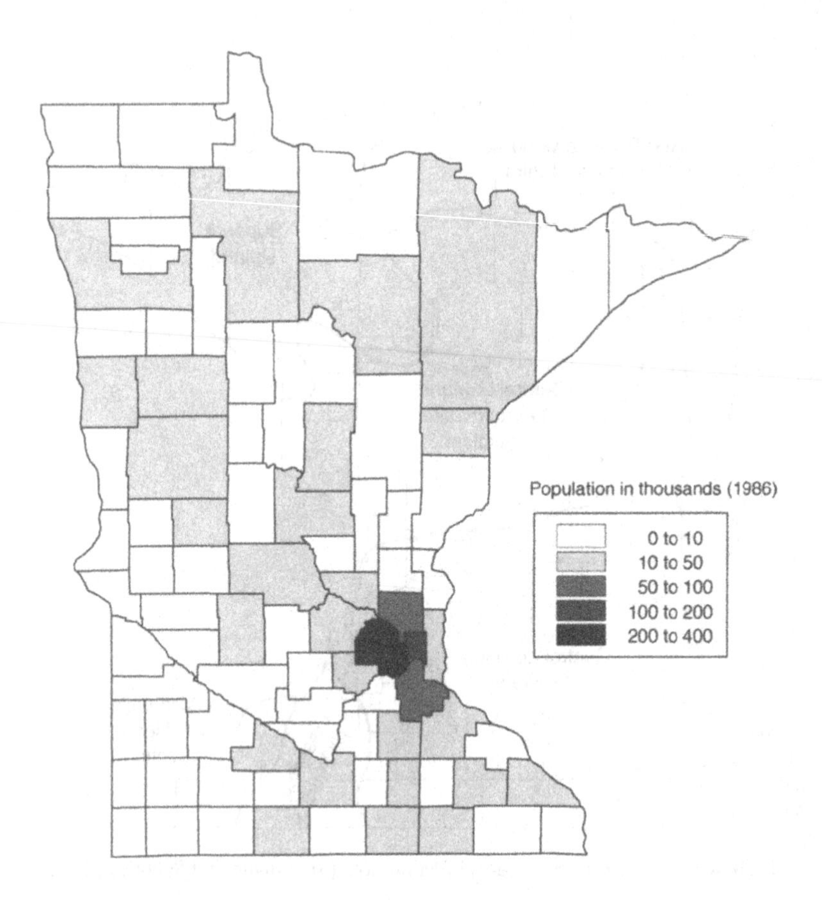

FIGURE 3.4. Population map of Minnesota.

Investigations

Our goal is to provide estimates for the proportion of houses in the state that exceed the EPA recommended action levels. Below are two sets of suggestions for making these estimates. The first is sample-based and the second model-based. It may be interesting to compare the two techniques.

The Sample-Based Method

- Use the number of houses in the sample with radon levels that exceed 4 pCi/l to estimate the proportion of houses in the state that exceed this level. Keep in mind the sampling scheme when creating the estimate. That is, in some counties one sampled house represents 500 houses and in other counties one house represents 10,000 houses.

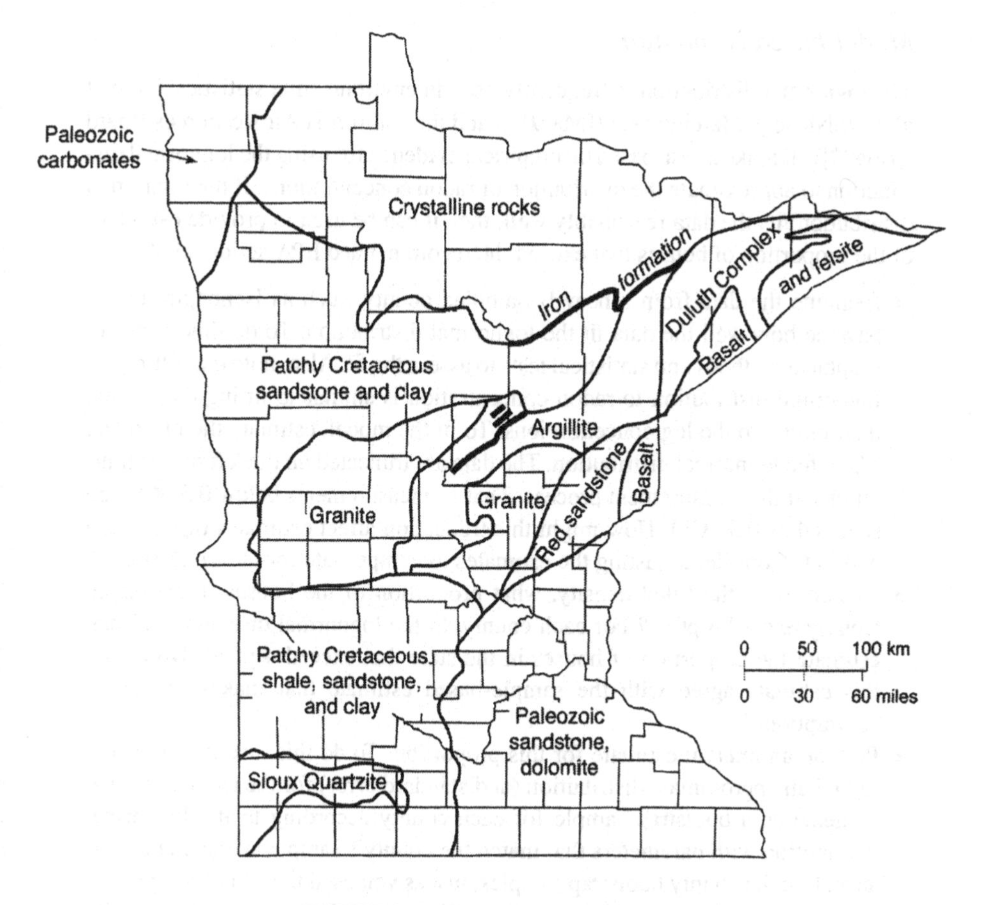

FIGURE 3.5. Bedrock map of Minnesota.

- Attach standard errors to your estimate, or provide an interval estimate for the proportion of houses in Minnesota that exceed 4 pCi/l.
- How would you use the sample survey results for each county to estimate the number of households in that county with radon levels that exceed 4 pCi/l?
- Many counties have as few as four or five observations. For a county with only four observations, the estimated proportion of houses in the county to exceed 4 pCi/l can only be either 0, 25, 50, 75, or 100%, according to whether 0, 1, 2, 3, or 4 houses exceed the level. Can you use the information for the entire state or for neighboring counties to help provide a more reasonable estimate for a county with a small sample?
- Using estimates for each county, color or shade a county map of Minnesota, indicating at least three levels of indoor radon concentration. What do you conclude about the geographical distribution of radon concentration in Minnesota? What factors limit the validity of your conclusions?

Model-based Estimation

The lognormal distribution is frequently used in environmental statistics. Nero et al. ([NSNR86]), Marcinowski ([Mar91]), and the California Air Resources Board ([Boa90]) all make a case based on empirical evidence for using the lognormal distribution to approximate the distribution of radon concentrations. If the lognormal distribution fits the data reasonably well, then it can be used to provide estimates of the proportion of houses that exceed the recommended EPA action level.

- Examine the data from a heavily sampled county, such as Hennepin, to determine how well the data fit the lognormal distribution. To do this, consider graphical methods and statistical tests to assess the fit. Also, note that fitting the lognormal distribution to radon concentration is similar to fitting the normal distribution to the log concentrations. To fit the model, estimate the mean and SD of the lognormal distribution. The data are truncated on the left due to limitations of the measurement process. That is, measurements below 0.5 pCi/l are reported as 0.5 pCi/l. How might this truncation affect estimates of the mean and SD? Consider adjusting the estimates to compensate for the truncation.
- According to the fitted density, what proportion of the houses in Hennepin County exceed 4 pCi/l? For each county, fit the lognormal distribution. Then estimate the proportion of houses in the state that exceed 4 pCi/l. How does this estimate agree with the sample-based estimate that makes no model assumptions?
- Provide an interval estimate for this proportion. To do this, use the bootstrap to find an approximate distribution (and standard error) for the estimator. That is, generate a bootstrap sample for each county according to the lognormal distribution with parameters that match the county's sample average and variance. Use the county bootstrap samples, just as you used the actual samples, to construct an estimate of the proportion of houses in the county and state with radon levels greater than 4 pCi/l. Repeat this process, taking more bootstrap samples, yielding more bootstrap estimates of the proportion. The distribution of the bootstrap proportions approximates the sampling distribution of the bootstrap estimator. It can be used to make an interval estimate from your original estimate of the proportion of houses with radon levels over 4 pCi/l.
- How does the procedure differ if the average and SD of the lognormal distribution are estimated directly from the sampled radon measurements, as opposed to estimating it indirectly through the average and SD of the logged measurements? Consider a simulation study that compares the properties of these estimates under various values of μ, σ, and n.

Theory

Stratified Random Sampling

The stratified random sample is a probability method for sampling from a population that is divided into subgroups called strata. A simple random sample is taken from each stratum. The strata are typically homogeneous groups of units. The benefits of this more complex sampling technique are that more information is obtained for a subgroup of the population and that a more accurate estimate for a population parameter is possible.

In this lab, the strata are counties. Within each county, a simple random sample was taken independently of the other counties. For example, in Hennepin County a simple random sample of 119 of the 392,500 houses was taken. Without regard to the sample from Hennepin County, a simple random sample of 3 houses from the 4600 houses in Yellow Medicine County was also taken. Similarly, simple random samples were independently taken from each of the remaining 85 counties. Altogether, 1003 houses were sampled. The sample sizes were chosen to reflect the size of the population in the county and the estimated radon levels for the county. That is, larger counties had larger samples, and counties that were thought to have high indoor radon concentrations had larger samples.

The Model

Consider a population with N units, where the units are grouped into J strata. Suppose there are N_1 units in the first stratum, N_2 in the second, and so on. Assign each unit in stratum #1 a number from 1 to N_1. Also assign each unit in stratum #2 a number from 1 to N_2. Do likewise for the rest of the strata.

The value of the characteristic for unit #i in stratum #j can be represented by $x_{j,i}$. That is, the #1 unit in stratum #1 has characteristic value $x_{1,1}$, the #2 unit in stratum #1 has value $x_{1,2}$, and so on. Then the average for stratum #j can be written as μ_j, for $j = 1, \ldots, J$, and the average for all units in the population is μ;

$$\mu_j = \frac{1}{N_j}\sum_{i=1}^{N_j} x_{j,i},$$

$$\mu = \frac{1}{N}\sum_{j=1}^{J}\sum_{i=1}^{N_j} x_{j,i}.$$

Notice that

$$\mu = \frac{N_1}{N}\mu_1 + \ldots + \frac{N_J}{N}\mu_J.$$

That is, the population average can be represented as a weighted combination of the stratum averages, where the weights are the relative sizes of the strata.

In stratified sampling, the sample size n is divided J ways into $n_1, \ldots, n_J$, where

$$n_1 + \ldots + n_J = n.$$

From the first stratum, a simple random sample of n_1 units is taken. From the second stratum, a simple random sample of n_2 units is taken, and so on. Each of the stratum samples are taken without regard to the others (i.e., they are independent samples). If we let $I(j, i)$ represent the index of the ith unit selected at random from the jth stratum, for $i = 1, \ldots, n_j$, and $j = 1, \ldots, J$ then the sample averages $\bar{x}_1, \ldots, \bar{x}_J$ can be combined as follows to estimate μ:

$$\bar{x}_j = \frac{1}{n_j}\sum_{i=1}^{n_j} x_{I(j,i)},$$

$$\tilde{x} = \frac{N_1}{N}\bar{x}_1 + \ldots + \frac{N_J}{N}\bar{x}_J.$$

This estimator is unbiased because each of the stratum sample averages is unbiased for its respective stratum population average.

To compute its variance, we use the fact that the strata samples are independent, so the strata sample averages are independent. We also use the fact that within each stratum a simple random sample was taken, which means

$$\text{Var}(\bar{x}_j) = \frac{\sigma_j^2}{n_j}\frac{N_j - n_j}{N_j - 1},$$

where σ_j^2 is the variance of the units in the jth stratum, namely,

$$\sigma_j^2 = \frac{1}{N_j}\sum_{i=1}^{N_j}(x_{j,i} - \mu_j)^2.$$

Using these facts, we find that

$$\begin{aligned}\text{Var}(\tilde{x}) &= \sum_{j=1}^{J} w_j^2 \text{Var}(\bar{x}_j) \\ &= \sum_{j=1}^{J} w_j^2 \frac{\sigma_j^2}{n_j}\frac{N_j - n_j}{N_j - 1},\end{aligned}$$

where $w_j = N_j/N$, $j = 1, \ldots J$.

Optimal Allocation

Those strata with more units and more variation across units contribute more to the variance of $\tilde{x}$ than smaller and less variable strata. This observation leads to the question of how best to choose strata sample sizes $\{n_j\}$ in order to minimize the variance of the estimator.

To answer this question, we begin by looking at the case of two strata (i.e., $J = 2$). The optimal allocation follows from the inequality:

$$\frac{w_1^2\sigma_1^2}{n_1} + \frac{w_2^2\sigma_2^2}{n_2} \geq \frac{(w_1\sigma_1 + w_2\sigma_2)^2}{n}.$$

(We ignore the finite population correction factor for now.) To see this, notice that when the right side is subtracted from the left side the difference is:

$$\frac{(n_2 w_1\sigma_1 - n_1 w_2\sigma_2)^2}{n_1 n_2 n},$$

which is never negative. It then follows that the left side is minimized when

$$n_2 w_1 \sigma_1 = n_1 w_2 \sigma_2,$$

or, equivalently, when

$$n_1 = n \frac{w_1 \sigma_1}{w_1 \sigma_1 + w_2 \sigma_2}.$$

In general, the optimal allocation is

$$n_i = n \frac{w_i \sigma_i}{\sum_{j=1}^{J} w_j \sigma_j},$$

and the variance of $\bar{x}$ is

$$\frac{1}{n}\left(\sum_{j=1}^{J} w_j \sigma_j\right)^2.$$

The proof for $J > 2$ is left as an exercise.

As expected, the optimal allocation takes a larger sample from the larger and the more variable strata. If the stratum variances are all the same then the optimal allocation coincides with proportional allocation. *Proportional allocation* takes the stratum sample size to be proportional to the stratum population size (i.e., $n_j/n = N_j/N$).

When the stratum variances are unknown, the method of proportional allocation can be used. Table 3.4 compares the variances of the sample average from a simple random sample, a stratified random sample with proportional allocation, and a stratified random sample with optimal allocation. The finite population correction factor is ignored in each of the calculations. Note that $\bar{\sigma} = \sum w_j \sigma_j$, which is in general different from the population standard deviation σ.

TABLE 3.4. Summary of variances for different sampling techniques (ignoring the finite population correction factor).

Sampling method	Variance	Difference
Simple	$\frac{1}{n}\sigma^2$	
		Simple–Proportional $\frac{1}{n}\sum_{j=1}^{J} w_j(\mu_j - \mu)^2$
Proportional	$\frac{1}{n}\sum_{j=1}^{J} w_j \sigma_j^2$	
		Proportional–Optimal $\frac{1}{n}\sum_{j=1}^{J} w_j(\sigma_j - \bar{\sigma})^2$
Optimal	$\frac{1}{n}(\sum_{j=1}^{J} w_j \sigma_j)^2$	

An Example

For the moment, treat Minnesota as a state with only three counties: Hennepin, Ramsey, and St. Louis (i.e., counties #27, #62, and #69). Their population sizes (in hundreds of houses) are 3925, 1809, and 81, and their sample sizes are 119, 42, and 122, respectively. The total population in this mini-state is 5815 hundred houses, and the total number of houses sampled is 283. Then the weights for these three counties are

$$w_H = 3925/5815 = 0.675, \quad w_R = 0.311, \quad w_S = 0.014,$$

where the subscripts denote the county — H for Hennepin, R for Ramsey, and S for St. Louis. The sample means for these three counties are

$$\bar{x}_H = 4.64, \quad \bar{x}_R = 4.54, \quad \bar{x}_S = 3.06.$$

Using the weights, we estimate the population average radon concentration to be

$$\bar{x} = (0.675 \times 4.64) + (0.311 \times 4.54) + (0.014 \times 3.06) = 4.59.$$

The sample SDs for each of the counties are

$$s_H = 3.4, \quad s_R = 4.9, \quad s_S = 3.6,$$

and if we plug these estimates into the formula for the Var($\bar{x}$) then we have an estimate for the variance:

$$\left(w_H^2 \frac{s_H^2}{119}\right) + \left(w_R^2 \frac{s_R^2}{42}\right) + \left(w_S^2 \frac{s_S^2}{122}\right) = 0.10.$$

In comparison, if proportional allocation had been used to take the samples, then the variance of the estimator would be about

$$\frac{1}{283}[(w_H \times s_H^2) + (w_R \times s_R^2) + (w_S \times s_S^2)] = 0.05\ .$$

The estimate of the variance under optimal allocation is also 0.05 to 2 decimal places.

The Lognormal Distribution

The geometric mean of a set of nonnegative numbers $x_1, x_2, \ldots x_n$ is defined to be

$$\left(\prod_{i=1}^{n} x_i\right)^{1/n}.$$

It can easily be shown that

$$\left(\prod_{i=1}^{n} x_i\right)^{1/n} = e^{\bar{y}},$$

where $y_i = \log x_i$ and $\bar{y} = n^{-1} \sum y_i$. This notion extends naturally to random variables: the geometric mean of a random variable X is $\exp(\mathbb{E}(\log X))$.

A nonnegative random variable X is said to have a lognormal distribution with parameters μ and σ^2 if log (base e) of X has a normal distribution with mean μ and variance σ^2 (Crow and Shimiza [CS88]). Note that μ and σ^2 are not the mean and variance of X but the mean and variance of $\log X$. The density function of the lognormal can be found via the change of variable formula for the transformation $X = \exp(Z)$, where Z has a normal distribution with mean μ and SD σ. For $x > 0$,

$$f(x) = \frac{1}{x\sigma\sqrt{2\pi}} \exp\left[-\frac{1}{2\sigma^2}\left(\log\frac{x}{\gamma}\right)^2\right],$$

where $\gamma = \exp(\mu)$ is the geometric mean of the lognormal distribution.

Then the mean and variance of the lognormal are

$$\mathbb{E}(X) = \gamma \exp(\sigma^2/2)$$

and

$$\mathrm{Var}(X) = \gamma^2[\exp(2\sigma^2) - \exp(\sigma^2)].$$

Note that for σ small, the geometric mean and expected value are close.

Parametric Bootstrap

In Chapter 2, we saw how to use the bootstrap to "blow up" the sample to make a bootstrap population and use this bootstrap population to approximate (through repeated sampling) the variance of the sample average.

In this lab, we consider the possibility that the survey results look as though they are random samples from the lognormal distribution. We can again use the bootstrap method to determine the variance of the sample average. This time, however, we use a parametric bootstrap that is based on the lognormal distribution. Rather than blowing up the sample in each county to make the county bootstrap population, we suppose the bootstrap population values in each county follow the lognormal distribution, with parameters determined by the county sample. Then by resampling from each county's bootstrap lognormal distribution, we generate bootstrap samples for the counties. Each county's bootstrap sample is used to estimate the lognormal parameters for the county, and then a bootstrap estimate of the proportion of houses in the state with radon levels exceeding 4 pCi/l is obtained. By repeated resampling from the counties, we can approximate the distribution of the bootstrap proportion. The bootstrap sampling procedure imitates stratified random sampling from each of the counties, with the number of bootstrap units sampled from a county matching the actual sample size taken.

Exercises

1. Consider the following population of six units:

$$x_1 = 1, x_2 = 2, x_3 = 2, x_4 = 4, x_5 = 4, x_6 = 5.$$

Suppose units 2, 3, 4, and 5 are in one stratum and units 1 and 6 are in a second stratum. Take a simple random sample of 2 units from the first stratum and a simple random sample of 1 unit from the second stratum.

a. Find the exact distribution of the stratified estimator for the population average.
b. Use the exact distribution to compute the expectation and standard deviation of the estimator.
c. Compare your calculations to the numeric results obtained from plugging the values for N_j, n_j, μ_j, and σ_j into the formulas derived in this chapter for the expected value and standard error of a stratified estimator.

2. The following results were obtained from a stratified random sample.

Stratum 1:	$N_1 = 100,$	$n_1 = 50,$	$\bar{x}_1 = 10,$	$s_1 = 50.$
Stratum 2:	$N_2 = 50,$	$n_2 = 50,$	$\bar{x}_2 = 20,$	$s_2 = 30.$
Stratum 3:	$N_3 = 300,$	$n_3 = 50,$	$\bar{x}_3 = 30,$	$s_3 = 25.$

a. Estimate the mean for the whole population.
b. Give a 95% confidence interval for the population mean.

3. Allocate a total sample size of $n = 100$ between two strata, where $N_1 = 200,000$, $N_2 = 300,000$, $\mu_1 = 100$, $\mu_2 = 200$, $\sigma_1 = 20$, and $\sigma_2 = 16$.

a. Use proportional allocation to determine n_1 and n_2.
b. Use optimal allocation to determine n_1 and n_2.
c. Compare the variances of these two estimators to the variance of the estimator obtained from simple random sampling.

4. For the strata in Exercise 3, find the optimal allocation for estimating the difference of means of the strata: $\mu_1 - \mu_2$. Ignore the finite population correction factor in your calculation.
5. It is not always possible to stratify a sample before taking it. In this case, when a simple random sample is taken, it can be stratified after the units are surveyed and it is known to which group each sampled unit belongs. That is, the sample is divided into strata after sampling and treated as a stratified random sample. This technique is called *poststratification*. For J strata, with strata sizes N_j and weights $w_j = N_j/N$, the poststratified estimator is

$$w_1\bar{x}_1 + \cdots + w_J\bar{x}_J.$$

Because we have poststratified, the number of units in the jth stratum, n_j, is random.

The variance of the poststratified estimator is approximately

$$\frac{N-n}{n(N-1)}\sum_j w_j\sigma_j^2 + \frac{1}{n^2}\frac{N-n}{N-1}\sum_j \frac{N-N_j}{N}\sigma_j^2,$$

where μ_j and σ_j^2 are the mean and standard deviation for stratum j. The first term in the variance is the same as the variance under proportional allocation,

TABLE 3.5. Numbers of female and male students in the sample of statistics students (Chapter 2) who did/did not play video games in the week prior to the survey, and who do/do not own a PC.

		Female	Male
Played	Yes	9	25
	No	29	28
Owns a PC	Yes	27	40
	No	11	13
	Total	38	53

with the finite population correction factor included. The second term is due to there being a random number of units in each stratum.
Poststratify the sample of statistics students (Chapter 2) by sex, and provide a 95% confidence interval for the proportion of students who played video games in the week prior to the survey (Table 3.5). Of the 314 students in the statistics class, 131 were women.

6. Consider the poststratification method introduced in Exercise 5. Show that the expected value for n_j equals the sample size for proportional allocation.
7. Prove that for J strata, the optimal choice for strata sample sizes $\{n_j\}$, where $\sum n_j = n$, is

$$n_j = n \frac{w_j \sigma_j}{\sum_{i=1}^{J} w_i \sigma_i}.$$

8. Suppose the cost of an observation varies from stratum to stratum, where the cost of sampling a unit from stratum j is c_j. Suppose there is a fixed start-up cost c_o to run the survey, so the total cost is

$$c_o + c_1 n_1 + \cdots + c_J n_J.$$

Now instead of fixing the total sample size n, we fix total survey cost C. Find the allocation of $n_1, ..., n_J$ that minimizes the variance of the estimator for the population mean. You may ignore the finite population correction factor.

9. *The delta method.* A second-order Taylor expansion of a function g around y_o is

$$g(y) \approx g(y_o) + (y - y_o) \times \frac{\partial}{\partial y} g(y)\Big|_{y_o} + \frac{1}{2}(y - y_o)^2 \times \frac{\partial^2}{\partial y^2} g(y)\Big|_{y_o}.$$

We can use this approximation to estimate the mean of $g(Y)$ by taking expectations of the right side, for $y_o = \mu$. Use a second-order expansion to approximate the expected value of a lognormal random variable and compare this approximation to the exact expression. Remember that if Y has a normal distribution with mean μ and variance σ^2, then e^Y has a lognormal distribution with parameters μ and σ^2.

10. Use a first-order Taylor expansion to estimate the variance of a lognormal random variable. How does your estimate compare to the exact expression?

11. Suppose the values of two characteristics are recorded on each unit in a simple random sample. Let the pair (x_i, y_i), $i = 1, \ldots, N$ represent the values of these pairs of characteristics for all members of the population. Suppose the population average for x is known and is μ_x, and the goal is to estimate μ_y, the population average for y. An alternative to $\bar{y}$ for an estimator of μ_y is the following ratio estimator:

$$\hat{\mu}_r = \bar{y}\frac{\mu_x}{\bar{x}}.$$

a. Show that this estimator is equivalent to

$$\frac{\sum y_{I(i)}}{\sum x_{I(i)}}\mu_x.$$

b. Explain why the expectation of $\hat{\mu}_r$ is not typically μ_y.

c. Show that the bias of the estimator is

$$-\text{Cov}(\bar{y}/\bar{x}, \bar{x}).$$

d. Use the delta method to show that $\hat{\mu}_r$ is approximately unbiased; that is, use a first-order approximation to the function $f(\bar{x}, \bar{y}) = \bar{y}/\bar{x}$,

$$f(\bar{x}, \bar{y}) \approx \mu_y/\mu_x + (\bar{y} - \mu_y)/\mu_x - (\bar{x} - \mu_x)\mu_y/\mu_x^2.$$

e. Use the same approximation as in part (d) to show the approximate variance of $\hat{\mu}_r$ is

$$\text{Var}(\bar{y} - \bar{x}\mu_y/\mu_x) = \frac{1}{n}\frac{N-n}{N-1}\text{Var}(y_{I(1)} - x_{I(1)}\mu_y/\mu_x).$$

Notice that if the pairs (x_i, y_i) roughly fall on a line through the origin of slope μ_y/μ_x then the expected sum of squares $\sum \mathbb{E}(y_i - x_i\mu_y/\mu_x)^2$ will be smaller than $\sum \mathbb{E}(y_i - \mu_y)^2$.

12. Use the ratio estimator introduced in Exercise 11 and the data in Table 3.5 to estimate the proportion of students who own PCs in the class, where it is known that 131 of the 314 students in the class are women (Chapter 2). Supply a 95% confidence interval for the population proportion. Do you think this ratio estimator is a better estimator than the proportion of students who own PCs in the sample? Explain.

Notes

Tony Nero and Philip Price of the Lawrence Berkeley Laboratory were very helpful in providing the data and in answering questions on radon and on the Minnesota survey.

The description of the sampling methodology is from a Minnesota Department of Health publication (Tate [Tat88]). The U.S. Department of the Interior publications (Nero [Ner88] and Otton [Ott92]) were the primary sources for the description of radon and how it enters houses. The geographic description of Minnesota was summarized from Schumann and Schmidt [SS88]. Pauling [Pau70] was the source for the description of the uranium series.

References

[Boa90] California Air Resources Board. Survey of residential indoor and outdoor radon concentrations in California, California Environmental Protection Agency, Sacramento, 1990.

[CS88] E.L. Crow and K. Shimiza. *Lognormal Distribution: Theory and Applications*. Marcel Dekker, New York, 1988.

[Mar91] F. Marcinowski. Nationwide survey of residential radon levels in the U.S. U.S. Environmental Protection Agency, Washington, D.C., 1991.

[Ner88] A.V. Nero. Controlling indoor air pollution. *Sci. Am.*, **258**(5):42–48, 1988.

[NSNR86] A.V. Nero, M.B. Schwehr, W.W. Nazaroff, and K.L. Revzan. Distribution of airborne radon-222 concentrations in U.S. homes. *Science*, **234**:992–997, 1986.

[Ott92] J.K. Otton. *The Geology of Radon*. U.S. Government Printing Office, Washington, D.C., 1992.

[Pau70] L. Pauling. *General Chemistry*. Dover, New York, 1970.

[SS88] R.R. Schumann and K.M. Schmidt. Radon potential of Minnesota. U.S. Geological Survey, Washington, D.C., 1988.

[Tat88] E.E. Tate. Survey of radon in Minnesota homes. Unpublished manuscript, Minnesota Department of Health, Minneapolis, 1988.

4
Patterns in DNA

[1]

TUESDAY, AUGUST 1, 1995 ★★★★★ New York Times

First Sequencing of Cell's DNA Defines Basis of Life

Feat is milestone in study of evolution

By Nicholas Wade

Life is a mystery, ineffable, unfathomable, the last thing on earth that might seem susceptible to exactdescription. Yet now, for the first time, a free-living organism has been precisely defined by the chemical identification of its complete genetic blueprint.

The creature is just a humble bacterium known as Hemophilus influenzae, but it nonetheless possesses all the tools and tricks required for independent existence. For the first time, biologists can begin to see the entire parts list, asit were, of what a living cell needs to grow, survive and reproduce itself.

Hemophilus -- no relation to the flu virus -- colonizes human tissues, where in its virulent form it can cause earaches and meningitis. Knowledge of its full genome has already given biologists a deeper insight into its genetic survivalstrategies.

"I think it's a great moment in science," said Dr. James D. Watson, codiscoverer of the structure of DNA and a former director of the Federal project to sequence the human genome. "With a thousand genes identified, we are beginning to see what a cell is," he said. ...

---

[1]Reprinted by permission.

Introduction

The human cytomegalovirus (CMV) is a potentially life-threatening disease for people with suppressed or deficient immune systems. To develop strategies for combating the virus, scientists study the way in which the virus replicates. In particular, they are in search of a special place on the virus' DNA that contains instructions for its reproduction; this area is called the origin of replication.

A virus' DNA contains all of the information necessary for it to grow, survive and replicate. DNA can be thought of as a long, coded message made from a four-letter alphabet: A, C, G, and T. Because there are so few letters in this DNA alphabet, DNA sequences contain many patterns. Some of these patterns may flag important sites on the DNA, such as the origin of replication. A complementary palindrome is one type of pattern. In DNA, the letter A is complementary to T, and G is complementary to C, and a complementary palindrome is a sequence of letters that reads in reverse as the complement of the forward sequence (e.g., GGGCATGCCC).

The origin of replication for two viruses from the same family as CMV, the herpes family, are marked by complementary palindromes. One of them, Herpes simplex, is marked by a long palindrome of 144 letters. The other, the Epstein–Barr virus, has several short palindromes and close repeats clustered at its origin of replication. For the CMV, the longest palindrome is 18 base pairs, and altogether it contains 296 palindromes between 10 and 18 base pairs long. Biologists conjectured that clusters of palindromes in CMV may serve the same role as the single long palindrome in Herpes simplex, or the cluster of palindromes and short repeats in the Epstein–Barr virus' DNA.

To find the origin of replication, DNA is cut into segments and each segment is tested to determine whether it can replicate. If it does not replicate, then the origin of replication must not be contained in the segment. This process can be very expensive and time consuming without leads on where to begin the search. A statistical investigation of the DNA to identify unusually dense clusters of palindromes can help narrow the search and potentially reduce the amount of testing needed to find the origin of replication. In practice, the CMV DNA was examined statistically for many different kinds of patterns. However, for this lab, the search will be restricted to looking for unusual clusters of complementary palindromes.

Data

Chee et al. ([CBB$^+$90]) published the DNA sequence of CMV in 1990. Leung et al. ([LBBK91]) implemented search algorithms in a computer program to screen the sequence for many types of patterns. Altogether, 296 palindromes were found that were at least 10 letters long. The longest ones found were 18 letters long. They occurred at locations 14719, 75812, 90763, and 173863 along the sequence.

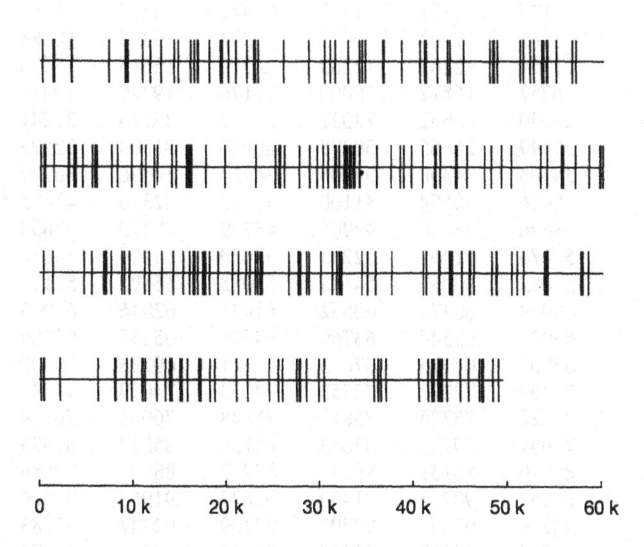

FIGURE 4.1. Diagram of the 296 palindrome locations for the CMV DNA (Chee et al. [CBB+90]).

Palindromes shorter than 10 letters were ignored, as they can occur too frequently by chance. For example, the palindromes of length two — AT, TA, GC and CG — are quite common.

Altogether, the CMV DNA is 229,354 letters long. Table 4.1 contains the locations of the palindromes in the DNA that are at least 10 letters long. Notice that the very first palindrome starts at position 177, the second is at position 1321, the third at position 1433, and the last at position 228953. Each palindrome is also located on a map of the DNA sequence in Figure 4.1. In this figure, a palindrome is denoted by a vertical line; clusters of palindromes appear as thick lines on the map.

Background

DNA

In 1944, Avery, MacLeod, and McCarty showed that DNA was the carrier of hereditary information. In 1953, Franklin, Watson, and Crick found that DNA has

TABLE 4.1. CMV palindrome locations for the 296 palindromes each at least ten base pairs long (Chee et al. [CBB+90]).

177	1321	1433	1477	3248	3255
3286	7263	9023	9084	9333	10884
11754	12863	14263	14719	16013	16425
16752	16812	18009	19176	19325	19415
20030	20832	22027	22739	22910	23241
25949	28665	30378	30990	31503	32923
34103	34398	34403	34723	36596	36707
38626	40554	41100	41222	42376	43475
43696	45188	47905	48279	48370	48699
51170	51461	52243	52629	53439	53678
54012	54037	54142	55075	56695	57123
60068	60374	60552	61441	62946	63003
63023	63549	63769	64502	65555	65789
65802	66015	67605	68221	69733	70800
71257	72220	72553	74053	74059	74541
75622	75775	75812	75878	76043	76124
77642	79724	83033	85130	85513	85529
85640	86131	86137	87717	88803	89586
90251	90763	91490	91637	91953	92526
92570	92643	92701	92709	92747	92783
92859	93110	93250	93511	93601	94174
95975	97488	98493	98908	99709	100864
102139	102268	102711	104363	104502	105534
107414	108123	109185	110224	113378	114141
115627	115794	115818	117097	118555	119665
119757	119977	120411	120432	121370	124714
125546	126815	127024	127046	127587	128801
129057	129537	131200	131734	133040	134221
135361	136051	136405	136578	136870	137380
137593	137695	138111	139080	140579	141201
141994	142416	142991	143252	143549	143555
143738	146667	147612	147767	147878	148533
148821	150056	151314	151806	152045	152222
152331	154471	155073	155918	157617	161041
161316	162682	162703	162715	163745	163995
164072	165071	165883	165891	165931	166372
168261	168710	168815	170345	170988	170989
171607	173863	174049	174132	174185	174260
177727	177956	178574	180125	180374	180435
182195	186172	186203	186210	187981	188025
188137	189281	189810	190918	190985	190996
191298	192527	193447	193902	194111	195032
195112	195117	195151	195221	195262	195835
196992	197022	197191	198195	198709	201023
201056	202198	204548	205503	206000	207527
207788	207898	208572	209876	210469	215802
216190	216292	216539	217076	220549	221527
221949	222159	222573	222819	223001	223544
224994	225812	226936	227238	227249	227316
228424	228953				

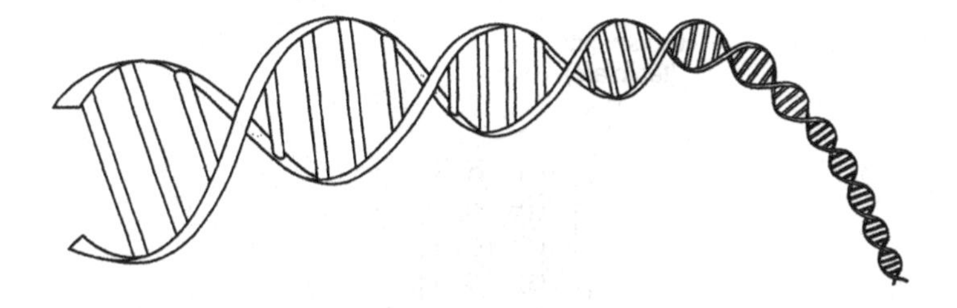

FIGURE 4.2. Paired ribbons of DNA forming the double helix structure.

a double helical structure (Figure 4.2) composed of two long chains of nucleotides. A single nucleotide has three parts: a sugar, a phosphate, and a base. All the sugars in DNA are deoxyribose — thus the name deoxyribose nucleic acid, or DNA. The bases come in four types: adenine, cytosine, guanine, and thymine, or A, C, G, T for short. As the bases vary from one nucleotide to another, they give the appearance of a long, coded message.

The two strands of nucleotides are connected at the bases, forming complementary pairs. That is, the bases on one strand are paired to the other strand: A to T, C to G, G to C, and T to A. Therefore, one strand "reads" as the complement of the other. This pairing forms a double helix out of the two strands of complementary base sequences.

The CMV DNA molecule contains 229,354 complementary pairs of letters or base pairs. In comparison, the DNA of the *Hemophilus influenzae* bacterium has approximately 1.8 million base pairs, and human DNA has more than 3 billion base pairs.

Viruses

Viruses are very simple structures with two main parts: a DNA molecule wrapped within a protein shell called a capsid. The DNA stores all the necessary information for controlling life processes, including its own replication. The DNA for viruses typically ranges up to several hundred thousand base pairs in length. According to *The Cartoon Guide to Genetics* ([GW91]), the replication of the bacteria *E. coli* happens as follows:

> In *E. coli* replication begins when a "snipping" enzyme cuts the DNA strand apart at a small region called the *origin*. In the neighborhood are plenty of free nucleotides, the building blocks for the new strands. When a free nucleotide meets its complementary base on the DNA, it sticks, while the "wrong" nucleotides bounce away. As the snipping enzyme opens the DNA further, more nucleotides are added, and a clipping enzyme puts them together.

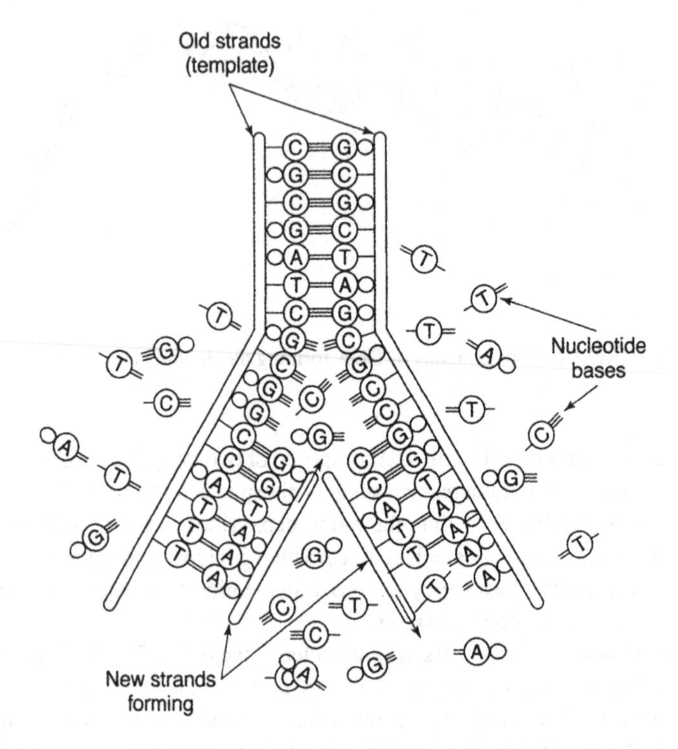

FIGURE 4.3. A sketch of DNA replication.

Figure 4.3 illustrates the replication process. The *origin* described in Gonick and Wheelis ([GW91]), where the snipping enzyme starts to cut apart the DNA strands, is the object of the search in this lab.

Human Cytomegalovirus

CMV is a member of the Herpes virus family. The family includes Herpes simplex I, chicken pox, and the Epstein–Barr virus. Some Herpes viruses infect 80% of the human population; others are rare but debilitating. As for CMV, its incidence varies geographically from 30% to 80%. Typically, 10 – 15% of children are infected with CMV before the age of 5. Then the rate of infection levels off until young adulthood, when it again increases ([Rya94, pp. 512–513]). While most CMV infections in childhood and adulthood have no symptoms, in young adults CMV may cause a mononucleosis-like syndrome.

Once infected, CMV typically lays dormant. It only becomes harmful when the virus enters a productive cycle in which it quickly replicates tens of thousands of copies. In this production cycle, it poses a major risk for people in immune-depressed states such as transplant patients who are undergoing drug therapy to suppress the immune system or people with Acquired Immune Deficiency Syndrome (AIDS). For these people, if the virus is reactivated, it can cause serious infections in internal organs. For example, CMV pneumonia is the leading cause of death among patients receiving bone marrow transplants. In AIDS patients, CMV infection often leads to neurological disorders, gastrointestinal disease and pneumonia. In addition, CMV is the most common infectious cause of mental retardation and congenital deafness in the United States.

Locating the origin of replication for CMV may help virologists find an effective vaccine against the virus. Research on the DNA for other Herpes viruses has uncovered the origin of replication for Herpes simplex I and Epstein–Barr. As stated earlier, the former is marked by one long palindrome of 144 base pairs, and the latter contains several short patterns including palindromes and close repeats. In earlier research, Weston ([Wes88]) found that a cluster of palindromes in the CMV DNA in the region 195,000 to 196,000 base pairs (see Figure 4.1) marked the site of another important function, called the enhancer.

Genomics

Recent advances in recombinant DNA and in machines that automate the identification of the bases have led to a burgeoning new science called genomics (Waterman [Wat89]). Genomics is the study of living things in terms of their full DNA sequences. Discoveries in genomics have been aided by advances in the fields of computer science, statistics, and other areas of mathematics, such as knot theory. For example, computer algorithms are being designed to search long sequences of DNA for patterns, information theory is facing the challenge of how to compress and manage these large databases, statistics and probability theory are being developed for matching sequences and identifying nonrandom structure in sequences, and knot theory has provided insights into the three-dimensional structure and molecular dynamics of DNA.

Investigations

How do we find clusters of palindromes? How do we determine whether a cluster is just a chance occurrence or a potential replication site?

- *Random scatter.*
  To begin, pursue the point of view that structure in the data is indicated by departures from a uniform scatter of palindromes across the DNA. Of course, a random uniform scatter does not mean that the palindromes will be equally spaced as milestones on a freeway. There will be some gaps on the DNA where no palindromes occur, and there will be some clumping together of palindromes. To look for structure, examine the locations of the palindromes, the spacings between palindromes, and the counts of palindromes in nonoverlapping regions of the DNA. One starting place might be to see first how random scatter looks by using a computer to simulate it. A computer can simulate 296 palindrome

sites chosen at random along a DNA sequence of 229,354 bases using a pseudo-random number generator. When this is done several times, by making several sets of simulated palindrome locations, then the real data can be compared to the simulated data.

- *Locations and spacings.* Use graphical methods to examine the spacings between consecutive palindromes and sums of consecutive pairs, triplets, etc., spacings. Compare what you find for the CMV DNA to what you would expect to see in a random scatter. Also, consider graphical techniques for examining the locations of the palindromes.
- *Counts.* Use graphical displays and more formal statistical tests to investigate the counts of palindromes in various regions of the DNA. Split the DNA into nonoverlapping regions of equal length to compare the number of palindromes in an interval to the number that you would expect from uniform random scatter. The counts for shorter regions will be more variable than those for longer regions. Also consider classifying the regions according to their number of counts.
- *The biggest cluster.* Does the interval with the greatest number of palindromes indicate a potential origin of replication? Be careful in making your intervals, for any small, but significant, deviation from random scatter, such as a tight cluster of a few palindromes, could easily go undetected if the regions examined are too large. Also, if the regions are too small, a cluster of palindromes may be split between adjacent intervals and not appear as a high-count interval. These issues are discussed in more detail in the Extensions section of this lab.

How would you advise a biologist who is about to start experimentally searching for the origin of replication? Write your recommendations in the form of a memo to the head biologist of a research team of which you are a member.

Theory

The Homogeneous Poisson Process

The homogeneous Poisson process is a model for random phenomena such as the arrival times of telephone calls at an exchange, the decay times of radioactive particles, and the positions of stars in parts of the sky. This model was first developed in the time domain. For the phone call example, it seems reasonable to think of phone calls coming from subscribers who are acting independently of one another. It is unlikely that any one subscriber will make a call in a particular time interval, but there are many subscribers, and it is likely that a few will make calls in a particular interval of time.

The process arises naturally from the notion of points haphazardly distributed on a line with no obvious regularity. The characteristic features of the process are that:

- The underlying rate (λ) at which points, called hits, occur doesn't change with location (homogeneity).
- The number of points falling in separate regions are independent.
- No two points can land in exactly the same place.

These three properties are enough to derive the formal probability model for the homogeneous Poisson process.

The Poisson process is a good reference model for making comparisons because it is a natural model for uniform random scatter. The strand of DNA can be thought of as a line, and the location of a palindrome can be thought of as a point on the line. The uniform random scatter model says: palindromes are scattered randomly and uniformly across the DNA; the number of palindromes in any small piece of DNA is independent of the number of palindromes in another, nonoverlapping piece; and the chance that one tiny piece of DNA has a palindrome in it is the same for all tiny pieces of the DNA.

There are many properties of the homogeneous Poisson process that can be used to check how well this reference model fits the DNA data. A few of these properties are outlined here.

Counts and the Poisson Distribution

One way to summarize a random scatter is to count the number of points in different regions. The probability model for the Poisson process gives the chance that there are k points in a unit interval as

$$\frac{\lambda^k}{k!}e^{-\lambda}, \qquad \text{for } k = 0, 1, \ldots .$$

This probability distribution is called the Poisson distribution (as it is derived from the Poisson process). The parameter λ is the rate of hits per unit area. It is also the expected value of the distribution.

The Rate λ

Usually the rate λ is unknown. When this is the case, the empirical average number of hits per unit interval can be used to estimate λ. It is a reasonable estimate because λ is the expected number of hits per unit interval. This technique for estimating λ by substituting the sample average for the expected value is called the *method of moments*. Another technique, more widely used to estimate an unknown parameter such as λ, is the method of *maximum likelihood*. For the Poisson distribution, both methods yield the same estimate for λ — the sample average. These two methods of parameter estimation are discussed in greater detail later in this chapter.

Goodness-of-Fit for Probability Distributions

We often hypothesize that observations are realizations of independent random variables from a specified distribution, such as the Poisson. We do not believe

TABLE 4.2. Palindrome counts in the first 57 nonoverlapping intervals of 4000 base pairs of CMV DNA (Chee et al. [CBB+90]).

Palindrome counts									
7	1	5	3	8	6	1	4	5	3
6	2	5	8	2	9	6	4	9	4
1	7	7	14	4	4	4	3	5	5
3	6	5	3	9	9	4	5	6	1
7	6	7	5	3	4	4	8	11	5
3	6	3	1	4	8	6			

that the data are exactly generated from such a distribution but that for practical purposes the probability distribution does well in describing the randomness in the outcomes measured. If the Poisson distribution fits the data reasonably well, then it could be useful in searching for small deviations, such as unusual clusters of palindromes.

In our case, we would want to use the homogeneous Poisson process as a reference model against which to seek an excess of palindromes. This only makes sense if the model more or less fits. If it doesn't fit well— for example if there is a lot of heterogeneity in the locations of the palindrome — then we would have to try another approach. A technique for assessing how well the reference model fits is to apply the chi-square goodness-of-fit test.

For example, divide the CMV DNA into 57 nonoverlapping segments, each of length 4000 bases, and tally the number of complementary palindromes in each segment (Table 4.2). There is nothing special about the number 4000; it was chosen to yield a reasonable number of observations (57). Notice that 7 palindromes were found in the first segment of the DNA, 1 in the second, 5 in the third, etc. The distribution of these counts appears in Table 4.3. It shows that 7 of the 57 DNA segments have 0, 1, or 2 palindromes in them, 8 segments have 3 palindromes, 10 segments have 4 palindromes each, ..., and 6 segments have at least 9 palindromes. However, these segments cover only the first 228,000 base pairs, excluding the last 1354, which include 2 palindromes. Hence we are now considering only a total of 294 palindromes.

The last column in Table 4.3 gives the expected number of segments containing the specified number of palindromes as computed from the Poisson distribution. That is, the expected number of intervals with 0, 1, or 2 palindromes is 57 $\times$ the probability of 0, 1, or 2 hits in an interval:

$$57\ \mathbb{P}(0,\ 1 \text{ or } 2 \text{ palindromes in an interval of length } 4000) = 57e^{-\lambda}[1 + \lambda + \lambda^2/2].$$

The rate λ is not known. There are 294 palindromes in the 57 intervals of length 4000, so the sample rate is 5.16 per 4000 base pairs. Plugging this estimate into the calculation above yields 0.112 for the chance that an interval of 4000 base pairs has 0, 1, or 2 palindromes. Then the approximate expected number of segments containing 0, 1, or 2 palindromes is 57×0.112, or 6.4. This is approximate because

TABLE 4.3. Distribution of palindrome counts for intervals of 4000 base pairs for the 294 palindromes in the first 228,000 base pairs of CMV DNA (Chee et al. [CBB+90]).

Palindrome	Number of intervals	
count	Observed	Expected
0–2	7	6.4
3	8	7.5
4	10	9.7
5	9	10.0
6	8	8.6
7	5	6.3
8	4	4.1
9+	6	4.5
Total	57	57

we are using an estimated value of λ. The remaining expectations are calculated in a similar fashion.

To compare the observed data to the expected, we compute the following *test statistic*:

$$\frac{(7-6.4)^2}{6.4}+\frac{(8-7.5)^2}{7.5}+\frac{(10-9.7)^2}{9.7}+\frac{(9-10.0)^2}{10.0}+$$
$$\frac{(8-8.6)^2}{8.6}+\frac{(5-6.3)^2}{6.3}+\frac{(4-4.1)^2}{4.1}+\frac{(6-4.5)^2}{4.5}=1.0\,.$$

If the random scatter model is true, then the test statistic computed here has an approximate *chi-square distribution* (also written χ^2) with six *degrees of freedom.* The size of the actual test statistic is a measure of the fit of the distribution. Large values of this statistic indicate that the observed data were quite different from what was expected. We use the χ^2 distribution to compute the chance of observing a test statistic at least as large as ours under the random scatter model:

$$\mathbb{P}(\chi_6^2 \text{ random variable} \geq 1.0) \;=\; 0.98\,.$$

From this computed probability, we see that deviations as large as ours (or larger) are very likely. It appears that the Poisson is a reasonable initial model. In this case, the observed values and expected values are so close that the fit almost seems too good. See the Exercises for a discussion of this point.

The *hypothesis test* performed here is called a *chi-square goodness-of-fit test.* In general, to construct a hypothesis test for a discrete distribution a distribution table is constructed from the data, where m represents the number of categories or values for the response and N_j stands for the number of observations that appear in category j, $j=1,\ldots,m$. These counts are then compared to what would be expected; namely,

$$\mu_j = np_j,\ \text{ where }\ p_j = \mathbb{P}(\text{an observation is in category j}).$$

Note that $\sum p_j = 1$ so $\sum \mu_j = n$. Sometimes a parameter from the distribution needs to be estimated in order to compute the above probabilities. In this case, the data are used to estimate the unknown parameter(s). The measure of discrepancy between the sample counts and the expected counts is

$$\sum_{j=1}^{m} \frac{(j\text{th Sample count} - j\text{th Expected count})^2}{j\text{th Expected count}} = \sum_{j=1}^{m} \frac{(N_j - \mu_j)^2}{\mu_j}.$$

When the statistic computed in this hypothesis test (called the *test statistic*) is large, it indicates a lack of fit of the distribution. Assuming that the data are generated from the hypothesized distribution, we can compute the chance that the test statistic would be as large, or larger, than that observed. This chance is called the *observed significance level*, or *p-value.*

To compute the p-value, we use the χ^2 distribution. If the probability model is correct, then the test statistic has an approximate chi-squared distribution with $m - k - 1$ degrees of freedom, where m is the number of categories and k is the number of parameters estimated to obtain the expected counts. The χ^2_{m-k-1} probability distribution is a continuous distribution on the positive real line. Its probability density function has a long right tail (i.e., it is skewed to the right) for small degrees of freedom. As the degrees of freedom increase, the density becomes more symmetric and more normal looking. See Appendix C for tail probabilities of the χ^2.

A rule of thumb for the test statistic to have an approximate χ^2 distribution is that the expected counts for each bin, or category, should be at least five. This means that some bins may need to be combined before performing the test. In the example above, the expected counts for the last two categories are not quite five. However, because the probability distribution is unimodal, it is okay to have one or two extreme bins with expected counts less than five. Also, for testing the goodness-of-fit of the uniform distribution, this rule of thumb can be ignored. (You may want to check these rules of thumb with a simulation study.)

If the χ^2 test gives a small p-value, then there is reason to doubt the fit of the distribution. When this is the case, a residual plot can help determine where the lack of fit occurs. For each category, plot the *standardized residual*

$$\frac{\text{Sample count} - \text{Expected count}}{\sqrt{\text{Expected count}}} = \frac{N_j - \mu_j}{\sqrt{\mu_j}}.$$

The denominator transforms the residuals (i.e., the differences between the observed and expected counts) in order to give them approximately equal variances. The square root allows for meaningful comparisons across categories. Note that the sum of the residuals is zero, but the standardized residuals do not necessarily sum to zero. Values of the standardized residuals larger than three (in absolute value) indicate a lack of fit. Figure 4.4 is such a plot for the entries in Table 4.3.

Locations and the Uniform Distribution

Under the Poisson process model for random scatter, if the total number of hits in an interval is known, then the positions of the hits are uniformly scattered across

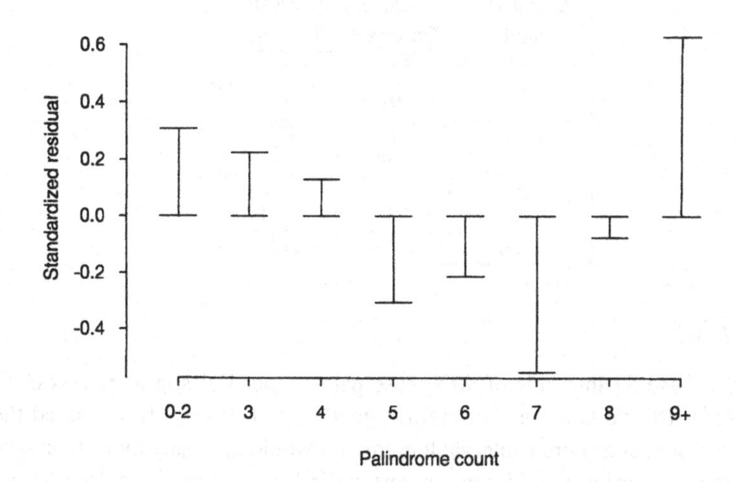

FIGURE 4.4. Standardized residual plot for the goodness-of-fit test of the Poisson distribution to the distribution of palindrome counts in 4000 base pair intervals of CMV DNA (Chee et al. [CBB+90]).

the interval. In other words, the Poisson process on a region can be viewed as a process that first generates a random number, which is the number of hits, and then generates locations for the hits according to the uniform distribution.

For the CMV DNA, there are a total of 296 palindromes on the CMV DNA. Under the uniform random scatter model, the positions of these palindromes are like 296 independent observations from a uniform distribution. The locations of the palindromes can be compared to the expected locations from the uniform distribution. Also, if the DNA is split into 10 equal subintervals, then according to the uniform distribution, we would expect each interval to contain 1/10 of the palindromes. Table 4.4 contains these interval counts. A χ^2 goodness-of-fit test can compare these observed and expected counts. We leave this computation to the Exercises.

TABLE 4.4. Observed and expected palindrome counts for ten consecutive segments of CMV DNA, each 22,935 base pairs in length (Chee et al. [CBB+90]).

Segment	1	2	3	4	5	
Observed	29	21	32	30	32	
Expected	29.6	29.6	29.6	29.6	29.6	
Segment	6	7	8	9	10	Total
Observed	31	28	32	34	27	296
Expected	29.6	29.6	29.6	29.6	29.6	296

TABLE 4.5. Distribution of palindrome counts for intervals of 400 base pairs for the 296 palindromes in CMV DNA (Chee et al. [CBB+90]).

Palindrome count	Number of intervals Observed	Expected
0	355	342
1	167	177
2	31	46
3	16	8
4	3	1
≥ 5	1	0.1
Total	573	573

Which Test?

Why did we use 57 intervals of 4000 base pairs in our Poisson goodness-of-fit test but only 10 intervals for the uniform goodness-of-fit test? If we based the Poisson test on much shorter interval lengths, we would get many more intervals, but a larger proportion would contain zero palindromes. For example, with an interval length of 400 base pairs (Table 4.5), 522 of the 573 intervals have 0 or 1 palindromes. The distribution of counts is now highly skewed, and the test is uninformative because a large proportion of the counts are in two categories (0 or 1 palindromes).

Alternatively, why not use larger intervals for the Poisson goodness-of-fit test? Suppose we divide the genome into 10 large, equal-sized intervals, as in Table 4.4. If we do this, we have hardly enough data to compare observed and expected numbers of intervals for a particular palindrome count. Our sample size here is 10, but the 10 intervals have 8 different palindrome counts. Previously, we had 57 intervals, which was enough to see the same palindrome counts for an interval many times (e.g., 10 of the intervals had a count of 4 palindromes).

Now with only 10 intervals, we change our approach from examining the goodness-of-fit of the Poisson distribution to the goodness-of-fit of the uniform distribution. In Table 4.3, we compared the *observed numbers of intervals* with a given count (or range of counts) of palindromes to the Poisson distribution. In Table 4.4, we are directly comparing the *observed counts of palindromes* in each of our 10 intervals with the uniform distribution, so we can use the many properties of the Poisson process to test the data in different ways. Be careful of the distinction between the Poisson process and the Poisson distribution; the Poisson distribution quantifies the frequencies of hits in fixed intervals of a Poisson process.

Spacings and the Exponential and Gamma Distributions

Another property that can be derived from the Poisson process is that the distance between successive hits follows an exponential distribution. That is,

$$\mathbb{P}(\text{the distance between the first and second hits} > t)$$

$$= \mathbb{P}(\text{No hits in an interval of length } t)$$
$$= e^{-\lambda t},$$

which implies that the distance between successive hits follows the exponential distribution with parameter λ. Similarly, it can be shown that the distance between hits that are two apart follows a gamma distribution with parameters 2 and λ. The exponential distribution with parameter λ is a special case of the gamma distribution for parameters 1 and λ. The χ_k^2 distribution is also a special case of the gamma distribution for parameters $k/2$ and $1/2$. The χ_k^2 distribution can be used in place of the gamma$(k/2, \lambda)$ distribution in gamma-quantile plots because λ is a scale parameter and only affects the slope of the plot.

Maximum Number of Hits

Under the Poisson process model, the numbers of hits in a set of nonoverlapping intervals of the same length are independent observations from a Poisson distribution. This implies that the greatest number of hits in a collection of intervals behaves as the maximum of independent Poisson random variables. If we suppose there are m such intervals, then

$$\begin{aligned}
&\mathbb{P}(\text{maximum count over } m \text{ intervals} \geq k) \\
&\qquad = 1 - \mathbb{P}(\text{maximum count} < k) \\
&\qquad = 1 - \mathbb{P}(\text{all interval counts} < k) \\
&\qquad = 1 - [\mathbb{P}(\text{first interval count} < k)]^m \\
&\qquad = 1 - \left[\frac{\lambda^0}{0!}e^{-\lambda} + \ldots + \frac{\lambda^{k-1}}{k-1!}e^{-\lambda}\right]^m .
\end{aligned}$$

From this expression, with an estimate of λ, we can find the approximate chance that the greatest number of hits is at least k. If this chance is unusually small, then it provides evidence for a cluster that is larger than that expected from the Poisson process. In other words, we can use the maximum palindrome count as a test statistic, and the computation above gives us the p-value for the test statistic.

Parameter Estimation

Suppose we have an independent sample $x_1, \ldots, x_n$ from a Poisson(λ) distribution where λ is unknown. The *method of moments* is one technique for estimating the parameter λ. It proceeds as follows:

1. Find $\mathbb{E}(X)$, where X has a Poisson(λ) distribution.
2. Express λ in terms of $\mathbb{E}(X)$.
3. Replace $\mathbb{E}(X)$ by $\bar{x}$ to produce an estimate of λ, called $\hat{\lambda}$.

For the Poisson distribution, the method of moments procedure is very simple because $\mathbb{E}(X) = \lambda$, so the estimate $\hat{\lambda}$ is $\bar{x}$.

Sometimes higher moments need to be taken; for example, $\mathbb{E}(X^2)$ is computed and $\sum x_i^2/n$ replaces $\mathbb{E}(X^2)$ to estimate an unknown parameter. See the Exercises for an example of such a problem.

The method of *maximum likelihood* is a more common procedure for parameter estimation because the resulting estimators typically have good statistical properties. The method searches among all Poisson distributions to find the one that places the highest chance on the observed data. For a Poisson(λ) distribution, the chance of observing $x_1, \ldots x_n$ is the following:

$$\frac{\lambda^{x_1}}{x_1!}e^{-\lambda} \times \cdots \times \frac{\lambda^{x_n}}{x_n!}e^{-\lambda} = \frac{\lambda^{\sum x_i}}{\Pi x_i!}e^{-n\lambda} = L(\lambda).$$

For the given data $x_1, \ldots, x_n$, this is a function of λ, called the *likelihood function*, and denoted by $L(\cdot)$. The maximum likelihood technique estimates the unknown parameter by the λ-value that maximizes the likelihood function L. Since the log function is monotonically increasing, the log likelihood function, which is typically denoted by l, is maximized at the same λ as the likelihood function. In our example, to find the maximum likelihood estimator, we differentiate the log likelihood as follows:

$$\begin{aligned} \frac{\partial l}{\partial \lambda} &= \frac{\partial}{\partial \lambda}\left[\sum x_i \log(\lambda) - n\lambda - \sum \log(x_i!)\right] \\ &= \sum x_i/\lambda - n, \\ \frac{\partial^2 l}{\partial \lambda^2} &= -\sum x_i/\lambda^2. \end{aligned}$$

By setting the first derivative to 0 and solving for λ, we find that the log likelihood is maximized at $\hat{\lambda} = \bar{x}$. The second derivative shows that indeed a maximum has been found because the counts x_i are always nonnegative.

Maximum likelihood for continuous distributions is similar. For example, suppose $x_1, \ldots, x_n$ form a sample from an exponential distribution with unknown parameter θ. Now the likelihood function given the observed data $x_1, \ldots, x_n$ is

$$L(\theta) = f_\theta(x_1, \ldots, x_n),$$

where f_θ is the joint density function for the n observations. The observations were taken independently, so the likelihood reduces to

$$L(\theta) = \theta^n e^{-\theta \sum x_i},$$

and the log likelihood is

$$l(\theta) = n \log(\theta) - \theta \sum x_i.$$

We leave it as an exercise to show that the maximum likelihood estimate $\hat{\theta}$ is $1/\bar{x}$.

Properties of Parameter Estimates

To compare and evaluate parameter estimates, statisticians use the mean square error:

$$\begin{aligned}\text{MSE}(\hat{\lambda}) &= \mathbb{E}(\hat{\lambda} - \lambda)^2 \\ &= \text{Var}(\hat{\lambda}) + [\mathbb{E}(\hat{\lambda}) - \lambda]^2.\end{aligned}$$

Note that this is the sum of the variance plus the squared bias of the estimator. Many of the estimators we use are unbiased, though sometimes an estimator with a small bias will have a small MSE (see the Exercises).

For independent identically distributed (i.i.d.) samples from a distribution f_λ, maximum likelihood estimates of λ often have good properties. Under certain regularity conditions on the sampling distribution, as the sample size increases, the maximum likelihood estimate (MLE) approaches λ, and the MLE has an approximate normal distribution with variance

$$\frac{1}{nI(\lambda)},$$

where $I(\lambda)$, called the *information*, is

$$\begin{aligned}I(\lambda) &= \mathbb{E}\left(\frac{\partial \log f_\lambda(X)}{\partial \lambda}\right)^2 \\ &= -\mathbb{E}\left(\frac{\partial^2 \log f_\lambda(X)}{\partial \lambda^2}\right).\end{aligned}$$

That is, $\sqrt{nI(\lambda)}(\hat{\lambda}-\lambda)$ has an approximate standard normal distribution for n large. The normal distribution can be used to make approximate confidence intervals for the parameter λ; for example,

$$\hat{\lambda} \pm 1.96/\sqrt{nI(\lambda)}$$

is an approximate 95% confidence interval for λ. The asymptotic variance for the MLE is also a lower bound for the variance of any unbiased parameter estimate.

Hypothesis Tests

The χ^2 goodness-of-fit test and the test for the maximum number of palindromes in an interval are examples of hypothesis tests. We provide in this section another example of a hypothesis test, one for parameter values. We use it to introduce the statistical terms in testing.

An Example

In Hennepin County, a simple random sample of 119 households found an average radon level of 4.6 pCi/l with an SD of 3.4 pCi/l. In neighboring Ramsey County, a simple random sample of 42 households had an average radon level of 4.5 pCi/l with an SD of 4.9 pCi/l (Chapter 3). It is claimed that the households in these two counties have the same average level of radon and that the difference observed in the sample averages is due to chance variation in the sampling procedure.

To investigate this claim, we conduct a hypothesis test. We begin with the probability model. Let $X_1, \ldots, X_{119}$ denote the radon levels for the sampled households from Hennepin County, and let $Y_1, \ldots, Y_{42}$ denote those from Ramsey County. Also, let μ_H and μ_R denote the average radon levels for all households in the respective counties, and σ_H^2, σ_R^2 denote the population variances. The *null hypothesis* is that the two counties have the same population means; namely,

$$H_0 : \mu_H = \mu_R,$$

and the *alternative hypothesis* is

$$H_A : \mu_H \neq \mu_R.$$

In making a hypothesis test, we assume that H_0 is true and find out how likely our data are under this model. In this example, $\bar{X}$ and $\bar{Y}$ are independent, their sampling distributions are approximately normal, and, under H_0, the difference $\bar{X} - \bar{Y}$ has mean 0. The *test statistic*,

$$Z = \frac{\bar{X} - \bar{Y}}{\sqrt{\sigma_H^2/119 + \sigma_R^2/42}},$$

has a *null distribution* that is approximately standard normal. We call this test statistic the *z statistic* because it is based on a normal approximation. (In Chapter 3, we saw that it might be appropriate to take logarithms of X_i and Y_i and then proceed with computing the test statistic because the data were thought to be approximately lognormal in distribution.)

Using estimates for σ_H and σ_R, the survey results produced an *observed test statistic* of 0.12, and the chance that $|Z|$ could be as large or larger than 0.12 is 0.90. The probability 0.90 is called the *p-value*. Notice that the p-value is two-sided — i.e., $0.90 = \mathbb{P}(|Z| \geq 0.12)$ — because the alternative hypothesis is that the two population averages are unequal. If the alternative was $\mu_H > \mu_R$ then the test would be one-sided and the p-value would be 0.45. Either way, we conclude that we have observed a typical value for the difference between sample averages, and the data support the null hypothesis.

If the p-value were very small, then we would conclude that the data provide evidence against the null hypothesis, and we would reject the null hypothesis in favor of the alternative. The typical levels at which the null hypothesis is rejected are 0.05 and 0.01. These cutoffs are called *significance levels* or α-levels. A test statistic that yields a p-value less than 0.05 is said to be *statistically significant*, and one that is less than 0.01 is *highly statistically significant*.

The p-value is *not* the chance that the null hypothesis is true; the hypothesis is either true or not. When we reject the null hypothesis, we do not know if we have been unlucky with our sampling and observed a rare event or if we are making the correct decision. Incorrectly rejecting the null hypothesis is a Type I error. A Type II error occurs when the null hypothesis is not rejected and it is false. We define α to be the chance of a Type I error and β to be the chance of a Type II error. Typically α is set in advance, and $1-\beta$, the *power* of the hypothesis test, is computed for various values of the alternative hypothesis. Power is the chance that we correctly reject the null hypothesis, so we want the power to be high for our test. For example, the power of the test in this example for $\alpha = 0.05$ and $\mu_H - \mu_R = 0.5$ is

$$\begin{aligned}\mathbb{P}\left(\frac{|\bar{X}-\bar{Y}|}{0.81} \geq 1.96\right) &= \mathbb{P}\left(|\bar{X}-\bar{Y}| \geq 1.96 \times 0.81\right)\\ &= \mathbb{P}\left(\frac{\bar{X}-\bar{Y}-0.5}{0.81} \geq 1.34\right)\\ &\quad + \mathbb{P}\left(\frac{\bar{X}-\bar{Y}-0.5}{0.81} \leq -2.58\right)\\ &= 0.09\,.\end{aligned}$$

That is, the chance that we would reject the null hypothesis of no difference, given an actual difference of 0.5, is about 1 in 10. This test is not very powerful in detecting a difference of 0.5 in the population means. A larger sample size in each county would have given a more powerful test.

Exercises

1. For the 91 students in the video survey (Chapter 2), do their expected grades for the course fit the "curve" of 20% As, 30% Bs, 40% Cs, and 10% Ds and Fs? Provide a χ^2 goodness-of-fit test of the data to the "curve."

Grade	A	B	C	D	F	Total
Count	31	52	8	0	0	91

2. For the data in Table 4.4, provide a goodness-of-fit test for the uniform distribution.
3. The negative binomial distribution is often used as an alternative to the Poisson distribution for random counts. Unlike the Poisson distribution, the variance of the negative binomial is larger than the mean. Consider the following parameterization for the negative binomial, $j = 0, 1, \ldots$,

$$\mathbb{P}(j) = \left(1 + \frac{m}{k}\right)^{-k} \frac{\Gamma(k+j)}{j!\,\Gamma(k)} \left(\frac{m}{m+k}\right)^{j}.$$

a. Establish the recursive relation:

$$\mathbb{P}(0) = \left(1 + \frac{m}{k}\right)^{-k},$$

$$\mathbb{P}(j) = \frac{(k+j-1)m}{j(m+k)}\mathbb{P}(j-1)\,.$$

b. The mean of the negative binomial is m, and the variance is $m + (m^2/k)$. Use these moments to find method of moment estimates of m and k.

c. Use the method of moment estimates of m and k and the recursive relationship for the probabilities to make a χ^2 goodness-of-fit test of the negative binomial to the palindrome counts in Table 4.2.

4. Unusually large values of the test statistic in a χ^2 goodness-of-fit test indicate a lack of fit to the proposed distribution. However, unusually small values may also indicate problems. Suspicion may be cast on the data when they are too close to the expected counts. What is the smallest value obtainable for a χ^2 test statistic? Why would very small values indicate a potential problem? For the χ^2 test for Table 4.3, the p-value is 0.98. This means that the chance of a test statistic smaller than the one observed is only $1/50$. How does the p-value change when intervals of length 1000 base pairs rather than 4000 base pairs are used? You will need to use the computer to summarize the palindrome counts.

5. Perform a simulation study on the sensitivity of the χ^2 test for the uniform distribution to expected cell counts below 5. Simulate the distribution of the test statistic for 40, 50, 60, and 70 observations from a uniform distribution using 8, 10, and 12 equal-length bins.

6. Let $\bar{x}$ be the average radon level for a simple random sample of n households in a county in Wisconsin that neighbors Washington County, Minnesota. Consider a 95% confidence interval for μ, the average radon level for the county. Explain why if the confidence interval were to contain the value 5, then the hypothesis test for $\mu = 5$ would not be rejected at the $\alpha = 0.05$ level. Also explain why if the confidence interval does not contain 5, then the hypothesis would be rejected at the $\alpha = 0.05$ level. You may use 4 pCi/L for the SD, the sample SD from neighboring Washington County in Minnesota (Chapter 3).

7. For the two-sided hypothesis test in Exercise 6 that $\mu = 5$ at the $\alpha = 0.05$ level, compute the power of the test against the alternatives that $\mu = 4, 4.5, 5, 5.5$, and 6. Take n to be 100. Use these computations to sketch the power curve. How does the power curve change when the alternative hypothesis is now one-sided: $\mu > 5$?

8. Suppose $X_1, \ldots, X_n$ are independent binomial(m, p) distributed random variables. Find the method of moments estimate of p and show that it is the same as the maximum likelihood estimate of p.

9. Suppose we observe $(x_1, \ldots, x_m)$ from a multinomial($100, p_1, \ldots, p_m$) distribution. Use LaGrange multipliers to find the maximum likelihood estimate of the probabilities, $p_1, \ldots, p_m$.

10. Suppose $X_1, \ldots, X_n$ are independent exponential(θ) distributed random variables.

a. Find the method of moments estimate of θ.

b. Show that it is the same as the maximum likelihood estimate.

c. Compute the information for the exponential(θ) distribution.

11. Suppose $X_1, \ldots, X_n$ are independent uniform(0,θ) distributed random variables.

a. Find the method of moments estimate of θ.
b. Find the maximum likelihood estimate of θ.
c. Find the MSE of the maximum likelihood estimate of θ for the uniform(0,θ) distribution. Compare it to the MSE of the method of moment estimate of θ.

12. Find the maximum likelihood estimate of α for n independent observations from the distribution with density

$$\alpha(1-x)^{\alpha-1}, \qquad \text{for } 0 \le x \le 1.$$

13. Find the maximum likelihood estimate of $\theta > 0$ for n independent observations from the Rayleigh distribution with density

$$\theta^{-2} x \exp^{-x^2/2\theta^2}, \qquad 0 \le x < \infty.$$

14. Compute the information for the Poisson(λ) distribution.

15. Find the maximum likelihood estimate for $\theta > 0$ given n independent observations from the Pareto distribution,

$$f(x) = \theta x_o^{\theta} x^{-\theta-1}, \qquad x \ge x_o,$$

where x_o is known. Also find the asymptotic variance of the maximum likelihood estimate.

16. For a homogeneous Poisson process in time, with rate λ per hour, show that the total number of hits in two nonoverlapping hours has a Poisson distribution with parameter 2λ. *Hint:*

$$\begin{aligned}&\mathbb{P}(n \text{ hits in two hours})\\ &= \sum_{k=0}^{n} \mathbb{P}(k \text{ hits in the first hour}, n-k \text{ hits in the second hour}).\end{aligned}$$

You may also need to use the formula

$$\sum_{k=0}^{n} \frac{n!}{k!(n-k)!} = 2^n.$$

17. Suppose we have two independent Poisson-distributed random variables, X and Y, one with rate λ and one with rate μ. Show that the distribution of $X+Y$ is Poisson with rate $\lambda+\mu$. *Hint:* Proceed as in Exercise 16 and use the formula

$$\sum_{k=0}^{n} \frac{n!}{k!(n-k)!} x^k y^{n-k} = (x+y)^n.$$

18. Suppose we have two independent Poisson random variables, X and Y, with rates λ and μ, respectively. Use the result from Exercise 17 to show that the distribution of X, given $X+Y=n$, is binomial(n, p) with $p = \lambda/(\lambda+\mu)$.

Extensions

One of the later parts of the Theory section examined the interval with the greatest number of palindromes. There it was noted that a tight cluster of palindromes can be split between two intervals. Then the corresponding interval counts are not very high, and the cluster remains hidden.

To circumvent this problem, we could slide a window 500 base pairs long along the DNA sequence one letter at a time and find the interval with the greatest number of palindromes out of all possible intervals 500 base pairs long.

To help us in this search, we can make a sliding bin plot (Figure 4.5). We calculate palindrome counts for overlapping intervals. These counts are plotted at the interval midpoints with connecting lines to illustrate areas of high density so, for example, if we choose an interval length of 1000 base pairs and an overlap of 500 base pairs, then (from Table 4.1) the intervals and their counts would be as in Table 4.6, and the sliding bin plot appears in Figure 4.5. To keep the figure simple, we use an overlap of 500 base pairs. An overlap of 1 base pair would find the interval of length 1000 base pairs with the greatest number of palindromes.

Try searching short overlapping regions, say of length 250, 500, or 1000, for clusters of palindromes.

Once potential cluster regions are found, you must decide whether the cluster is typical of what you may find among 296 palindromes scattered randomly across a DNA strand of 229,354 base pairs. That is, say you find an interval of 500 base pairs with 6 palindromes in it. Table 4.7 gives the probability that a random scatter

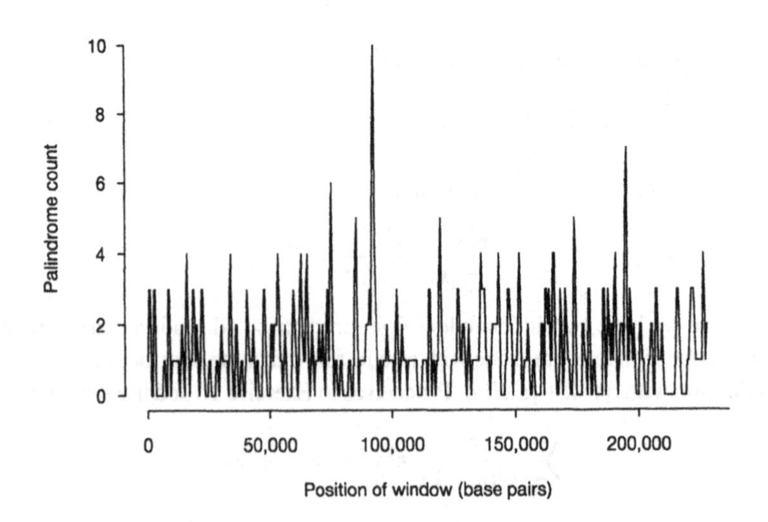

FIGURE 4.5. Sliding bin plot for the 296 palindromes in the CMV DNA, with intervals of 1000 base pairs and overlap of 500 base pairs (Chee et al. [CBB+90]).

TABLE 4.6. Sample computations of the bin counts for a sliding bin plot of the CMV DNA (Chee et al. [CBB+90]).

Start point	End point	Palindrome count
1	1000	1
501	1500	3
1001	2000	3
1501	2500	0
2001	3000	0
2501	3500	3
	etc.	

TABLE 4.7. Probabilities of the maximum palindrome count for a sliding bin plot with intervals of length 250, 500, and 1000 base pairs, for the CMV DNA (calculated from Leung et al.[LSY93] and Naus [Nau65]).

Interval	Maximum palindrome count							
length	4	5	6	7	8	9	10	11
250	0.67	0.08	0.006	0.0003	—	—	—	—
500	—	0.61	0.12	0.013	0.001	0.0001	—	—
1000	—	—	—	0.34	0.07	0.012	0.002	0.0002

of 296 palindromes across a DNA sequence of 229,354 base pairs would contain an interval of 500 base pairs with at least 6 palindromes in it. From the table, we see that this chance is about 12%. Probabilities for intervals of length 250, 500, and 1000 base pairs appear in the table. Note that this probability is different from the chance that the maximum of 458 nonoverlapping intervals of 500 base pairs contains 6 or more palindromes, which is only .03. These probabilities have been computed using the algorithms found in Leung et al. ([LSY93]) and Naus ([Nau65]).

Notes

Ming-Ying Leung introduced us to these data and the associated problem. Her solution appears in Leung et al. ([LSY93]) and is the basis for the material in the Extensions section. The probabilities used in this section were determined by Naus ([Nau65]). Hwang ([Hwa77]) provided an algorithm for approximating this chance. Table 4.7 lists these probabilities computed for the special case of 296 palindromes in a sequence of 229,354 base pairs. They are calculated from an example in Leung et al. ([LSY93]). Masse et al. ([MKSM91]) provides another analysis of the CMV DNA.

We recommend Gonick and Wheelis ([GW91]) as a source for the background material on DNA. The material on CMV comes from Anders and Punterieri

([AP90]), Ryan ([Rya94]), and Weston ([Wes88]). Pitman ([Pit93]) and Kingman ([Kin93]) were sources for the development of the Poisson process.

References

[AP90] D.G. Anders and S.M. Punterieri. Multicomponent origin of cytomegalovirus lytic-phase DNA replication. *J. Virol.*, **65**:931–937, 1990.

[CBB+90] M.S. Chee, A.T. Bankier, S. Beck, R. Bohni, C.M. Brown, R. Cerny, T. Hosnell, C.A. Hutchinson III, T. Kourzarides, J.A. Martignetti, E. Preddie, S.C. Satchwell, P. Tomlinson, K.M. Weston, and B.G. Barell. Analysis of the protein coding content of human cytomegalovirus strain ad169. *Curr. Top. Microbiol. Immunol.*, **154**:126–169, 1990.

[GW91] L. Gonick and M. Wheelis. *The Cartoon Guide To Genetics*. Harper Perennial, New York, 1991.

[Hwa77] F.K. Hwang. A generalization of the Karlin-McGregor theorem on coincidence probabilities and an application to clustering. *Ann. Probab.*, **5**:814–817, 1977.

[Kin93] J.F.C. Kingman. *Poisson Processes*. Oxford University Press, Oxford, 1993.

[LBBK91] M.Y. Leung, B.E. Blaisdell, C. Burge, and S. Karlin. An efficient algorithm for identifying matches with errors in multiple long molecular sequences. *J. Mol. Biol.*, **221**:1367–1378, 1991.

[LSY93] M.Y. Leung, G. Schachtel, and H.S. Yu. Scan statistics and DNA sequence analysis: The search for an origin of replication in a virus. University of Texas at San Antonio, 1993. Preprint.

[MKSM91] M.J. Masse, S. Karlin, G.A. Schachtel, and E.S. Mocarski. Human cytomegalovirus origin of DNA replication (orilyt) resides within a highly complex repetitive region. *Proc. Natl. Acad. Sci. USA*, **89**:5246–5250, 1991.

[Nau65] J.I. Naus. The distribution of the size of the maximum cluster of points on a line. *J. Am. Stat. Assoc.*, **60**:532–538, 1965.

[Pit93] J.P. Pitman. *Probability*. Springer-Verlag, New York, 1993.

[Rya94] K.J. Ryan. *Sherris Medical Microbiology, 3rd edition*. Appleton and Lange, Norwalk, CT, 1994.

[Wat89] M. Waterman. *Mathematical Methods for DNA Sequences*. CRC Press, Boca Raton, 1989.

[Wes88] K. Weston. An enhancer element in the short unique region of human cytomegalovirus regulates the production of a group of abundant immediate early transcripts. *Virology*, **162**:406–416, 1988.

5

Can She Taste the Difference?

MONDAY, APRIL 14, 1997 ★★★★★ San Francisco Chronicle[1]

Wake Up and Smell Health Benefits of Fresh Coffee

By Charles Petit

The distinctive aroma of freshly brewed coffee is not only pleasant, says a University of California chemist, it might be chock full of things that are good for you.

The molecules wafting up from a steaming cup of coffee, he has discovered, combine to form potent anti-oxidants. In principle, they should have cancer- and age-fighting effects similar to other anti-oxidants, including vitamin C and vitamin E.

Of course, just waking up and smelling the coffee won't do much good. The nose cannot possibly absorb enough aroma molecules to make an appreciable difference to health. You have to drink it.

But if initial calculations are correct, there is as much anti-oxidant capacity in the aromatic compounds of a cup of fresh coffee as in three oranges, said Takayuki Shibamoto, a professor of environmental toxicology at UC Davis....

Because the compounds are light and escape rapidly into the air, "you have to drink it in about 20 minutes after it is brewed," he said. In other words, the smell of fresh coffee is from the good stuff evaporating into the air.

Shibamoto emphasized that all he has so far is a hypothesis. Much more research will be needed to show whether coffee -- despite its ability to cause stomach aches and send nerve-rattling caffeine through your arteries -- is actually a health tonic....

And it appears that the health effects, if they are there, should be the same for caffeine-free coffee as for regular coffee....

[1]Reprinted by permission.

Introduction

One of R.A. Fisher's first designed experiments was prompted by an English lady's claim that she could tell the difference between a cup of tea prepared by adding the milk to the cup before the tea rather than after. She preferred the milk first. Fisher found it hard to believe that she could distinguish between the two preparations, and he designed an experiment to put the lady's skill to the test.

In this lab, you will conduct an experiment similar to Fisher's to determine if a subject can tell the difference between decaffeinated and regular coffee. We all know someone who claims they can tell the difference between the two. Here you will test the sensory discrimination of the supposed expert. Unlike the other labs, you will design and carry out an experiment in order to produce the data to analyze.

The essential ideas in this experiment appear in many other experimental designs, including ones used to test the skills of touch therapists (Rosa et al. [Rosa98]) and to test subjects purported to have extrasensory perception (Tart [Tart76]).

Data

The data for this lab are to be collected by you. They are the results from an experiment that you design and conduct. In your experiment, a subject will be asked to distinguish, by taste alone, between cups of regular and decaffeinated coffee.

For each cup of coffee tasted by the subject, record whether the coffee is decaffeinated (D) or regular (R), and record the subject's classification of the coffee as decaffeinated (D) or regular (R). It may be a good idea to also keep track of the order in which the cups were prepared, and the order in which they were served to the subject. This could be accomplished with identification numbers, as shown in Table 5.1.

Background

Rothamsted Experimental Station

In 1834, John Bennet Lawes, a chemistry student at Oxford University, left school to return with his widowed mother to the family manor of Rothamsted in England. At Rothamsted, he found that fertilizing the soil with bone meal did not improve his crop yields. Although bone meal worked well on sand, peat, and limestone, it was ineffective on the clay soil at the manor. Lawes used his training as a chemist to develop a better fertilizer. He turned a barn into a chemical laboratory, and in this laboratory he discovered that the treatment of the soil with acid improved his crop yields.

Lawes supported and expanded his laboratory with funds from a patent on his fertilizer, and in 1889 he created a trust for the endowment of a continuing research

TABLE 5.1. Example observations and data description for results from a taste testing experiment.

Order poured	1	2	3
Order served	5	7	1
Type	R	D	R
Opinion	R	D	D

Variable	Description
Order prepared	Each cup of coffee is assigned a number 1, ..., N according to the order in which it was poured.
Order served	A number 1, ..., N used to denote the order in which the coffee was served to the subject.
Type	The type of coffee: D=decaffeinated; R=regular.
Opinion	Subject's classification: D=decaffeinated; R=regular.

station at Rothamsted. Fisher joined the staff at Rothamsted in 1919, where he quickly made an impact at tea time. According to his daughter [Box78]:

> Already, quite soon after he had come to Rothamsted, his presence had transformed one commonplace tea time to an historic event. It happened one afternoon when he drew a cup of tea from the urn and offered it to the lady beside him, Dr. B. Muriel Bristol, an algologist [someone who studies algae]. She declined it, stating that she preferred a cup into which the milk had been poured first. "Nonsense," returned Fisher, smiling, "Surely it makes no difference." But she maintained, with emphasis, that of course it did. From just behind, a voice suggested, "Let's test her." It was William Roach who was not long afterward to marry Miss Bristol. Immediately, they embarked on the preliminaries of the experiment, Roach assisting with the cups and exulting that Miss Bristol divined correctly more than enough of those cups into which tea had been poured first to prove her case.
>
> Miss Bristol's personal triumph was never recorded, and perhaps Fisher was not satisfied at that moment with the extempore experimental procedure. One can be sure, however, that even as he conceived and carried out the experiment beside the trestle table, and the onlookers, no doubt, took sides as to its outcome, he was thinking through the questions it raised: How many cups should be used in the test? Should they be paired? In what order should the cups be presented? What should be done about chance variations in the temperature, sweetness, and so on? What conclusion could be drawn from a perfect score or from one with one or more errors?
>
> Probably this was the first time he had run such an experiment, for it was characteristic of him, having conceived an idea in one context, to revert to that context in expounding the idea later, rather than to select a new example from innumerable possibilities. And, of course, when he came to write *The Design of Experiments* (1935) more than a dozen years later, the "lady with

FIGURE 5.1. Various coffee makers: French press (right), filter cones (middle two), and automatic drip (left).

the tea-cups" took her rightful place at the initiation of the subject. ... In the subsequent pages [of the book] he considered the questions relevant to designing this particular test as a prime example, for the same questions arise, in some form, in all experimental designs.

Brewing Coffee

We asked the staff at Peet's Coffee how to brew a good cup of coffee and received the following instructions for three types of coffee makers: French press, filter cone, and automatic drip (Figure 5.1).

For all methods, start with fresh, cold water. If your water is heavily chlorinated, hard, or tastes funny, then use bottled or filtered water. Ideally, the water should have between 50 and 150 parts per million of dissolved solids.

Use two level tablespoons (10 grams) of ground coffee for each six ounces (180 milliliters) of water. Keep coffee (whole or ground beans) in an airtight container in the refrigerator or freezer. The fineness of the grind required depends on the method used for making the coffee (see below). In general, too fine a grind will cause bitterness, and too coarse a grind will yield watery coffee.

For the French press (right picture in Figure 5.1), heat the water to just below the boiling point (200° to 205° Fahrenheit, or 93° to 96° Celsius). Rinse the coffee pot with hot water. Add medium to coarsely ground coffee to the bottom of the preheated container. Pour water over the grounds, stir, wait a minute, stir again, and then push the plunger down. Do not brew for more than three minutes.

For filter cones (middle pictures in Figure 5.1), heat the water and preheat the container as described for the french press. Use a clean nylon or gold filter. Place the filter in the cone, and add medium to finely ground coffee. Wet the coffee grounds, then fill the cone with water. Continue to add water to keep the infusion going, remove grounds before the last few drops of coffee pass through the filter, and stir before serving.

For automatic drip machines (left picture in Figure 5.1), add cold water to the machine, and rinse the paper filter with hot water before adding a medium grind of beans. To avoid overextraction, make only the amount of coffee that can be brewed in four to six minutes. As with the filter cone, remove the grounds before the last few drops of coffee pass through the filter, and stir the coffee before serving.

Finally, coffee can be kept warm for only about 20 minutes before it starts to turn bitter. Do not reheat coffee.

Decaffeinated Coffee

There are two basic techniques for removing caffeine from coffee beans. For both techniques, the caffeine is extracted from green beans, before the beans are roasted.

The Swiss water process uses warm water under pressure to extract caffeine. First, the process creates an extraction of green coffee water solubles. The beans used to create this extraction are discarded, and the resulting liquid is used to remove the caffeine from other batches of beans. To do this, the extract is filtered to remove the caffeine, and then the decaffeinated liquid is continuously heated and circulated through the beans. Since the extract contains all green coffee solubles except caffeine, the beans give up only their caffeine to the liquid.

The direct contact method of decaffeination uses methylene chloride to remove the caffeine and wax from the beans. The temperatures used in this process are kept low so as not to destroy the chlorophyll in the beans. To remove the methylene chloride from the beans, they are washed and dried under a vacuum, which also keeps the temperature low. The U.S. Food and Drug Administration limits the amount of methylene chloride to 10 parts per million (ppm) in green coffee. Some processing plants guarantee the level of this chemical to be less than 5 ppm, with typical amounts less than 1 ppm. Roasting further reduces the level of this chemical.

Investigations

In this lab, you will design, conduct, and analyze results from your own experiment to test the sensory discrimination of a subject who claims to be able to taste the difference between regular and decaffeinated coffee. First you will need to find such a subject for your experiment. Then to proceed with designing your experiment, use the questions posed by Fisher's daughter — the questions that she says were on his mind as he carried out the impromptu experiment over tea at Rothamsted.

- ***What should be done about chance variations in the temperature, sweetness, and so on?*** Before conducting your experiment, carefully lay out the procedure for making the cups of coffee. Ideally you would want to make all cups of coffee identical, except for the caffeine. But it is never possible to control all of the ways in which the cups of coffee can differ from each other. Some must always be dealt with by randomization.

- *How many cups should be used in the test? Should they be paired? In what order should the cups be presented?* Fisher [Fisher66] suggests that the experiment should be designed such that, "if discrimination of the kind under test is absent, the result of the experiment will be wholly governed by the laws of chance." Also keep in mind that the number and ordering of the cups should allow a subject ample opportunity to prove his or her abilities and keep a fraud from easily succeeding at correctly discriminating the type of coffee in all the cups served.
- *What conclusion could be drawn from a perfect score or from one with one or more errors?* For the design you are considering, list all possible results of the experiment. For each possible result, decide in advance what action you will take if it occurs. In determining this action, consider the likelihood that someone with no powers of discrimination could wind up with that result. You may want to make adjustments in your design to increase the sensitivity of your experiment. For example, if someone is unable to discriminate between decaffeinated and regular coffee, then by guessing alone, it should be highly unlikely for that person to determine correctly which cups are which for all of the cups tasted. Similarly, if someone possesses some skill at differentiating between the two kinds of coffee, then it may be unreasonable to require the subject to make no mistakes in order to distinguish his or her abilities from a guesser.
- Write out an instruction sheet for your experimental process. Conduct a "dress-rehearsal" to work out the kinks in the process. After your practice run, you may want to make changes in your instruction sheet to address any problems that arose.
- You should now be ready to conduct your experiment. Record your results carefully, and note any unusual occurrences in the experimental process. Use a form similar to the one shown in Table 5.1 to record the successes and failures of the subject.
- Summarize your results numerically. Do they support or contradict the claim that the subject possesses no sensory discrimination? Use your list of all possible events and subsequent actions to come to a conclusion. Discuss the reasons behind the decision that you have made.
- What changes would you make to your experimental process if you had the opportunity to do it over again?

To help you in designing your experiment, here are some pitfalls that one student discovered in his experimental procedures:

> "The subject simply didn't like the brand of coffee that I bought. I chose the brand that I did because it had a decaffeinated and a caffeinated version. If I were to do the experiment over, I might try to buy a kind of coffee that I knew the expert would enjoy. This has its own problems because if you choose the expert's favorite coffee, he would probably be able to identify the version that he normally drinks (whether it be caffeinated or decaffeinated) and deduce the other one.

The expert was used to drinking coffee with sugar, and I served the coffee black. Whether this is an important factor is debatable. I think that if I were to do the experiment over again, I would sugar the coffee, to simulate as closely as possible the expert's usual experience.
Another problem was that after six or seven sips of two different coffees, the expert's ability to differentiate was diminished. Both "mistakes" made by our expert were made in the second half of the tastings. If I were to re-design the experiment, I would leave some time between each tasting. In order to make sure each cup of coffee was the same, I could make a single pot and use a thermos to keep it hot throughout the tasting period."

Theory

For this lab, we observe simple counts such as how many cups of regular coffee were classified as regular by the subject. These counts can be summarized and presented in a 2 × 2 table as shown in Table 5.2. The four numbers a, b, c, and d reported in the cells of the table correspond to the four possible categories: regular cup and subject classifies it as regular; decaffeinated cup and subject classifies it as regular; regular cup and subject classifies it as decaffeinated; and decaffeinated cup and subject classifies it as decaffeinated, respectively. The total number of cups of coffee made is

$$n = a + b + c + d.$$

From the table, we see that $a + c$ cups of regular coffee are prepared and the subject classifies $a + b$ of them as regular. Ideally, if the subject can taste the difference, then the counts b and c should be small. Conversely, if the subject can't really distinguish between the two types of coffee, then we would expect a and c to be about the same.

In this section, we propose several ways of testing the subject's skill, and we derive exact and approximate tests for the hypothesis that the subject cannot distinguish between the two types of coffee. For more on hypothesis tests, see Chapter 4.

TABLE 5.2. Counts of cups of coffee properly labeled as decaf.

		Coffee Prepared		
		Regular	Decaf	
Subject says	Regular	a	b	$a+b$
	Decaf	c	d	$c+d$
		$a+c$	$b+d$	n

The Hypergeometric Distribution

Suppose that to test the subject, 8 cups of coffee are prepared, 4 regular and 4 decaf, and the subject is informed of the design (i.e., that there are 4 cups of regular and 4 decaf). Also suppose that the cups of coffee are presented to the subject in random order. The subject's task is to identify correctly the 4 regular coffees and the 4 decafs.

This design fixes the row and column totals in Table 5.2 to be 4 each; that is,

$$a + b = a + c = c + d = b + d = 4.$$

With these constraints, when any one of the cell counts is specified, the remainder of the counts are determined. That is, given a,

$$b = 4 - a, \quad c = 4 - a, \quad \text{and} \quad d = a.$$

In general, for this experimental design, no matter how many cups of regular coffee are served, the row total $a + b$ will equal $a + c$ because the subject knows how many of the cups are regular. Then for n cups of coffee, of which $a + c$ are known to be regular, once a is given, the remaining counts are specified.

We can use the randomization of the cups to judge the skill of the subject. Begin by formulating the null hypothesis. We take the position that the subject has no sensory discrimination. Then the randomization of the order of the cups makes the 4 cups chosen by the subject as regular coffee equally likely to be any 4 of the 8 cups served.

There are $\binom{8}{4} = 70$ possible ways to classify 4 of the 8 cups as regular. If the subject has no ability to discriminate between decaf and regular, then by the randomization, each of these 70 ways is equally likely. Only one of them leads to a completely correct classification. Hence a subject with no discrimination has chance $1/70$ of correctly discriminating all 8 cups of coffee.

To evaluate the results of the experiment, we need only consider the five possibilities: the subject classifies 0, 1, 2, 3, or 4 of the regular cups of coffee correctly. The chance of each of these possible outcomes is shown in Table 5.3. These probabilities are computed from the hypergeometric distribution:

$$\mathbb{P}(a) = \frac{\binom{4}{a}\binom{4}{4-a}}{\binom{8}{4}} \qquad a = 0, 1, 2, 3, 4.$$

With these probabilities, we can compute the p-value for the test of the hypothesis that the subject possesses no sensory discrimination. Recall that the p-value is the chance of observing a result as extreme or more extreme than the one observed, given the null hypothesis. If the subject makes no mistakes, then the p-value is $1/70 \approx 0.014$, and if the subject makes one mistake, then the p-value is

$$\mathbb{P}(1 \text{ or fewer mistakes}) = \frac{1}{70} + \frac{16}{70} \approx 0.24.$$

Making one or fewer mistakes could easily occur by chance if the subject had no sensory discrimination. Only when the subject performs perfectly would we reject this hypothesis.

TABLE 5.3. Hypergeometric probabilities for the number of regular cups of coffee that have been correctly determined by the subject, assuming no sensory discrimination. (For 4 cups of regular and 4 cups of decaf).

Number of mistakes	0	1	2	3	4
Probability	$\frac{1}{70}$	$\frac{16}{70}$	$\frac{36}{70}$	$\frac{16}{70}$	$\frac{1}{70}$

This test is known as *Fisher's exact test.* See Exercise 2 for an example of Fisher's exact test when there are an unequal number of regular and decaf cups of coffee. Notice that with this test there are only a finite number of possible outcomes, and as a result there are only a finite number of possible p-values. This means the critical values that correspond exactly to the traditional significance levels of 0.05 and 0.01 may not exist.

Two Alternative Designs

As an alternative, we could serve 8 cups of coffee, where for each cup a coin is flipped: if it lands heads, regular is served; and if it lands tails, decaf is served. For this design, there are 2^8 possible ways in which the subject can classify the cups of coffee. Under the hypothesis that the subject possesses no sensory discrimination, each of these 256 possibilities is equally likely, and the chance of making no mistakes is $1/256$. This probability and others can be determined from the binomial distribution.

If B and C are the random variables used to denote the number of decaf coffees classified as regular by the subject and the number of regular coffees classified as decaf, respectively, then the chance of $b + c$ mistakes is

$$\mathbb{P}(b+c) = \binom{8}{b+c}\frac{1}{256}, \qquad (b+c) = 0, 1, \ldots, 8.$$

Notice that this design constrains only the overall total n in Table 5.2, which means that it is possible that the coffees served are all decaf or all regular. In these cases, the experimental design denies the subject the advantage of judging by comparison.

As another alternative, we could serve the cups of coffee in pairs, where in each pair one cup is regular and one is decaf. As in the first design, the subject would choose 4 cups as regular and 4 as decaf, and we need only keep track of mistakes of one type — regular coffee said to be decaf. However, the pairing makes it easier for a person with no discriminating abilities to guess correctly:

$$\mathbb{P}(b) = \binom{4}{b}\frac{1}{16}, \qquad b = 0, \ldots, 4.$$

With this design, more than 8 cups of coffee need to be served to make the probability of no mistakes small. Also for this design, we would not use Table 5.2 to summarize the results of the experiment because it hides the pairing that is present in the data. All that needs to be reported are the number of correctly and incorrectly discriminated pairs.

An Approximate Test

With a large number of counts, it can be cumbersome to compute p-values for Fisher's exact test. Instead, we can use the normal approximation to the hypergeometric distribution to test the hypothesis. (See the Exercises and Appendix B for an example of this approximation.)

Let A, B, C, D be the random variables underlying the counts in the 2×2 table. We have already seen that for Fisher's design, A has a hypergeometric distribution. It can be shown (see the Exercises) that

$$\mathbb{E}(A) = (a+b)\frac{a+c}{n}$$

$$\text{Var}(A) = (a+b)\frac{(a+c)}{n}\frac{(b+d)}{n}\frac{(c+d)}{n-1}.$$

Use the expected value and standard deviation of A to standardize the cell count:

$$z = \frac{a - \mathbb{E}(A)}{\text{SD}(A)} \approx \frac{a - \frac{(a+b)(a+c)}{n}}{\sqrt{\frac{(a+b)(a+c)(b+d)(c+d)}{n^3}}}$$

$$= \frac{\sqrt{n}(ad - bc)}{\sqrt{(a+b)(a+c)(b+d)(c+d)}}.$$

Note that $n - 1$ was approximated by n in the standardization. This z statistic has an approximate standard normal distribution. That is, we can approximate Fisher's exact test with a *z test*. For intermediate sample sizes, we would use a continuity correction in the approximation; that is, we would substitute $a \pm 0.5$ for a in z. See Exercise 3 for an example of the continuity correction.

We note that

$$z^2 = \frac{n(ad - bc)^2}{(a+b)(a+c)(b+d)(c+d)}. \tag{5.1}$$

We will refer to this representation throughout the remainder of this section.

Contingency Tables

Two-by-two tables that cross-classify subjects according to two dichotomous characteristics are called *contingency tables*. We present here two additional models for A, B, C, and D, the random variables underlying the cell counts in a 2×2 table. For each model, we show that the counts can be analyzed using the test statistic z^2 in equation (5.1).

The Multinomial

In Chapter 2, a simple random sample of statistics students found that the men in the sample appeared more likely to enjoy playing video games than the women (Table 5.4). Ignoring the slight dependence between observations that arises from sampling without replacement, we can think of the 4 counts in the table as an

TABLE 5.4. Counts of students according to their sex and whether they like to play video games (Chapter 2).

		Sex		
		Male	Female	
Like to play	Yes	43	26	69
	No	8	12	20
		51	38	89

observation from a multinomial distribution. That is, (a, b, c, d) is an observation of a multinomial with parameters $(n, \pi_A, \pi_B, \pi_C, \pi_D)$. The probability π_A is the chance that the student selected will be a male who likes to play video games, π_B is the chance that the student will be a female who likes to play video games, and so on. In our case $a = 43$, $b = 26$, $c = 8$, $d = 12$, and $n = 89$.

Suppose we want to test the hypothesis that sex and attitude toward video games are independent. Here we would use the chi-square goodness-of-fit test that was introduced in Chapter 4. Recall that the chi-square test compares the observed counts to the expected counts from the multinomial. For m categories, the test statistic is

$$\sum_{j=1}^{m} \frac{(j\text{th Sample count } - j\text{th Expected count})^2}{j\text{th Expected count}}.$$

In our case $m = 4$.

To determine the expected counts for our example, note that the null hypothesis of independence between sex and attitude implies that

$$\pi_A = \alpha\beta, \quad \pi_B = \alpha(1-\beta), \quad \pi_C = (1-\alpha)\beta, \quad \pi_D = (1-\alpha)(1-\beta),$$

where α is the probability of choosing a student who likes video games, and β is the chance of choosing a male student. Then the expected counts are

$$\mathbb{E}(A) = n\alpha\beta, \qquad \mathbb{E}(B) = n\alpha(1-\beta),$$
$$\mathbb{E}(C) = n(1-\alpha)\beta, \quad \mathbb{E}(D) = n(1-\alpha)(1-\beta).$$

The probabilities α and β need to be estimated in order to obtain numeric values for these expected counts. A natural estimate to use for α is the proportion of students in the sample who like video games (i.e., $(a+b)/n = 69/89$.) Similarly, $(a+c)/n = 51/89$ can be used to estimate β. (Exercise 12 shows that they are the maximum likelihood estimates.) We use these estimates to find

$$\begin{aligned}\mathbb{E}(A) &\approx n\frac{(a+b)}{n}\frac{(a+c)}{n} \\ &= 89 \times \frac{69}{89} \times \frac{51}{89} \\ &= 39.5\,.\end{aligned}$$

Similarly, the other approximate expected cell counts are 11.5, 29.5, and 8.5, and the test statistic is

$$\frac{(43-39.5)^2}{39.5}+\frac{(26-29.5)^2}{29.5}+\frac{(8-11.5)^2}{11.5}+\frac{(12-8.5)^2}{8.5}=3.2\,.$$

To find the p-value — the chance of observing a test statistic as large as or larger than 3.2 — we compute

$$\mathbb{P}(\chi_1^2 > 3.2) = 0.08.$$

The p-value indicates we have observed a typical value for our statistic; we would not reject the null hypothesis of independence between sex and attitude toward video games.

Recall that the degrees of freedom in the chi-square test are calculated as follows: from 4, the number of categories, subtract 1 for the constraint that the counts add to 89 and subtract 2 for the parameters estimated.

This special case of the chi-square test is called the *chi-square test of independence*. This chi-square test statistic,

$$\frac{(a-n\frac{(a+b)}{n}\frac{(a+c)}{n})^2)}{n\frac{(a+b)}{n}\frac{(a+c)}{n}}+\cdots+\frac{(d-n\frac{(c+d)}{n}\frac{(b+d)}{n})^2)}{n\frac{(c+d)}{n}\frac{(b+d)}{n}}, \tag{5.2}$$

is identical to z^2 in equation (5.1). We leave it as an exercise to show this equivalence.

Independent Binomials

Suppose for the moment that the male and female students in the previous example were sampled independently. In other words, a simple random sample of $a + c = 51$ male students was taken, and a simple random sample of $b + d = 38$ females students was taken independently of the sample of males. For this design, ignoring the dependence that occurs with sampling from a finite population, A has a binomial$(51, \gamma_A)$ distribution and, independent of A, the random variable B has a binomial$(38, \gamma_B)$ distribution. Also, $C = 51 - A$ and $D = 38 - B$.

As in the previous example, we may be interested in whether males and females have similar attitudes toward video games. Assume that the chance that a male chosen for the sample likes video games is the same as the chance that a female chosen for the second sample likes video games. This assumption of *homogeneity* means that $\gamma_A = \gamma_B = \gamma$, say.

To test this null hypothesis of homogeneity, we compare estimates of γ_A and γ_B. To estimate γ_A we would use the fraction of males who like video games, namely $a/(a + c) = 43/51$. Similarly, we estimate γ_B by $b/(b + d) = 26/38$. Under the null hypothesis, the difference

$$\frac{A}{(a+c)}-\frac{B}{(b+d)}$$

has expected value 0 and variance

$$\gamma(1-\gamma)\left[\frac{1}{(a+c)}+\frac{1}{b+d)}\right].$$

The observed standardized difference,

$$z=\frac{\frac{a}{(a+c)}-\frac{b}{(b+d)}}{\sqrt{\gamma(1-\gamma)\left[\frac{1}{(a+c)}+\frac{1}{(b+d)}\right]}},$$

has an approximate standard normal distribution for large sample sizes. A test using this standardized difference is called a *two-sample z test*. We estimate the variance of the difference by plugging in the pooled sample variance for $\gamma(1-\gamma)$:

$$\frac{(a+b)}{n}\times\frac{(c+d)}{n}\left[\frac{1}{(a+c)}+\frac{1}{(b+d)}\right]=\frac{69}{89}\frac{20}{89}\left[\frac{1}{51}+\frac{1}{38}\right]=0.008.$$

Then our observed test statistic is $(43/51-26/38)/\sqrt{0.008}=1.8$, and the p-value is

$$\mathbb{P}(|Z|>1.8)=0.08.$$

Notice that this p-value is the same as in the previous example for the chi-square test of homogeneity. This equality is not a fluke: Z^2 has a χ_1^2 distribution, and the square of 1.8 is 3.2. The two analyses lead to the same test statistic — the square of this two-sample test statistic equals the quantity in equation (5.1). We leave it as an exercise to prove this result.

Why Are All These Statistics the Same?

Three different models for the 2 × 2 contingency table all led to the same test statistic. The reason for this lies in the relations between the hypergeometric, multinomial, and binomial distributions. Both the hypergeometric and the binomial models can be derived from the multinomial by placing conditions on the distribution.

If (A, B, C, D) has a multinomial distribution with parameters n, π_A, π_B, π_C, and π_D, then given $A+C=a+c$ we can show that A is binomial$(a+c, \gamma_A)$, where $\gamma_A=\pi_A/(\pi_A+\pi_C)$; B is binomial$(b+d, \gamma_B)$, where $\gamma_B=\pi_B/(\pi_B+\pi_D)$; and A and B are independent.

Further conditioning on $A+B=a+b$, for the special case when $\gamma_A=\gamma_B=\gamma$, the distribution of A becomes hypergeometric, where

$$\mathbb{P}(A=a)=\frac{\binom{a+b}{a}\binom{c+d}{b}}{\binom{n}{a+b}}.$$

Note that A has the hypergeometric distribution found in Fisher's exact test. We leave the proof of these relationships to the Exercises.

$I \times J$ Contingency Tables

A cross-classification of subjects where there are I categories in the first classification and J categories in the second can be represented in a contingency table with I rows and J columns. Call the count in the ith row and the jth column of the table c_{ij}, for $i = 1, \ldots, I$ and $j = 1, \ldots, J$. We can generalize the multinomial and independent binomial models of the 2×2 table to the $I \times J$ table.

Multinomial

Suppose the IJ counts form one observation from a multinomial distribution with parameters n and π_{ij}, $i = 1, \ldots, I$ and $j = 1, \ldots J$. The assumption of independence between the two classifications means that $\pi_{ij} = \alpha_i \beta_j$, where the $0 < \alpha_i < 1$, $\sum \alpha_i = 1$, $0 < \beta_j < 1$, and $\sum \beta_j = 1$. To test the hypothesis of independence, if α and β are unknown we would use the statistic

$$\sum_{i=1}^{I} \sum_{j=1}^{J} \frac{(c_{ij} - n\hat{\alpha}_i \hat{\beta}_j)^2}{n\hat{\alpha}_i \hat{\beta}_j},$$

where $\hat{\alpha}_i = \sum_j c_{ij}/n$ and $\hat{\beta}_j = \sum_i c_{ij}/n$. Under the null hypothesis, provided the cell counts are all at least 5, the test statistic has an approximate χ^2 distribution with $(I-1)(J-1)$ degrees of freedom. The degrees of freedom are calculated by taking IJ cells and subtracting 1 for the constraint that they add to n and $I - 1 + J - 1$ for the parameters that were estimated. This test is called the *chi-square test of independence*.

Independent Multinomials

To generalize the test of the null hypothesis of homogeneous binomial proportions, suppose I samples are taken of size $n_1, \ldots, n_I$, respectively. If, for each of these samples, units are classified according to one of J categories, then for $i = 1, \ldots, I$, $(c_{i1}, \ldots, c_{iJ})$ form one observation from a multinomial with parameters n_i, γ_{ij}, $j = 1, \ldots, J$, where $0 < \gamma_{ij} < 1$ and $\sum_j \gamma_{ij} = 1$. The I multinomials are independent. The null hypothesis of homogeneity states that $\gamma_{1j} = \gamma_{2j} = \cdots = \gamma_{Ij} = \gamma_j$, for $j = 1, \ldots, J$. To test this hypothesis here, we use the *chi-square test of homogeneity*,

$$\sum_{i=1}^{I} \sum_{j=1}^{J} \frac{(c_{ij} - n_i \hat{\gamma}_j)^2}{n_i \hat{\gamma}_j},$$

where $\hat{\gamma}_j = \sum_i c_{ij}/n$. Again, provided the counts are large enough, the test statistic has an approximate χ^2 distribution with $(I-1)(J-1)$ degrees of freedom. The degrees of freedom are calculated by taking IJ cells and subtracting I for the constraints that each row adds to n_i and $J-1$ for the parameters that are estimated.

As in the 2×2 contingency table, the chi-square test for homogeneity and the chi-square test for independence yield the same test statistic. The proof is left as an exercise.

Exercises

1. Suppose 10 cups of coffee, 5 decaf and 5 regular, are served according to Fisher's design. Find the p-value for Fisher's exact test when one cup of regular coffee is incorrectly categorized. What is the p-value when no mistakes are made? What about two mistakes?
2. Suppose 6 of the 10 cups of coffee in the experiment described in Exercise 1 are regular and the remaining 4 are decaf. How do the p-values change? Which design, $5 + 5$ or $6 + 4$, is preferred?
3. Suppose 20 cups of coffee, 10 decaf and 10 regular, are served according to Fisher's design.

    a. Find the p-value for Fisher's exact test when two cups of regular coffee are incorrectly categorized.
    b. Compare this p-value to the one obtained using the z test.
    c. Compute the p-value for the z test using the continuity correction. That is, find the chance that at most 2.5 cups are incorrectly labeled. Does this correction improve the p-value?

4. Suppose 8 cups of coffee are served to a subject, where for each cup a coin is flipped to determine if regular or decaf coffee is poured. Also suppose that the subject correctly distinguishes 7 of the 8 cups. Test the hypothesis that the subject has no sensory discrimination. What is your p-value?
   Suppose instead that the cups were presented to the subject in pairs, where each pair had one cup of decaf and one regular, and the order was determined by the flip of a coin. How many cups of coffee would you need to serve the subject to obtain a p-value for one mistake that is roughly the same size as in the previous design?
5. Table 5.5 gives a finer categorization of the cross-classification of students in the video survey. Here the students are categorized according to whether they like video games very much, somewhat, or not really. Test the hypothesis that sex and attitude are independent. Do your conclusions from the test differ from the conclusions using the data in Table 5.4?
6. Table 5.6 cross-classifies students in the video survey according to the grade they expect in the class and whether they like to play video games. Test the hypothesis that these two characteristics are independent.

TABLE 5.5. Counts of students according to their sex and whether they like to play video games (Chapter 2).

		Like to play			
		Very	Somewhat	No	
Sex	Male	18	25	8	51
	Female	5	21	12	38
		23	46	20	89

TABLE 5.6. Counts of students according to their expected grade and whether they like to play video games (Chapter 2).

		Like to play			
		Very	Somewhat	No	
Expected grade	A	10	14	6	30
	B or C	13	32	14	59
		23	46	20	89

7. In Stearns County, Minnesota, it was found that 15 of 27 houses sampled had radon concentrations exceeding 4 pCi/l. In neighboring Wright county, a sample of 14 houses found 9 with radon concentrations over 4 pCi/l. Test the hypothesis that the proportion of houses in Stearns County with radon levels over 4 pCi/l is the same as the proportion for Wright County.
8. Consider a simple random sample of n subjects from a population of N subjects, where M of the subjects in the population have a particular characteristic.

   a. Show that T, the number of subjects in the sample with the particular characteristic, has a hypergeometric distribution, where

   $$\mathbb{P}(T=k)=\frac{\binom{M}{k}\binom{N-M}{n-k}}{\binom{N}{n}},$$

   for $k=0,1,\ldots,n$. (We assume that $n<\min(M,N-M)$.)
   b. Use the expected value and variance from Chapter 2 for totals from simple random samples to derive the expected value and variance of a hypergeometric distribution.

9. Consider Fisher's exact test. Use the results from Exercise 8 to show:

$$\mathbb{E}(A)=(a+b)\frac{a+c}{n},$$

$$\text{Var}(A)=(a+b)\frac{(a+c)}{n}\frac{(b+d)}{n}\frac{(c+d)}{n-1}.$$

10. Show that the estimated expected counts for the independent binomial model and the homogeneous parameter model equal the expected counts for the hypergeometric model:

$$\mathbb{E}(A)=n\frac{(a+b)}{n}\frac{(a+c)}{n},\quad \mathbb{E}(B)=n\frac{(a+b)}{n}\frac{(b+d)}{n},$$

$$\mathbb{E}(C)=n\frac{(c+d)}{n}\frac{(a+c)}{n},\quad \mathbb{E}(D)=n\frac{(c+d)}{n}\frac{(b+d)}{n}.$$

11. Show that the square of the two-sample z statistic

$$\frac{\frac{a}{(a+b)}-\frac{c}{(c+d)}}{\sqrt{\frac{(a+c)}{n}\times\frac{(b+d)}{n}\left[\frac{1}{(a+b)}+\frac{1}{(c+d)}\right]}}$$

reduces to equation (5.1).

12. Suppose (A, B, C, D) follows a multinomial with parameters n, π_A, π_B, π_C, and π_D. Show that for observations (a, b, c, d), the maximum likelihood estimates of π_A, π_B, π_C, and π_D are $a/n, b/n, c/n$, and d/n, respectively. Show that in the case of independence — that is, when $\pi_A = \alpha\beta$, $\pi_B = \alpha(1-\beta)$, $\pi_C = (1-\alpha)\beta$, and $\pi_D = (1-\alpha)(1-\beta)$ — that the maximum likelihood estimates of α and β are $(a+b)/n$ and $(a+c)/n$, respectively.
13. Consider the chi-square test of independence for a 2×2 contingency table. Show that the test statistic (5.2) is equivalent to

$$\frac{n(ad - bc)^2}{(a+b)(a+c)(b+c)(b+d)}.$$

14. Suppose (A, B, C, D) has a multinomial distribution with parameters n, π_A, π_B, π_C, and π_D. Given $A + C = a + c$, show that A has a binomial distribution with parameters $a+c$ and $\gamma_A = \pi_A/(\pi_A + \pi_C)$; B has a binomial distribution with parameters $b+d$ and $\gamma_B = \pi_B/(\pi_B + \pi_D)$; and A and B are independent.
15. Continue with Exercise 14 and further suppose that A and B are independent binomials with parameters $a+c, p$ and $b+d, p$, respectively. Given $A + B = a + b$, prove that A has the hypergeometric distribution in Fisher's exact test.

Extensions

For a final example of a model for a 2×2 contingency table, suppose the table corresponds to a cross-classification by sex (female, male) and residence (urban, rural) of the number of cases of a particular cancer in a given state in a given year. (In practice, we would classify by age too, but this is a simplified example.) Then our counts are a, b, c, and d, where a is the number of cancer cases that are female and urban, b the number of male and urban cases, c the number of female and rural cases, and d the number of male and rural cases.

For reasons to be explained in Chapter 6, we might begin our analysis supposing that the underlying random variables A, B, C, and D are independent Poisson, with parameters (means) α, β, γ, and δ. To estimate these parameters, we use the method of maximum likelihood. For observations a, b, c, and d, the likelihood function is

$$L(\alpha, \beta, \gamma, \delta) = \frac{e^{-\alpha}\alpha^a}{a!}\frac{e^{-\beta}\beta^b}{b!}\frac{e^{-\gamma}\gamma^c}{c!}\frac{e^{-\delta}\delta^d}{d!}.$$

By maximizing L with respect to α, β, γ, and δ, we find that (see the Exercises) if $a > 0, b > 0, c > 0$, and $d > 0$,

$$\hat{\alpha} = a, \quad \hat{\beta} = b, \quad \hat{\gamma} = c, \quad \hat{\delta} = d.$$

An alternative model is the *multiplicative model*. If we think of urban females as a baseline category and suppose that the male relative risk of getting this cancer (in this state in this year) is λ and the relative risk of rural inhabitants is μ, then a simple model for these data would have expected values

$$\mathbb{E}(A) = \alpha, \qquad \mathbb{E}(B) = \lambda\alpha,$$
$$\mathbb{E}(C) = \mu\alpha, \qquad \mathbb{E}(D) = \lambda\mu\alpha;$$

in other words, $\alpha\delta = \beta\gamma$. The use of relative risk parameters such as λ and μ simplifies discussion of cancer rates, but does it fit the data?

For the multiplicative model, $\delta = \beta\gamma/\alpha$. As a result, only three parameters need to be estimated. The likelihood function in this case is

$$L(\alpha, \beta, \gamma) = \frac{e^{-\alpha}\alpha^a}{a!}\frac{e^{-\beta}\beta^b}{b!}\frac{e^{-\gamma}\gamma^c}{c!}\frac{e^{-\beta\gamma/\alpha}(\beta\gamma/\alpha)^d}{d!}.$$

We leave it to the Exercises to show that the maximum likelihood estimates of the parameters are, for $a + b > 0$, $a + c > 0$, $b + c > 0$, and $c + d > 0$,

$$\hat{\alpha} = \frac{(a+b)(a+c)}{n}, \quad \hat{\beta} = \frac{(a+b)(b+d)}{n},$$
$$\hat{\gamma} = \frac{(c+d)(a+c)}{n}, \quad \hat{\delta} = \frac{(c+d)(b+d)}{n}.$$

To test how well this model fits our data, we compare the observed counts to their expected counts in a chi-square test. Since the expected counts are the same as in the previous examples for the hypergeometric, multinomial, and binomial distributions, we will get exactly the same test statistic:

$$\frac{n(ad - bc)^2}{(a+b)(a+c)(b+c)(b+d)}.$$

Also, as before, there is one degree of freedom. This degree of freedom comes from 4 freely varying cells (n is not fixed) less 3 estimated parameters.

We saw earlier that the relations among the various distributions (multinomial-binomial and multinomial-hypergeometric) explained how the test statistics all reduced to the same quantity. The same reasoning underlies this model. For A, B, C, D independent Poisson random variables, when we condition on $A+B+C+D = n$, then (A, B, C, D) has a multinomial distribution with parameters n, π_A, π_B, π_C, and π_D, where $\pi_A = \alpha/(\alpha + \beta + \gamma + \delta)$, and so forth. For the multiplicative model, these probabilities further reduce to

$$\pi_A = \frac{1}{(1+\lambda)(1+\mu)}, \quad \pi_B = \frac{\lambda}{(1+\lambda)(1+\mu)},$$
$$\pi_C = \frac{\mu}{(1+\lambda)(1+\mu)}, \quad \pi_D = \frac{\lambda\mu}{(1+\lambda)(1+\mu)}.$$

We see that the multiplicative Poisson model reduces to a multinomial model, which in turn is an independent binomial model with two probabilities, $1/(1 + \lambda)$ and $1/(1 + \mu)$, that need to be estimated.

Exercises

1. Suppose A, B, C, and D are independent Poisson random variables, with parameters α, β, γ, and δ. Show that for the given observations a, b, c, and d, the maximum likelihood estimates for α, β, γ, and δ are a, b, c, and d, respectively, provided a, b, c, and d are positive.
2. Suppose A, B, C, and D are independent Poisson random variables, with parameters α, β, γ, and $\delta = \beta\gamma/\alpha$. Derive the maximum likelihood estimates of α, β, γ, and δ given observations a, b, c, and d.
3. Suppose A, B, C, and D are independent Poisson random variables with means that satisfy the multiplicative model. Given $A + B + C + D = n$, prove that (A, B, C, D) has a multinomial distribution with parameters n, π_A, π_B, π_C, and π_D, where $\pi_A = \pi_\lambda \pi_\mu$, $\pi_B = (1 - \pi_\lambda)\pi_\mu$, $\pi_C = \pi_\lambda(1 - \pi_\mu)$, $\pi_D = (1 - \pi_\lambda)(1 - \pi_\mu)$, and $\pi_\lambda = 1/(1 + \lambda)$, $\pi_\mu = 1/(1 + \pi_\mu)$.

Notes

We chose an experiment to test a subject's ability to discriminate between decaffeinated and regular coffee because it seemed the modern American version of tea tasting. Other comparisons that we considered, and that you may wish to use, were based on comparing brands of food, such as peanut butter and soft drinks. You may also want to collect data according to the multinomial and binomial models. A sample of students cross-classified by sex and handedness would provide an example for the multinomial model. Two samples, one of fraternity students and one of dormitory students with regard to their smoking habits, would provide an example of the independent binomial model.

Cathy, the manager of Peet's Coffee shop on Domingo Avenue in Berkeley, California, was very helpful in providing information on brewing and decaffeinating coffee. Much of this information was paraphrased from their brochure, *Peet's Coffee: Recommendations & Descriptions*, and from their web site—www.peets.com. Peet's recommends the French press for brewing coffee and the direct-contact method for removing caffeine from the coffee beans. To mail order Peet's coffee call 1-800-999-2132 or visit their web site.

References

[Box78] J.F. Box. *R.A. Fisher: The Life of a Scientist.* John Wiley & Sons, Inc., New York, 1978.

[Fisher66] R.A. Fisher. *The Design of Experiment, 8th edition.* Hafner Publishing Company, New York, 1966.

[Rosa98] L. Rosa, E. Rosa, L. Sarner, and S. Barrett. A close look at therapeutic touch. *J. Am. Med. Assoc.*, **279**(13):1005–1010, April, 1998.

[Tart76] C. Tart. *Learning to Use Extrasensory Perception.* University of Chicago Press, Chicago, 1976.

6

HIV Infection in Hemophiliacs

MONDAY, MAY 31, 1999 ★★★★★· San Francisco Chronicle [1]

Genetic Bloodhound

New test from rival Bay Area biotech firms can sniff out viral infections in donations

By Tom Abate

With 14 million pints of blood collected from U.S. donors each year, screening blood-borne diseases is a daunting task. Each pint is tested within 24 hours for HIV and other viruses.

Right now, the tests theoretically miss just 1.5 HIV and 10 hepatitus C viruses in every million parts, but safety regulators say that isn't good enough.

In response, U.S. blood banks launched a new genetic test this spring that promises to make the slim odds of infection even slimmer.

This new technique, called nucleic acid testing, or NAT, is supposed to be sensitive enough to detect the earliest traces of HIV or hepatitis C viruses that can slip through current blood tests.

Although current tests already have made infection from donated blood exceedingly rare, NAT could reduce the risk 80 to 90 percent for hepatitis and somewhat less for HIV. ...

But some experts have questioned whether the cost of the new test -- it could add 5 percent to the overall cost of a pint of blood -- is worth the slim improvement in safety margins.

Jim AuBuchon, a pathology professor at Dartmouth Medical Center in New Hampshire, says NAT should reduce the statistical probability of transfusion-caused HIV from 24 cases a year at present to 16 a year. But he notes that there hasn't been a single real case of HIV-infected blood reported in five years -- which makes him wonder whether there actually would be an increased safety margin for HIV. "This is the least cost-effective medical procedure I have ever seen," he says.

[1] Reprinted by permission.

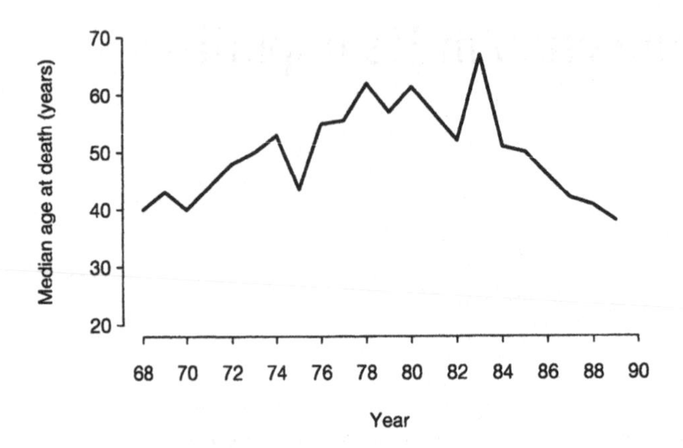

FIGURE 6.1. Median age at death for hemophiliacs in the U.S. (Chorba [Chorba94] and Cohen [Cohen94]).

Introduction

Major advances in hemophilia treatment were made during the 1960s and 1970s when new techniques were developed for concentrating blood clotting factors used in transfusions. As a result, the median age at death increased steadily (Figure 6.1). However in the 1980s, large numbers of people with hemophilia began to die from Acquired Immune Deficiency Syndrome (AIDS). Hemophiliacs are typically immunosuppressed and susceptible to bacterial infections, but in the 1980s people with hemophilia began to die from other types of infections, such as fungal infections, that are more associated with AIDS. Immune failure soon became the single most common cause of death in hemophiliacs (Harris [Harris95]), and the median age at death peaked in 1984 before dropping dramatically in the late 1980s.

The clotting factor supply was contaminated with the Human Immunodeficiency Virus (HIV) and it was thought that HIV infection caused AIDS because most or all of the deaths due to AIDS occurred in people who were infected with HIV. However, the implication of causality is difficult to prove in part because the definition of AIDS adopted by the U.S. Centers for Disease Control (CDC) requires patients to have seroconverted (i.e., become infected with HIV) in order to be diagnosed as having AIDS.

A few leading scientists have questioned whether HIV causes AIDS. Some contend that HIV infection is necessary for AIDS but that it alone is not sufficient to cause AIDS. Others, such as the retrovirologist Peter Duesberg at the University of California, Berkeley, have claimed that HIV is a harmless passenger virus that acts as a marker for the number of transfusions a patient with hemophilia has received.

According to Duesberg, AIDS is caused by the large quantity of contaminants people with hemophilia receive in their lifetime dosages of clotting factor.

In this lab, you will have the opportunity to examine this question of whether HIV causes AIDS using data from a large study of hemophiliacs. As Cohen ([Cohen94]) states:

> Hemophiliacs offer a unique window on the effects of HIV infection because there are solid data comparing those who have tested positive for antibodies to HIV — and are presumably infected — with those who have tested negative. In addition, the health status of hemophiliacs has been tracked for more than a century, providing an important base line. And unlike homosexual groups, hemophiliac cohorts are not riddled with what Duesberg thinks are confounding variables, such as illicit drug use.

Data

The data for this lab are from the Multicenter Hemophilia Cohort Study (MHCS), which is sponsored by the U.S. National Cancer Institute (NCI). The study followed over 1600 hemophilia patients at 16 treatment centers (12 in the U.S. and 4 in western Europe) during the period from January 1, 1978 to December 31, 1995 (Goedert et al. [GKA89], Goedert [Goedert95]). The MHCS is one of the two large U.S. epidemiological studies of hemophiliacs, the other being the Transfusion Safety Study Group of the University of Southern California.

Patients in the MHCS are classified according to age, HIV status, and severity of hemophilia. See Table 6.1 for a description of the data and sample observations. To determine severity of hemophilia, on each annual questionnaire patients indicate the amount of clotting-factor concentrate they have received in that year. Hemophiliacs are fairly consistent over time with respect to dose, except that low-dose users might have an occasional moderate year if they had surgery. The severity of hemophilia reported here is calculated from the average annual clotting-factor concentrate received in the 1978–84 period.

Each record in the data set corresponds to one stratum of hemophiliacs in one calendar year of the study. For example, the first column in Table 6.1 provides information on all severe hemophiliacs in 1983 who were HIV-negative and between the ages of 10 and 14 inclusive. In this group in 1983, the patients contributed 6.84 person years to the study, and none of them died. In general, a subject's person years are calculated from the beginning of the study, or their birth date if they were born after January 1, 1978, until the time of last contact or the end of the study, whichever is earlier. When a subject dies or drops out before the end of the study, then the last contact occurs at the time of death or withdrawal.

Figure 6.2 shows a time line for three hypothetical subjects for the 18 years of the study. Subject A is a severe hemophiliac who was 6 years old when he entered the study, seroconverted in 1985, and was alive as of December 31, 1995. He contributes 18 person years to the study. Subject B, also a severe hemophiliac,

TABLE 6.1. Sample observations and data description for 29067.22 person years in the Multicenter Hemophilia Cohort Study (Goedert et al.[GKA89], Goedert [Goedert95]).

Year	83	83	91	91	91	91	91	91	91	91
HIV	1	1	2	2	2	2	2	2	2	2
Severity	1	1	1	1	1	1	3	3	3	3
Age	3	4	5	6	7	8	2	3	4	5
Person years	6.84	9.04	30.14	33.74	28.51	18.92	7.47	14.64	14.83	18.57
Deaths	0	0	1	2	2	3	0	0	0	1

Variable	Description
Year	Calendar year: 78; ... ;95.
HIV	HIV status: 1=negative; 2=positive.
Severity	Average annual dose of clotting-factor concentrate: 1= over 50,000 units; 2= 20,001–50,000 units; 3= 1–20,000 units; 4=unknown; 5=none.
Age	Age in 5-year intervals: 1= under 5; 2= 5 to 9;... ; 14= 65 and over.
Person years	Total amount of time during the calendar year that the people in the stratum were alive and part of the study.
Deaths	Number of deaths in the calendar year for the stratum.

entered the study at age 11, seroconverted in 1984, and died on September 30, 1990. This subject contributes 12.75 person years to the study. Subject C, a mild hemophiliac, was born on July 1, 1985 and dropped out of the study on March 31, 1990, not having seroconverted at that point. He contributes 4.75 person years. Each subject's person years can be further allocated to strata within calendar years. Here are some examples: for 1983, subject A contributes one person year to the HIV-negative, 10–14 year old, severe hemophilia group; subject B contributes one person year to the HIV-negative, 15–19 year old, severe hemophilia group; and subject C has not yet entered the study and does not contribute anything. Later, in 1990: one person year is contributed to the HIV-positive, 15–19 year old, severe strata; 0.75 person years and one death to the HIV-positive, 20–24 year old, severe group; and 0.25 years to the HIV-negative, under 5, mild hemophiliacs.

Background

Hemophilia

There are twelve factors in plasma, numbered I through XII, that play a role in blood coagulation. A person whose factor VIII level is so low that it leads to uncontrolled bleeding is diagnosed with hemophilia A, or classical hemophilia.

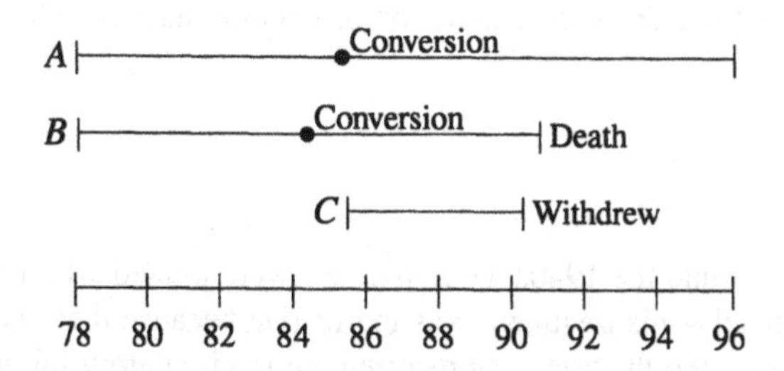

FIGURE 6.2. Diagrammatic representation of follow-up on three hypothetical subjects in the MHCS. (Subject A is a severe hemophiliac who was 6 years old at the start of the study; subject B, also a severe hemophiliac, was 11 when he entered the study; and subject C, born July 1, 1985, was a mild hemophiliac.)

Hemophilia B, also named Christmas disease after someone who had the disease, occurs when factor IX levels are too low.

Hemophilia is classified into three degrees of severity according to the quantity of the clotting factor present in the blood. Severe hemophilia A means the individual's plasma contains less than 1% of the factor VIII found in a healthy adult; this amount is 0.2 micrograms per milliliter of plasma. Moderate cases have 1–5%, and mild cases have 6–24% of the normal quantity of the clotting factor.

Most hemophiliacs (85%) are of Type A, and of these 70% are severe cases. Fourteen percent are of Type B, and the remaining 1% have factor disorders other than VIII and IX. Severe hemophiliacs spontaneously bleed without trauma several times a month.

Hemophiliacs do not have trouble with their platelets, meaning nose bleeds, cuts, and scratches are not a problem for them. Instead, bleeding occurs internally, most often in muscles and joints, especially ankles, knees, elbows, shoulders, and hips. The bleeding destroys the cartilage in the joints and leads to chronic pain and permanent disability. Internal bleeding occurs at other sites as well. Prior to the AIDS epidemic, approximately 1/4 of the deaths in patients with hemophilia were attributed to intracranial bleeding. Moderate and mild hemophiliacs rarely have unprovoked bleeding episodes in joints, and mild cases may go undetected into adulthood.

Hemophilia is a sex-linked genetic disease. That is, the genes coding for factors VIII and IX are located on the X-chromosome. A woman has two X-chromosomes, and if only one of the chromosomes has the specific mutation then she is called a carrier. Carriers generally have factor levels from 25% to 49% of the normal level and tend not to bleed. Males have one X-chromosome inherited from their mother, and one Y-chromosome from their father. If the X-chromosome has the specific mutation, then the individual will have hemophilia; about one in 5000 male births result in hemophilia. Hemophilia in females occurs in the very rare case when both X-chromosomes have mutated. There is no family history of the bleeding disorder for about one-third of hemophiliacs, as the disorder often results from a new genetic mutation. The U.S. population of hemophiliacs in 1994 was about 35,000.

Treatment

From the late 1800s into the 1940s, hemophiliacs were treated with blood transfusions. Unfortunately, this treatment was ineffective because there is very little factor VIII in blood, and the transfusions often led to circulatory failure. Plasma transfusions introduced in the 1950s improved the treatment, but the real change in treatment occurred in the 1960s when Pool discovered that the material formed in the bottom of a bag of fresh frozen plasma had high concentrations of the coagulation factors, and techniques for processing these crude concentrates were developed. To make these blood products, plasma from 2000 to 30,000 donors is pooled, purified, and stabilized in a concentrated form. These new concentrates dramatically improved the physical conditions and life expectancy of hemophiliacs. Patients were able to administer treatment at home, and bleeding episodes could be rapidly reversed. People with severe hemophilia require regular transfusions because the body constantly recycles these plasma factors. Within eight hours of a transfusion, half of the factor VIII received will be eliminated from the bloodstream.

In the early 1980s, the concentrates became contaminated from donors with the Human Immunodeficiency Virus (HIV), and by 1985 roughly two-thirds of the U.S. hemophiliac population was infected with HIV. Hemophiliacs began dying of AIDS. Deaths increased from 0.4 per million in 1978 to 1.3 per million in the 1979–89 period (Chorba [Chorba94]), and the median age at death dropped from over 60 years in the early 1980s (Figure 6.1) to 38 in 1989.

Starting in 1985, donors were screened for blood-borne viruses, and all blood products were tested for hepatitis and HIV. Later, new heat techniques and genetic processing were developed to better purify the concentrates. From 1985 to 1994, only 29 cases of HIV transmission were reported among recipients of blood products screened for HIV.

AIDS

Acquired Immune Deficiency Syndrome (AIDS) is the name given to a new medical syndrome: a fatal immune deficiency acquired by previously healthy patients. The immune deficiency in AIDS involves a particular kind of cell in blood and lymphatic tissues, called a T-lymphocyte. In the syndrome, a subset of these cells, the CD4+ cells, gradually disappear. These CD4+ cells help stimulate the immune system. A healthy adult has a CD4+ cell count between 800 and 1000. Under physical stress, injury, and chronic stress, the CD4+ count might drop to 500 and mild non-fatal infections may result. The CD4+ count of an adult with full-blown AIDS is under 200, and this count continues to decrease over time. Another T-lymphocyte cell is the CD8+. In a healthy adult, the CD8+ cell count is between

400 and 500. AIDS does not affect the CD8+ cell count, so the ratio CD4+/CD8+ is generally less than 1 for people with AIDS.

Those diagnosed with AIDS die of immune failure; they are subject to fatal opportunistic infections, such as the Cytomegalovirus (CMV) described in Chapter 4. People with low CD4+ counts had been dying from immune failure prior to the AIDS epidemic. However, these deaths were generally associated with cancer, malnutrition, tuberculosis, radiation, or chemotherapy. AIDS differs from these causes of immune deficiency in that it is "acquired," meaning that it is "picked up" by healthy people.

In 1993, the CDC diagnosed AIDS when the following symptoms are present: CD4+ count under 500 or a CD4+/CD8+ ratio under 1; HIV infection; and either a CD4+ count under 200 or opportunistic infection. Some medical researchers object to this definition because it includes HIV infection and therefore skirts the question as to whether HIV causes AIDS. Conservative skeptics would like HIV infection to be dropped from the CDC definition; other more extreme dissenters, such as Duesberg, advocate expanding the definition to include all those patients with CD4+ counts under 500 or a CD4+/CD8+ ratio under 1. This criterion would include a large number of HIV-free cases.

Does HIV Cause AIDS?

The risk groups for AIDS in western countries are illicit drug users, recipients of blood products, and male homosexuals. All people in these high-risk groups who have low and declining CD4+ counts have been infected with HIV (Harris [Harris95]). (French researcher Luc Montagnier first isolated the HIV virus in 1983 from a French patient who later died of AIDS.) This observation does not prove that HIV causes AIDS. However, according to Harris, the CDC reported after a massive search that it had found fewer than 100 cases without HIV infection that had CD4+ counts that were less than 300. These people were not in the usual AIDS risk groups. In addition, their CD4+ cell counts were often higher than 300 and did not progressively decrease.

Duesberg claims that HIV is nothing more than a benign passenger virus, and that AIDS is a result of lifestyle choices: illicit drug use; use of AZT, the first drug approved for treating AIDS; and contaminants in blood products. HIV skeptics argue that the lifetime dosages of illicit drugs, AZT, and factor concentrates need to be considered in epidemiological studies of the relationship between HIV infection and AIDS. They also point out that HIV does not satisfy Koch's postulates for proving that an organism causes disease.

Robert Koch, a 19th century physician, postulated simple rules that should be fulfilled in order to establish that an organism causes disease. One postulate states that the organism must be isolated in pure culture; that the culture be used to transmit disease; and that the organism must be isolated again from the diseased subject. This has not been confirmed with HIV.

Some attempts to argue that HIV satisfies Koch's postulate have been made. For example, animal experiments on a strain of HIV found in AIDS patients in

West Africa has caused AIDS in monkeys. However, experiments have not been successful in infecting animals with the strain found in the U.S. and many other countries. Also, three lab workers have been accidentally infected with the pure HIV virus. These workers did not belong to any of the high-risk groups, and all developed AIDS. Two of the three did not receive AZT treatment.

Drug Use

The earliest proven case of modern AIDS was in 1959 in the U.K., and the first case found in the U.S. was in 1968. Both of these individuals were victims of severe immune deficiency, and their tissues were preserved and later tested positive for HIV. Additionally, 4% of preserved serum samples from injectable drug users in 1971–72 in the U.S. have been found to be HIV-positive.

It appears that HIV was around long before the epidemic that began in the 1980s. There are two possible reasons for the dramatic increase in AIDS cases. In the late 1960s, illicit drug use became more widespread in the U.S., and the disposable plastic syringe made it possible for large-scale injectable-drug abuse. Also around this time, one faction of male homosexuals began to engage in the high infection risk of the "bath house" lifestyle. Ten years later, there was a large enough fraction of HIV-infected people to contaminate the blood supply.

Evidence that HIV causes AIDS was found in a Dutch study of illegal drug use (Cohen [Cohen94]). After controlling for lifetime dosages, it was found that the CD4+ counts of HIV-positive drug users were well outside the normal range, and the CD4+ counts for comparable HIV-negative users were in the normal range.

AZT

The Concorde study, a British and French study of HIV-positive subjects, found that the mortality of those who began immediate AZT treatment was not significantly different from the mortality of HIV-positive subjects who deferred their AZT treatment. These figures are in Table 6.2. According to Cohen [Cohen94], Duesberg claims that

> The Concorde data exactly prove my points: the mortality of the AZT-treated HIV-positives was 25% higher than that of the placebo [deferred] group.

This statement is examined in more detail in the Exercises.

TABLE 6.2. Observed deaths in HIV-positive subjects according to whether treatment with AZT was immediate or deferred. Data are from the Concorde study as reported in Cohen [Cohen94]).

	Immediate	Deferred
Total deaths	96	76
HIV-related deaths	81	69
Number of subjects	877	872

Factor VIII Contaminants

People with hemophilia represent a well-defined group whose mortality and morbidity have been studied over the long term. One long-term study of hemophiliacs is the U.K. National Haemophilia Register (NHR), established in 1976 to follow all diagnosed hemophiliacs living there.

From 1977 to 1991, over 6000 male hemophiliacs were living in the U.K., and in the 1979–86 period approximately two-thirds of them received transfusions that were potentially contaminated with HIV. As of January, 1993, 15% of these 6000+ individuals had died, 82% were alive, and 3% were lost to followup. Table 6.3 shows the number of deaths and death rates according to severity of hemophilia and HIV infection (Darby et al. [DEGDSR95]). In the 1978–84 period, the mortality rate among all severe hemophiliacs was 8 deaths per 1000 person years (8 d/1000 py). Over the next 8 year period, the mortality rate for HIV-negative severe hemophiliacs remained at 8 d/1000 py, but for HIV-positive severe hemophiliacs it increased to 50 d/1000 py. According to Darby et al.,

> During 1985–92, there were 403 deaths among all HIV seropositive patients, whereas only 60 would have been predicted from the rates in seronegatives, suggesting that 85% of the deaths in seropositive patients were due to HIV infection. Most of the excess deaths were certified as due to AIDS or to conditions recognized as being associated with AIDS.

Figure 6.3 shows mortality rates by calendar period for HIV-positive and HIV-negative patients. The vertical bars in the plot are 95% confidence intervals based on the normal approximation.

The rates in Table 6.3 and Figure 6.3 are weighted averages of rates in the age groups < 15, 15–24, 25–34, 35–44, 45–54, 55–64, and 65–84. The weights are based on the total number of person years at risk in the period 1985–92 for all HIV-positive patients of all degrees of severity. Standardization of rates is described in the Theory section of this chapter.

TABLE 6.3. Observed deaths and standardized death rates per 1000 person years in all hemophiliacs in the NHR by HIV status and severity of hemophilia (Darby et al. [DEGDSR95]).

	Severe hemophilia				Moderate or mild hemophilia			
	HIV-negative		HIV-positive		HIV-negative		HIV-positive	
Years	Deaths	Rate	Deaths	Rate	Deaths	Rate	Deaths	Rate
85–86	8	15.0	43	23.9	5	2.4	13	19.4
87–88	13	9.3	74	41.3	14	2.0	10	23.8
89–90	13	9.9	96	56.8	19	4.6	22	63.0
91–92	7	3.6	121	80.8	26	4.1	24	84.7
85–92	41	8.1	334	49.1	64	3.5	69	45.2

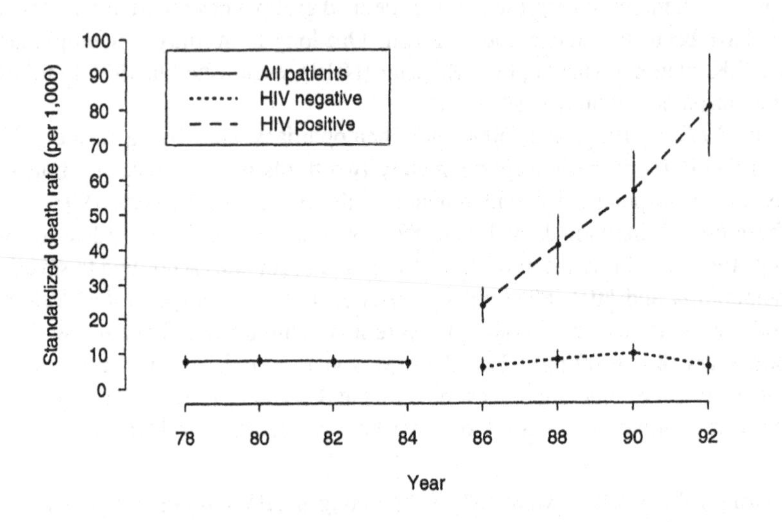

FIGURE 6.3. Death rates, with confidence intervals, for severe hemophiliacs per 1000 person years, by calendar year and HIV status (Darby et al. [DEGDSR95]). In this figure, the HIV-negative group includes those of unknown HIV status.

Investigations

The main question to be addressed here is how the mortality of HIV-positive hemophiliacs compares to that of HIV-negative hemophiliacs.

Table 6.4 shows that the crude mortality rate for the HIV-positive hemophiliacs in the MHCS is 34.1 deaths per 1000 person years, and the rate for HIV-negative hemophiliacs is only 1.6 deaths per 1000 person years. However, it may be inappropriate to make such crude comparisons because the HIV-positive and HIV-negative hemophiliacs may differ according to some factor that is associated with mortality, which confounds the results. For example, if the HIV-positive hemophiliacs tend to be older than the HIV-negative hemophiliacs, then the additional deaths in this group may be explained by differences in the age distribution between the two groups. If there is a difference in age distribution, then we should compare the rates within age categories.

- Begin by comparing the age distribution of person years for the HIV-negative and HIV-positive groups. Is the HIV-negative population younger or older than the HIV-positive population? From 1980 to 1985, the clotting-factor concentrates were contaminated with HIV, and it was during this time period that the hemophiliacs who received contaminated transfusions were infected. By 1985, methods were available for screening blood supplies, and HIV was virtually eliminated from blood products. How might this window of exposure affect the age distribution of the HIV-infected hemophiliacs over time? Is calendar year a potential confounder?

- According to Duesberg, it is not HIV that has caused AIDS among hemophiliacs, but the "other junk" in the transfusions the patient receives. Duesberg claims that these other contaminants are the real cause of AIDS, and that it is critical to take into account the quantities of clotting-factor concentrate that people with hemophilia receive in their lifetime for a fair comparison of mortality rates of seropositive and seronegative patients. Do you find evidence in the MHCS study to support Duesberg's claim?
- Consider various weighting schemes for standardizing the mortality rates. If the ratio of mortality rates for HIV-positive to HIV-negative patients within an age subgroup is roughly constant across age subgroups, then the age-specific mortality rates for an HIV group may be combined to provide a single rate. It is usually easier to compare these single rates. The HIV-positive mortality rates for each age group may be combined using a weighted average, where the weights reflect the relative size of each age group in some standard population. The same age adjustment would then be made to the HIV-negative rates. Alternatively, the rates may be combined with weights chosen to minimize the variability in the difference between these rates. In either case, these combined rates should be interpreted with caution, as changes in the weights can produce quite different summary rates.
- Provide interval estimates for the difference in mortality rates between HIV-positive and HIV-negative hemophiliacs according to severity of hemophilia and calendar year. Present your results graphically. Darby et al. (Figure 6.3) provide similar estimates using the NHR. These mortality curves give a picture of the progression of the AIDS epidemic through the hemophilia population in the U.K. The bars used to denote variability in the figure are based on the normal approximation. The normal approximation may not be entirely appropriate in this setting because of small counts, and a population rather than a random sample was observed. Nonetheless, the intervals are useful in providing a comparison of the seropositive and seronegative groups.

Write your report as a letter to the editor of *Nature* to follow up on Darby et al. [DEGDSR95]. Explain your findings and whether or not they corroborate the response of Darby et al. to what Cohen [Cohen94] in *Science* labels the "Duesberg phenomenon."

Theory

The data for this lab are from a *cohort study*. The cohort or group under observation is hemophiliacs. The population in the MHCS is *dynamic* rather than *fixed* as new subjects enter the study after the initial start date. A cohort study is *prospective*: subjects are followed through time, occurrence of disease or death are noted, and the rate at which people die or become diseased (i.e., the incidence of disease) is measured. In contrast, a *cross-sectional* study takes a snapshot of a population at a fixed point in time. With a cross-sectional study, the prevalence of disease, or proportion of subjects with a disease, is measured.

Proportions, Rates, and Ratios

Proportions, rates, and ratios are used to measure disease. A proportion measures prevalence; it is a fraction in which the numerator is included in the denominator, such as the proportion of hemophiliacs who are HIV-infected. A rate measures change in one quantity per unit change in another quantity, where the second quantity is usually time. The speed of a car in kilometers per hour is an instantaneous rate; the total distance traveled in a car divided by the total travel time is an average rate. In the MHCS, the mortality rate for hemophiliacs is an average rate; it is the ratio of the number of deaths (d) to the number of person years (py) observed in the study period. For the HIV-positive hemophiliacs, the mortality rate (Table 6.4) is

$$434 \text{ deaths}/12,724 \text{ person years} = 34.1 \text{ d}/1000 \text{ py}.$$

Person years provides a measure of exposure to HIV infection.

Ratios provide a means of comparing two proportions or two rates. The ratio of mortality rates, or *mortality ratio*,

$$\frac{34.1 \text{ HIV-negative d}/1000 \text{ py}}{1.6 \text{ HIV-positive d}/1000 \text{ py}} = 21,$$

indicates that the average mortality rate for HIV-positive hemophiliacs is 21 times that for HIV-negative hemophiliacs.

As a rule of thumb, a small number such as 0.5 is often added to each of the counts to stabilize the effect of small cell counts on the rates and ratios. The use of 0.5 is arbitrary; another value, such as 0.1, may be added to the counts and may produce different rates and ratios. In our example, the mortality ratio of HIV-positive to HIV-negative hemophiliacs is 21; when 0.5 is added to each of the death tallies, the ratio drops to 19. On the other hand, adding 0.1 does not change the ratio. Standard errors for these rates and ratios are useful for making confidence intervals and testing differences. This topic is discussed next.

TABLE 6.4. Observed deaths, person years, and crude death rates per 1000 person years for hemophiliacs in the MHCS by HIV status (Goedert et al. [GKA89]).

	Deaths	PY	Rate
HIV-negative	26	16,343	1.6
HIV-positive	434	12,724	34.1

Poisson Counts

In epidemiology (Kleinbaum et al. [KKM82]), the standard errors for rates and ratios of rates are based on the assumption that counts (e.g., of deaths in strata) have either binomial or Poisson distributions. The usual justification for the latter is what is sometimes called the *law of small numbers*; namely, the Poisson approximation to the binomial. This "law" asserts that the total number of successes in a large number n of Bernoulli trials, with small common probability p of success, is approximately Poisson distributed with parameter np.

For a large group of individuals, such as hemophiliacs, the individual chances of death in some short period are quite small, so there is some plausibility to the assumption that mortality experience is like a series of Bernoulli trials, where the total number of deaths can be approximated by the Poisson distribution. However, we are not dealing with perfectly homogeneous groups of people, rates are seldom completely stable over the relevant time periods, and deaths are not always independent, so the Bernoulli assumption will be violated to some extent. One consequence of this violation is a greater variance associated with the counts than would be the case with the binomial or Poisson distributions. This is sometimes dealt with by using models incorporating what is known as "over-dispersion" (McCullagh and Nelder [MN89]).

We will calculate variances of rates and ratios under the Poisson assumption in what follows, with the knowledge that in many cases these will be understated. If we let λ represent the mortality rate per 1000 person years, then the number of deaths D observed over $m \times 1000$ person years follows a Poisson($m\lambda$) distribution with expected value and variance,

$$\mathbb{E}(D) = m\lambda,$$
$$\text{Var}(D) = m\lambda.$$

The observed mortality rate is $R = D/m$. Its expectation is λ, and SE is $\sqrt{\lambda/m}$. Recall from Chapter 4 that R is the maximum likelihood estimate of λ. Provided $m\lambda$ is large, the distribution of R is well approximated by the normal. We can estimate $m\lambda$ by D, so the SE of R can be estimated by $\sqrt{D}/m$. Hence we find an approximate 95% confidence interval for λ,

$$R \pm 1.96\sqrt{D}/m.$$

Note that these intervals and estimates are conditional on the number of person years observed; that is, we treat m as a nonrandom quantity.

Comparing Rates

Most often, we would like to compare the mortality rates for different groups such as HIV-positive and HIV-negative hemophiliacs. Suppose for HIV-positive hemophiliacs we observe D_+ deaths over $m_+ \times 1000$ person years, and we have a Poisson model with parameter $m_+\lambda_+$. Take $R_+ = D_+/m_+$ as an estimate for λ_+.

Similarly, for HIV-negative hemophiliacs, D_- follows a Poisson($m_-\lambda_-$) distribution, and $R_- = D_-/m_-$ estimates λ_-. To compare rates for the two groups, we consider the ratio

$$\frac{R_+}{R_-} = \frac{m_- D_+}{m_+ D_-}.$$

The distribution of possible values for this ratio is often asymmetric; the ratio has to be nonnegative and is one when the rates are equal. Hence it is often misleading to use a normal approximation to provide confidence intervals.

One remedy is to create confidence intervals for the logarithm of the ratio $\log(R_+/R_-)$, which tends to have a more symmetric distribution. In the exercises of Chapter 3, we used the delta method to approximate variances of transformed random variables. We leave it as an exercise to derive the approximate variance

$$\mathrm{Var}[\log(R_+/R_-)] \approx \frac{1}{\lambda_+ m_+} + \frac{1}{\lambda_- m_-},$$

which we estimate by

$$\frac{1}{D_+} + \frac{1}{D_-}.$$

Then an approximate 95% confidence interval for the log of the ratios is

$$\log(R_+/R_-) \pm 1.96\sqrt{\frac{1}{D_+} + \frac{1}{D_-}}.$$

We take exponentials of the left and right endpoints of the interval to obtain an approximate 95% confidence interval for R_+/R_-:

$$\frac{R_+}{R_-}\exp\left(\pm 1.96\sqrt{\frac{1}{D_+} + \frac{1}{D_-}}\right).$$

Notice that the confidence interval is not symmetric about the ratio R_+/R_-.

Rates can also be compared via differences such as $(R_+ - R_-)$. Ratios have the advantage of always producing viable estimates for rates; negative rates may arise from estimates and confidence intervals based on differences.

When rates for more than two groups are to be compared, one group may be designated as the reference group. Typically it is the lowest-risk group. Then ratios are calculated with the reference group in the denominator.

Adjusting Rates

The figures 34.1 and 1.6 d/1000 py from Table 6.4 are crude mortality rates because they do not control for potential confounding factors such as age or severity of hemophilia. For instance, HIV is transmitted via contaminated blood products, and severe hemophiliacs receive more transfusions and have higher mortality rates than moderate or mild hemophiliacs. It seems likely that HIV-infected hemophiliacs are more often severe cases, so the comparison of these crude rates can be misleading.

TABLE 6.5. Observed deaths, person years, and crude death rates per 1000 person years in the MHCS by HIV status and severity of hemophilia (Goedert et al. [GKA89]).

	Severe hemophilia			Moderate/mild hemophilia		
	Deaths	PY	Rate	Deaths	PY	Rate
HIV-negative	4	1939	2.3	16	6387	2.6
HIV-positive	199	4507	44.2	138	3589	38.5

It would be more appropriate to compare mortality rates within subgroups of hemophiliacs that have the same severity of hemophilia. Table 6.5 provides these figures; the 434 deaths in the HIV-positive group are divided among 199 severe, 138 moderate/mild, and 97 of unknown severity. For those diagnosed with severe hemophilia, the mortality rates are 2.3 d/1000 py for the HIV-negative and 44.2 d/1000 py for the HIV-positive.

Controlling for age produces many subgroups, as shown in Table 6.6. To simplify the comparison of groups according to the main factor, such as HIV status, we can *standardize* the mortality rates by age. One way to do this is to apply the mortality rates for age groups to a standard population. For example, the distribution of person years from 1978 to 1980 for all hemophiliacs in the MHCS could be used as a standard hemophiliac population. Figure 6.4 displays the distribution of these person years.

The age-standardized mortality rate for HIV-positive hemophiliacs is computed as follows

$$(0.38 \times 7.9) + (0.29 \times 21.2) + \cdots + (0.02 \times 126.5) = 25 \text{ d}/1000 \text{ py}.$$

A similar calculation for HIV-negative hemophiliacs gives an age standardized rate of 1.2 d/1000 py. These rates are standardized to the same age distribution. Other factors may be simultaneously controlled in a similar manner. However, sparsity of data can make these comparisons difficult.

An alternative weighting scheme for standardizing ratios selects weights that minimize the variance of the estimator. These are known as *precision weights*. Suppose we want to compare the mortality rates of two groups, controlling for a second factor that has G groups. For simplicity, we will denote the two main

TABLE 6.6. Observed deaths, person years, and crude death rates per 1000 person years in the MHCS by HIV status and age (Goedert et al. [GKA89]).

		HIV-positive			HIV-negative		
		Deaths	PY	Rate	Deaths	PY	Rate
Age	<14	16	2034	7.9	1	6240	0.2
	15–24	84	3967	21.2	1	3764	0.3
	25–34	112	3627	30.9	3	3130	1.0
	35–44	117	2023	57.8	4	1713	2.3
	45–54	52	655	79.4	6	905	6.6
	55+	53	419	126.5	11	590	22.4

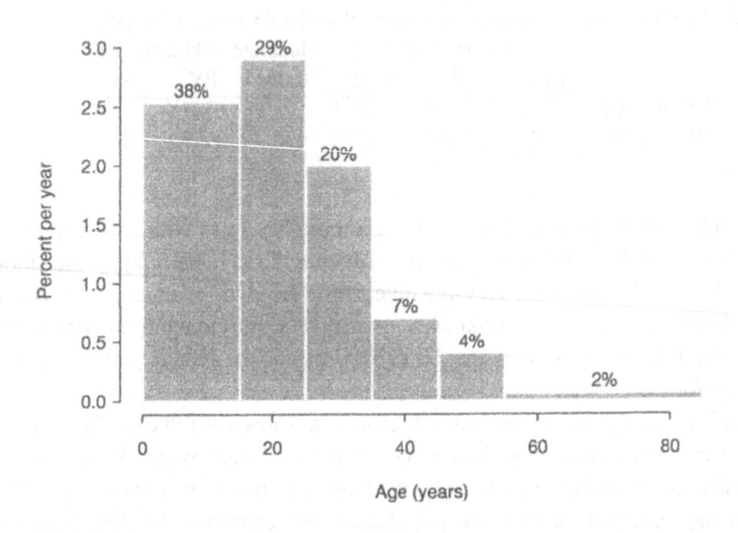

FIGURE 6.4. Histogram of the distribution of person years from 1978 to 1980 for all hemophiliacs in the MHCS. The area of each bar is provided to the nearest percent.

groups by + and −. Let D_{+g} be the number of deaths in the gth subgroup of the positive group and D_{-g} be the number of deaths in the corresponding subgroup of the negative group. Weights $w_1, \ldots, w_G$, called *precision weights*, are chosen to minimize the variance

$$\mathrm{Var}\left[\sum_{g=1}^{G} w_g \log(R_{+g}/R_{-g})\right].$$

Recall that the log transformation symmetrizes the distribution of the ratio of rates. We leave it to the Exercises to show that the minimizing weights are

$$w_g = \frac{1/\mathrm{Var}[\log(R_{+g}/R_{-g})]}{\sum_{i=1}^{G} 1/\mathrm{Var}[\log(R_{+i}/R_{-i})]},$$

which we would estimate by

$$\hat{w}_g = \frac{D_{+g}D_{-g}}{D_{+g} + D_{-g}} \left(\sum_{i=1}^{G} \frac{D_{+i}D_{-i}}{D_{+i} + D_{-i}}\right)^{-1}.$$

The precision weights are not viable when the number of deaths in stratum g is 0 (i.e., $D_{+g} = D_{-g} = 0$), as the weight $\hat{w}_g$ would not be defined. We could correct the problem by adding a small increment to each D. However, the weights may be sensitive to the size of the increment, especially if more than one stratum has zero weight.

Standardizing rates may be misleading when the ratios R_{+g}/R_{-g} vary greatly. For example, suppose the ratio for one subgroup is less than 1, which indicates a "protective" effect, and also suppose the ratio is greater than 1 for another subgroup, indicating a "detrimental effect." These two contradictory effects can be masked in weighted averages of rates. Standardizing is most appropriate when all ratios of rates in subgroups go in the same direction (i.e., are all greater than 1) and are all roughly the same size. See the Exercises for examples of this.

Mantel–Haenszel Test

The confidence intervals introduced earlier can be used to test a hypothesis that two rates are the same (i.e., that the ratio of rates is 1). If the 95% confidence interval does not include 1, then the hypothesis would be rejected at the 5% level.

An alternative test is based on comparing the observed to expected deaths under the hypothesis that the rates are the same. This test is a Mantel–Haenszel-type test. It is applied conditionally given the total deaths in the two groups, $n = D_+ + D_-$. In the Exercises, we show that given n, m_+, and m_-, the number of deaths in the positive group D_+ follows a binomial distribution with parameters n and

$$p = \frac{m_+\lambda_+}{m_+\lambda_+ + m_-\lambda_-}.$$

Under the hypothesis that $\lambda_+ = \lambda_-$, the probability p simplifies to $m_+/(m_+ + m_-)$, and

$$\mathbb{E}(D_+) = n\frac{m_+}{m_+ + m_-},$$

$$\mathrm{SD}(D_+) = \frac{\sqrt{nm_+m_-}}{m_+ + m_-}.$$

Additionally, the Mantel–Haenszel test statistic,

$$\left(\frac{D_+ - np}{\sqrt{np(1-p)}}\right)^2,$$

has an approximate χ_1^2 distribution under the null hypothesis.

The Mantel–Haenszel test can also be applied to a stratified analysis such as one that controls for age. For example, let D_{+g} have a Poisson$(m_{+g}\lambda_{+g})$ distribution, and similarly the D_{-g} have a Poisson$(m_{-g}\lambda_{-g})$ distribution, $g = 1 \ldots, G$. Given $n_g = D_{+g} + D_{-g}$, $g = 1, \ldots, G$, the D_{+g} are independent binomials. Under the hypothesis that $\lambda_{+g} = \lambda_{-g}$, for all g, the parameters for the binomial are n_g and $p_g = m_{+g}/(m_{+g} + m_{-g})$, and the test statistic

$$\left(\frac{\sum[D_{+g} - n_g p_g]}{\sqrt{\sum n_g p_g(1-p_g)}}\right)^2$$

has a χ_1^2 distribution. This test statistic is robust against small cell counts because the cell counts are summed before the ratio is taken.

TABLE 6.7. Events for hypothetical patients in a cohort study.

Subject	Entered	Exposed	Last contact	Outcome
A	1/1/80	7/1/85	12/31/95	alive
B	1/1/80	1/1/84	9/1/92	died
C	1/1/80	—	4/1/94	died
D	1/1/80	—	1/1/88	withdrew
E	1/1/80	—	12/31/95	alive
F	1/1/80	—	1/1/90	withdrew
G	1/1/82	1/1/84	9/1/91	died
H	7/1/85	—	9/1/92	died
I	9/1/83	7/1/83	12/31/95	alive
J	1/1/90	—	12/31/95	alive
K	1/1/81	1/1/85	1/1/94	withdrew
L	4/1/83	1/1/85	1/1/91	died

Exercises

1. Use the z statistic to conduct a test of the hypothesis that the proportion of deaths in the group of patients in the Concorde study (Table 6.2) who received immediate AZT treatment equals the proportion of deaths in the group that deferred treatment. Comment on Duesberg's analysis of the data as reported in Cohen ([Cohen94]) and on page 126.
2. Consider the time line (Table 6.7) for the hypothetical patients in a study of the effect of exposure to some agent on mortality. For each calendar year of the study, determine the number of person years and deaths in the exposed and unexposed groups.
3. Let D_1 be an observation from a Poisson(λ_1) distribution, and let D_2 be from a Poisson(λ_2). Given that we know the total $n = D_1 + D_2$, show that D_1 follows a binomial($n, \lambda_1/(\lambda_1 + \lambda_2)$) distribution. That is, show

$$\mathbb{P}(D_1 = k | D_1 + D_2 = n) = \binom{n}{k} \left(\frac{\lambda_1}{\lambda_1 + \lambda_2}\right)^k \left(\frac{\lambda_2}{\lambda_1 + \lambda_2}\right)^{n-k}.$$

4. Let D have a Poisson(λ) distribution. Consider the transformation $\log(D)$.

   a. Use a first-order Taylor series expansion (see Exercise 9 in Chapter 3) to show that for large λ

$$\mathrm{Var}(\log(D)) \approx \frac{1}{\lambda}.$$

   b. Use this result twice to derive the approximation

$$\mathrm{Var}[\log(R_1/R_2)] \approx \frac{1}{m_1\lambda_1} + \frac{1}{m_2\lambda_2},$$

   where $R_1 = D_1/m_1$, D_1 has a Poisson($m_1\lambda_1$) distribution, and D_2 is independent of D_1 and similarly defined.

5. Use the summary statistics in Table 6.4 to provide a 95% confidence interval for the mortality rate of HIV-positive hemophiliacs.
6. Use the information in Table 6.5 to make a 95% confidence interval for the ratio of mortality rates for HIV-positive and HIV-negative hemophiliacs.
7. **a.** Determine the precision weights for the age categories shown in Table 6.6.
   b. Use these weights to provide an age-adjusted estimate for the ratio of mortality rates of HIV-positive and HIV-negative hemophiliacs.
   c. Provide a confidence interval for the population ratio.
8. Use the weights from Exercise 7 to standardize the mortality rate for HIV-positive hemophiliacs displayed in Table 6.6. Explain the differences obtained.
9. Compare the ratios of mortality rates of HIV-positive to HIV-negative hemophiliacs for the age groups in Table 6.6. Do they all go in the same direction? Are they all roughly the same magnitude?
10. Add 0.5 to each of the counts in Table 6.6 and recalculate the age-adjusted rates for HIV-positive and HIV-negative hemophiliacs using the standard weights displayed in the histogram in Figure 6.4. Compare these rates to the 25 and 1.2 d/1000 py obtained without adding a small constant to each cell. Also compare them to the rates obtained by adding 0.1 to each count. How sensitive are the rates to the size of the addition?
11. Compute the Mantel–Haenszel test statistic to compare HIV-positive and HIV-negative deaths; adjust for severity of hemophilia using Table 6.5.
12. Test the hypothesis of no difference in mortality between HIV-positive and HIV-negative hemophiliacs. Use the Mantel–Haenszel test and adjust for age (Table 6.6).
13. Study the robustness of the Mantel–Haenszel test to small cell counts. Add small values such as 0.1, 0.5, and 0.8 to the cell counts in Table 6.6 and compute and compare the Mantel–Haenszel test statistics.
14. For uncorrelated D_g with variance σ_g^2, $g = 1, \ldots, G$, show that the w_g that minimize

$$\mathrm{Var}\left(\sum w_g D_g\right)$$

subject to the constraint that $\sum w_g = 1$ are

$$w_g = \frac{1/\sigma_g^2}{\sum 1/\sigma_j^2}.$$

Notes

Darby (1995) served as the impetus for studying this subject. Articles by Cohen (1994) and Harris (1995) were the source for much of the background material on the controversy over whether HIV causes AIDS. The background material on hemophiliacs was from Hoyer [Hoyer94] and Levine [Levine93]. Kleinbaum et al. [KKM82] provides further details on standardizing rates and the Mantel–Haenszel-type test. See especially Chapter 17, Sections 1 and 3. The data were made available by James J. Goedert and Philip S. Rosenberg of the Multicenter Hemophilia Cohort Study. Christine Chiang assisted in the preparation of the background material for this lab.

References

[Chorba94] T.L. Chorba, R.C. Holman, T.W. Strine, M.J. Clarke, and B.L. Evatt. Changes in longevity and causes of death among persons with hemophilia A. *Am. J. Hematol.*, **45**:112–121, 1994.

[Cohen94] J. Cohen. The Duesberg Phenomenon. *Science*, **266**:1642–1649, 1994.

[DEGDSR95] S.C. Darby, D.W. Ewart, P.L.F. Glangrande, P.J. Dolin, R.J. Spooner, and C.R. Rizza. Mortality before and after HIV infection in the complete UK population of haemophiliacs. *Nature*, **377**:79–82, 1995.

[GKA89] J.J. Goedert, C.M. Kessler, L.M. Aledort, R.J. Biggar, W.A. Andes, G.C. White, J.E. Drummond, K. Vaidya, D.L. Mann, M.E. Eyster, M.V. Ragni, M.M. Lederman, A.R. Cohen, G.L. Bray, P.S. Rosenberg, R.M. Friedman, M.W. Hilgartner, W.A. Blattner, B. Kroner, and M.H. Gail. A prospective study of human immunodeficiency virus type 1 infection and the development of AIDS in subjects with hemophilia. *N. Engl. J. Med.*, **321**:1141–1148, 1989.

[Goedert95] J.J. Goedert. Mortality and haemophilia. *Lancet*, **346**:1425-1426, 1995.

[Harris95] S.B. Harris. The AIDS heresies: A case study in skepticism taken too far. *Skeptic*, **3**:42–79, 1995.

[Hoyer94] L.W. Hoyer. Hemophilia A. *N. Engl. J. Med.*, **330**:38–43, 1994.

[KKM82] D.G. Kleinbaum, L.L. Kupper, and H. Morgenstern. *Epidemiologic Research: Principles and Quantitative Methods*. Lifetime Learning Publications, Belmont, CA, 1982.

[Levine93] P.H. Levine. HIV infection in hemophilia. *J. Clin. Apheresis*, **8**:120–125, 1993.

[MN89] P. McCullagh and J.A. Nelder. *Generalized linear models, 2nd edition*. Chapman and Hall, New York, 1989.

7

Dungeness Crab Growth

SUNDAY, OCTOBER 20, 1996 San Francisco Examiner [1]

Stripping the Seas

Fishermen are taking from the oceans faster than species can be replenished

By Jane Kay

Commercial fishing vessels are hauling so much seafood from the world's oceans that more than 100 species are in danger, scientists say.

Last week, an international science group issued a "red list" of imperiled marine fish, including some species of tuna, swordfish, shark and Pacific red snapper.

"The ocean cannot sustain the massive removal of wildlife needed to keep nations supplied with present levels of food taken from the sea," said Sylvia Earle, an internationally known marine biologist from Oakland.

Regulators and fishing groups, among them the National Marine Fisheries Service and the Pacific Fishery Management Council, consider the list an indicator of failing fish populations.

The World Conservation Union's 1996 list of 5,205 threatened animals includes 120 marine fish - a record number. It was based on the recommendations of 30 experts brought together by the Zoological Society of London.

Until now, the group has put only a handful of marine fish on the list, which began in the 1960s and is updated every three years. ...

The problems associated with overfishing began more than a half century ago when the annual catch of sea creatures was about 20 million tons.

In the 1940s, demand for fresh fish grew.

With advanced fish-locating technology, factory-size vessels, longer fishing lines and nets, the world catch exceeded 60 million tons by the 1960s. It peaked in 1989 at 86 million tons, then began to dwindle.

In 1993, the National Marine Fisheries Service announced that of 157 commercially valuable fish species in the United States, 36 percent were overfished and 44 percent were fished at the maximum level. ...

---

[1] Reprinted by permission.

Introduction

Dungeness crabs (*Cancer magister* Dana, Figure 7.1) are commercially fished between December and June along the Pacific coast of North America. In U.S. waters, nearly the entire adult male Dungeness crab population is fished each year. Female crabs are not fished in order to maintain the viability of the crab population. However, the question of fishing female crabs has been raised as a possible means of controlling the large fluctuations in yearly catches of crabs. To support the change in fishing law, it has been noted that the fishing industry in Canada allows female crabs to be fished and does not suffer from such large fluctuations in catches (Wickham [Wic88]). It has also been argued that the great imbalance in the sex ratio may have contributed to the decline in the crab population along the central California coast. According to biologists, the imbalance may have caused an increase in the parasitic ribbon worm population to the level where the worm now destroys 50–90% of the crab eggs each year (Wickham [Wic88]).

Size restrictions on male crabs are set to ensure that they have at least one opportunity to mate before being fished. To help determine similar size restrictions for female crabs, more needs to be known about the female crab's growth.

The lack of growth marks on crab shells makes it difficult to determine the age of a crab. This is because crabs molt regularly, casting off their old shell and growing a new one. Adult female Dungeness crabs molt in April and May, although they do not necessarily molt yearly. Biologists (Mohr and Hankin [MH89]) require size-specific information on molting to understand the female crab's growth pattern. Of particular interest is the size of the increase in the width of the shell having observed only the size of the shell after the crab molted; for example, for a female crab with a postmolt shell that measures 150 mm across, the scientists want to provide a prediction for how much the shell changed in size. In this lab, you will have the opportunity to study the growth patterns of female Dungeness crabs in order to assist biologists in developing recommendations for size restrictions on fishing female crabs.

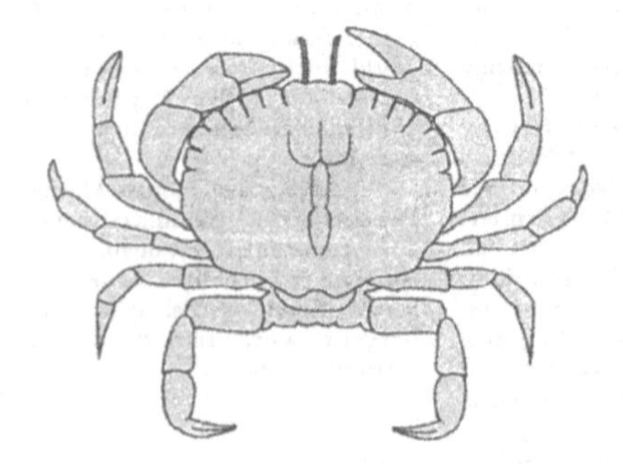

FIGURE 7.1. Dungeness Crab (*Cancer magister*).

TABLE 7.1. Sample observations and data description for the 472 Dungeness crabs collected in 1981, 1982, and 1992 (Hankin et al. [HDMI89]).

Premolt	113.6	118.1	142.3	125.1	98.2	119.5	116.2
Postmolt	127.7	133.2	154.8	142.5	120.0	134.1	133.8
Increment	14.1	15.1	12.5	17.4	21.8	14.6	17.6
Year	NA	NA	81	82	82	92	92
Source	0	0	1	1	1	1	1

Variable	Description
Premolt	Size of the carapace before molting.
Postmolt	Size of the carapace after molting.
Increment	postmolt–premolt.
Year	Collection year (not provided for recaptured crabs).
Source	1=molted in laboratory; 0=capture–recapture.

The Data

The data for this lab were collected as part of a study of the adult female Dungeness crab. The study was conducted by Hankin, Diamond, Mohr, and Ianelli ([HDMI89]) with assistance from the California Department of Fish and Game and commercial crab fishers from northern California and southern Oregon.

Two sets of data are provided. The first consists of premolt and postmolt widths of the carapaces (shells) of 472 female Dungeness crabs. These data are a mixture of some laboratory data and some capture–recapture data. They were obtained by scientists and commercial fisheries over three fishing seasons. The first two seasons were in 1981 and 1982. The third season was in 1992. The information available in the first data set is summarized in Table 7.1. The size measurements were made on the external carapace along the widest part of the shell, excluding spines. All measurements are in millimeters.

The capture–recapture data were obtained by tagging 12,000 crabs. These crabs were caught, measured, tagged with a unique identification number, and returned to the water. The crabs were tagged and released in January through March of each year, before the annual spring molting season. Commercial fisheries brought tagged crabs they caught in their traps to the laboratory for second measurements. Commercial traps have netting designed to catch the larger male crabs; female crabs caught with these traps were typically larger than 155 mm. For an incentive to return the tagged crabs, a lottery of the returned crab tags, with a $500 prize, was held at the end of each fishing season.

The laboratory data were collected during the molting season for female crabs. Crabs that were in a premating embrace were caught and brought to the laboratory. The premolt carapace width was measured when the crab was first collected, and the postmolt measurements were made three to four days after the crab left its old

TABLE 7.2. Sample observations and data description for the 362 female crabs caught in 1983 (Mohr and Hankin [MH89]).

Postmolt	114.6	116.8	128.7	139.3	142.4	148.9	150.8	155.8
Molt	0	1	0	1	1	0	0	1

Variable	Description
Postmolt	Size of the carapace after molting.
Molt classification	1=clean carapace; 0=fouled carapace.

shell to ensure that the new shell had time to harden. The postmolt measurements for all crabs were made in the laboratory, after which they were released.

The second set of data were collected in late May, 1983, after the molting season. The carapace width was recorded as well as information on whether the crab had molted in the most recent molting season or not. The crabs were collected in traps designed to catch adult female crabs of all sizes, and so it is thought that the sample is representative of the adult female Dungeness crab population. This sample consists of 362 crabs. Table 7.2 contains a description of this second set of data.

Background

The Dungeness crab (*Cancer magister*) is a large marine crustacean found off the Pacific coast of North America from the Aleutian Islands to Baja, California. It is one of the largest and most abundant crabs on the Pacific coast.

The crab has a broad, flattened hard shell, or carapace, that covers the back of the animal. The shell is an external skeleton that provides protection for the crab (Figure 7.1). To accommodate growth, the crab molts periodically, casting off its shell and growing a new one. To do this, the crab first takes in water to enlarge its body and split open its shell. The animal then withdraws from the shell and continues to take in water and swell. Then the exterior tissues of the crab harden to form the new shell. Once the shell has hardened, the water is released, and the animal shrinks to its original size, which creates room to grow in the new shell. The molting process is complete in four to five days. Immediately after the molting season, it is fairly easy to determine whether a crab has recently molted; its shell is clean, free of barnacles, and lighter in color.

Crabs mate in April and May when the female molts. The male crabs molt later in the year in July and August. During the female crabs' molting season male and female crabs enter shallow water; a male and female will embrace prior to the female's molting. When the female leaves her shell, the male deposits a sperm packet in the female. Once the female crab's shell has hardened, the male and female separate. The female stores the sperm for months as her eggs develop. In the fall, she extrudes her eggs and fertilizes them with the stored sperm. She carries her eggs on four pairs of bristly abdominal appendages until they hatch, after which they pass through several stages as free-swimming plankton and a final larval stage before they metamorphose into juvenile crabs. The juveniles molt every few months for about two years, until they reach maturity. This occurs when the carapaces (shells) are 90 to 100 mm in width.

Dungeness crabs have many kinds of predators in their different stages of development. The parasitic ribbon worm preys on crab eggs; many kinds of fish eat crab plankton; adult crabs, shorebirds, and fish prey on the juvenile crabs; and otters, sea lions, sharks, and large bottom fish consume adult crabs.

Investigations

In this lab, you will examine the relationship between premolt and postmolt carapace size and summarize your results both numerically and graphically. When you analyze the data, keep in mind that some crabs molted in the laboratory and others molted in the ocean. Studies suggest that crabs in captivity have smaller molt increments than those in the wild. Although the crabs in this study were held in captivity for only a few days, a comparison of the crabs caught by these two collection methods is advisable. Also keep in mind that the crabs were fished in the early 1980s and ten years later.

Figure 7.2 is a histogram for the size distribution of the sample of female crabs collected after the 1983 molting season. The shaded portion of each bar represents the crabs that molted, and the unshaded portion corresponds to those that did not molt. The goal of this lab is to create a similar histogram for the size distribution of the crabs before the molting season, with the shaded region representing the molted crabs. The data where both premolt and postmolt sizes are available can be used to determine the relationship between a crab's premolt and postmolt size, and this relationship can be used to develop a method for predicting a crab's premolt size from its postmolt size.

- Begin by considering the problem of predicting the premolt size of a crab given only its postmolt size. Develop a procedure for doing this, and derive an expression for the average squared error you expect in such a prediction.
- Examine a subset of the data collected, say those crabs with postmolt carapace width between 147.5 and 152.5 mm. Compare the predictions of premolt size for this subset with the actual premolt size distribution of the subset. Do this for one or two other small groups of crabs.
- Use your procedure to describe the premolt size distribution of the molted crabs collected immediately following the 1983 molting season. Make a histogram for the size distribution prior to the molting season of the crabs caught in 1983. Use shading to distinguish the crabs that molted from those that did not molt.

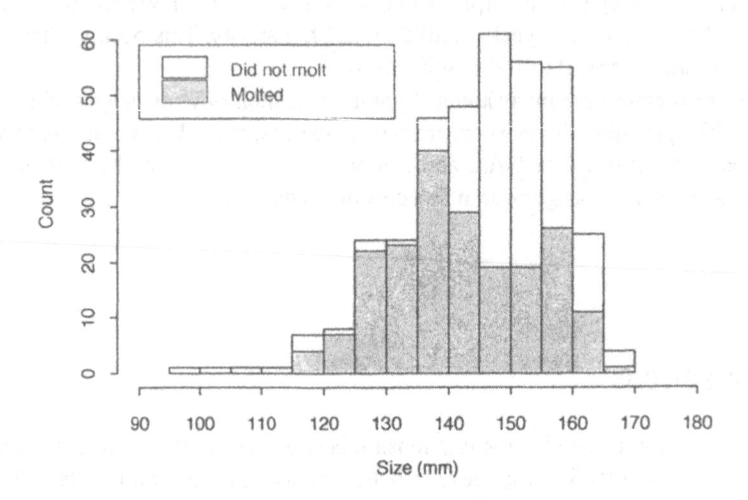

FIGURE 7.2. Size distribution of 362 adult female Dungeness crabs shortly after the 1983 molting season (Mohr and Hankin [MH89]).

Theory

A subset of the crab data will be used to illustrate the theoretical concepts in this lab. It includes those female Dungeness crabs that molted in the laboratory and that have premolt shells at least 100 mm wide.

Correlation Coefficient

From the scatter plot in Figure 7.3, we see that the pre- and postmolt sizes of the crab shells are highly linearly associated. That is, the points on the scatter plot in Figure 7.3 are closely bunched around a line. The linear association referred to here is more formally measured by the *correlation coefficient*. The correlation between premolt and postmolt size for the crabs collected in this study is 0.98. This correlation is very high; correlation coefficients range from -1 to $+1$. A correlation of exactly $+1$ or -1 indicates that all points in the scatter plot fall exactly on a line. The sign of the correlation determines whether the line has positive or negative slope.

Examples of other correlation coefficients are the correlation between the length and weight of babies in the CHDS, which is 0.72, and the correlation of the height and weight of the mothers in the CHDS, which is 0.44.

Positive correlation coefficients indicate that above average values in one variable, such as length, are generally associated with above average values in the second variable, such as weight. It also indicates that below average values in the

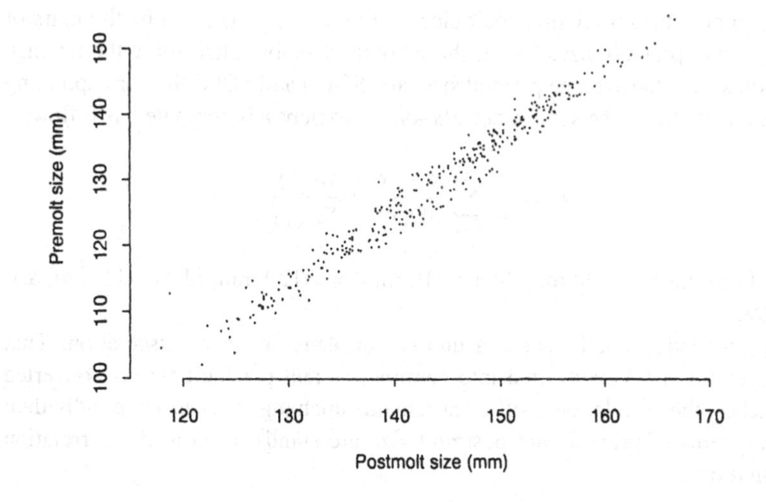

FIGURE 7.3. Scatter plot of premolt and postmolt carapace width for 342 adult female Dungeness crabs (Hankin et al. [HDMI89]).

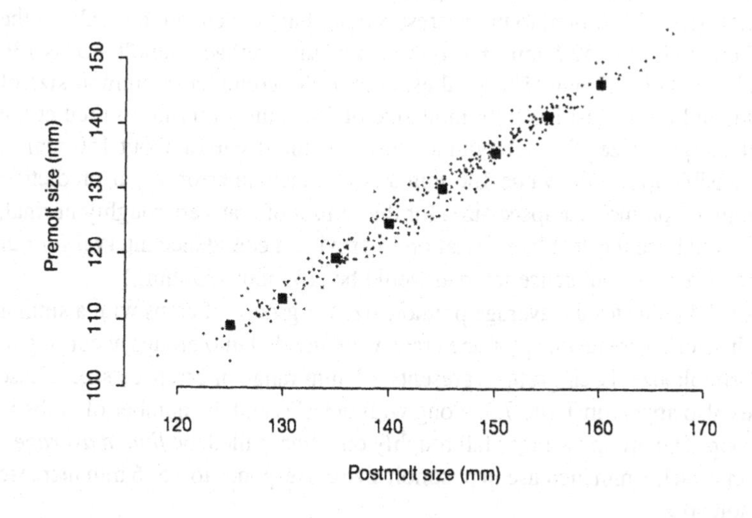

FIGURE 7.4. Scatter plot of premolt and postmolt carapace width for 342 adult female Dungeness crabs; also shown are the average premolt carapace widths at 5 mm increments of postmolt carapace width (Hankin et al. [HDMI89]).

first variable are typically associated with below average values in the second. Conversely, negative correlation coefficients indicate that above average values in one variable are generally associated with below average values in the other variable.

To compute the correlation coefficient, let $(x_1, y_1)\ldots, (x_n, y_n)$ be the pairs of postmolt and premolt sizes for all the laboratory crabs. Then for $\bar{x}$ the average postmolt size, $\bar{y}$ the average premolt size, and SD(x) and SD(y) the corresponding standard deviations, the sample correlation coefficient r is computed as follows:

$$r = \frac{1}{n}\sum_{i=1}^{n} \frac{x_i - \bar{x}}{\mathrm{SD}(x)} \times \frac{y_i - \bar{y}}{\mathrm{SD}(y)}.$$

In our example $\bar{x} = 144$ mm, SD(x)=10 mm, $\bar{y}$ =129 mm, SD(y)=11 mm, and r =0.98.

The correlation coefficient is a unitless measure of linear association. That is, if premolt size is converted into centimeters and postmolt size is converted into inches, the correlation coefficient remains unchanged because the individual measurements of premolt and postmolt size are standardized in the correlation computation.

Averages

From the scatter plot in Figure 7.3, we see that the crabs with a postmolt carapace of 150 mm have premolt carapaces of about 136 mm. We also see that there is natural variability in their premolt sizes. Altogether, there are 69 crabs in the sample with a postmolt size of 150 mm, to the nearest 5 mm; that is, there are 69 crabs in the range from 147.5 to 152.5 mm. For these crabs, the average premolt size is 136 mm and the SD is 2.7 mm. The smallest crab in the group has a premolt size of 130 mm, and the largest has a premolt size of 143 mm. In trying to predict the premolt carapace size of a crab with a known postmolt size of about 150 mm, it seems sensible to use the value 136 mm and to attach an error to our prediction of 2.7 mm. (If premolt carapace sizes for this subset of crabs are roughly normal, then we could use the standard deviation to provide a confidence interval for our estimate — a 68% confidence interval would be [133 mm,139 mm].)

Figure 7.4 indicates the average premolt size for groups of crabs with a similar postmolt size. To create this plot, the crabs were divided into groups according to their postmolt size. Each group represents a 5 mm range in carapace size. These averages also appear in Table 7.3, along with the SDs and the number of crabs in each group. The group averages fall roughly on a line, called the *line of averages*. Notice that each 5 mm increase in postmolt size corresponds to a 5–6 mm increase in premolt size.

The method of least squares can be used to find an equation for this line:

$$\text{premolt} = a + b \times \text{postmolt}.$$

TABLE 7.3. Summary statistics on premolt carapace size for groups of Dungeness crabs with similar postmolt size (Hankin et al. [HDMI89]).

Postmolt size (mm)	Premolt size (mm) Average	SD	Number of crabs
122.5 to 127.5	109	2.8	14
127.5 to 132.5	113	2.2	34
132.5 to 137.5	119	1.9	39
137.5 to 142.5	125	2.8	56
142.5 to 147.5	130	3.0	55
147.5 to 152.5	136	2.7	69
152.5 to 157.5	141	2.0	49
157.5 to 162.5	146	2.3	15

This method finds the line that minimizes the squared difference between the observed premolt size and the premolt size on the line:

$$\sum(\text{premolt} - a - b \times \text{postmolt})^2.$$

Minimizing the sum of squares above is usually done on the individual observations, not the grouped data. That is, the least squares method minimizes the following sum of squares with respect to a and b:

$$\sum_{i=1}^{n}(y_i - a - bx_i)^2,$$

for pairs (x_i, y_i) of postmolt and premolt sizes for n crabs. The minimizing values are denoted by $\hat{a}$ and $\hat{b}$, and the resulting line, $y = \hat{a} + \hat{b}x$, is called the ***regression line*** of premolt size on postmolt size.

The regression line for the crab data is drawn on the plot of averages in Figure 7.5. The equation for the line is

$$\text{premolt} = -29 + 1.1 \times \text{postmolt}.$$

Notice that the slope of this line corresponds to an average 5.5 mm increase in premolt size for each 5 mm increase in postmolt size.

The Regression Line

The regression line has a slope of

$$\hat{b} = r\frac{\text{SD}(y)}{\text{SD}(x)} = 1.1 \text{ mm/mm}.$$

(The derivation is left to the Exercises.) The slope is expressed in terms of the standard deviations and the correlation coefficient of the two variables. In our example, an increase of one SD in postmolt size corresponds, on average, to an increase of 0.98 SDs in premolt size. The linear association of premolt and postmolt size is so high that a one SD increase in postmolt size is associated with nearly a one SD increase in premolt size.

For any pair of variables, there are two regression lines. In Figure 7.5, we plot the regression line for determining premolt size in terms of postmolt size. There is also the regression line that describes postmolt size in terms of premolt size. This line is in a sense the more natural one to consider because the premolt size is measured before the crab molts and can be thought to determine the postmolt size. Its slope is

$$0.98 \text{ SD(postmolt)}/\text{SD(premolt)} = 0.88 \text{ mm/mm}.$$

However, in this lab, we are interested in the less "natural" regression line of premolt on postmolt because, for the crabs measured after the molting season, our task is to determine their premolt size.

An Example

Regression analysis can often be used to address questions involving biological populations that exhibit natural variability. The effect of the correlation on the slope of the regression line is more clearly seen in an example where the linear association is not as high as for the premolt and postmolt shell sizes. Take, for example, the height and weight of mothers in the CHDS study. There, the average height is 64 inches with an SD of 2.5 inches, the average weight is 128 pounds with an SD of 19 pounds, and the correlation is 0.44. Figure 7.6 shows the scatter plot of height and weight for the mothers. Two lines are also marked on the scatter plot; one is the regression line. Its slope is 3.5 pounds/inch (i.e., 0.44×19 pounds/2.5 inches). This line is much less steep then the SD line, the other line marked on the scatter plot. The SD line has a slope equal to the ratios of the SDs with the sign of r (that is, $r\text{SD}(y)/|r|\text{SD}(x)$), so here the slope is 19 pounds/2.5 inches = 7.6 pounds/inch. This effect is called ***regression to the mean*** because those mothers with height one SD below average have weights that are on average only about 0.44 SDs below the overall average (i.e., they are closer than one SD to the overall mean weight).

Residuals

Table 7.3 shows that the crabs' premolt sizes in each group vary about the group's premolt average, with an SD about 2.5 mm. Similarly, a crab's premolt size varies about the regression line. The residual is a name for the difference between a crab's actual premolt size and the regression line prediction of it. More concretely, for crab i, with premolt size y_i and postmolt size x_i, the residual is

$$r_i = y_i - (\hat{a} + \hat{b}x_i), \text{ or}$$
$$r_i = y_i - \hat{y}_i,$$

where the regression line prediction is $\hat{y}_i = \hat{a} + \hat{b}x_i$. Note that the residuals have an average of zero, so the SD and root mean square of the residuals are equal.

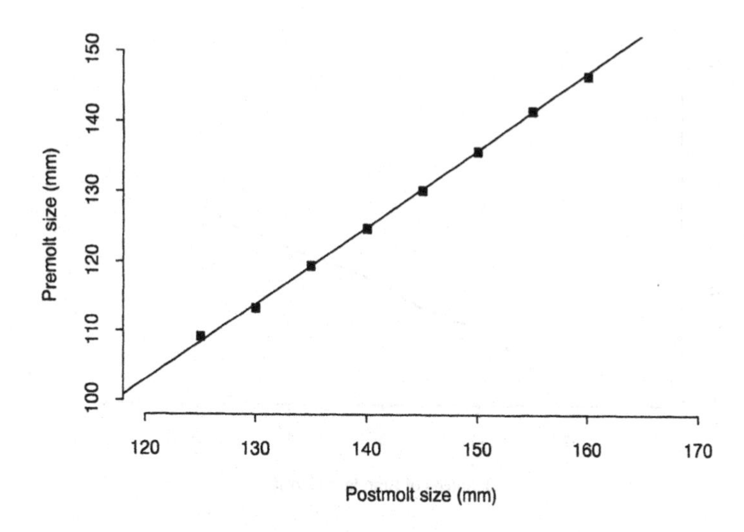

FIGURE 7.5. Regression line for premolt shell size in terms of postmolt shell size for 342 adult female Dungeness crabs; also shown are the average premolt sizes for 5 mm increments in postmolt size (Hankin et al. [HDMI89]).

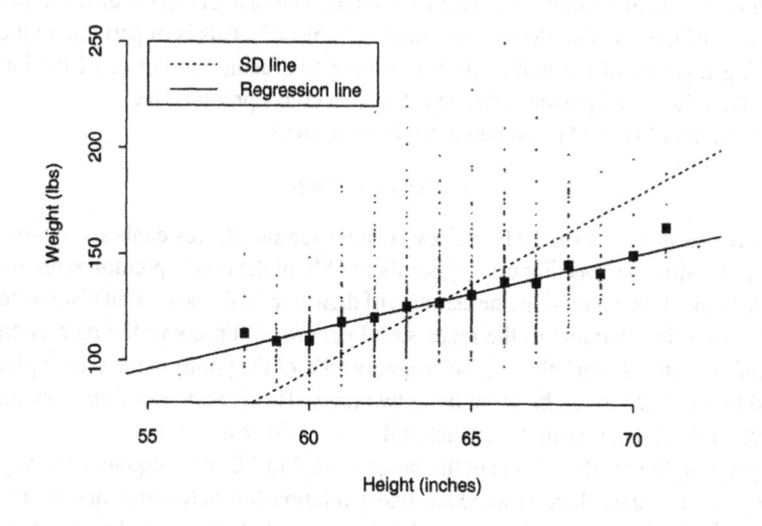

FIGURE 7.6. Scatter plot of heights and weights of 1197 mothers in the CHDS (Yerushalmy [Yer64]); also shown here are the regression line and SD line for weight as determined by height.

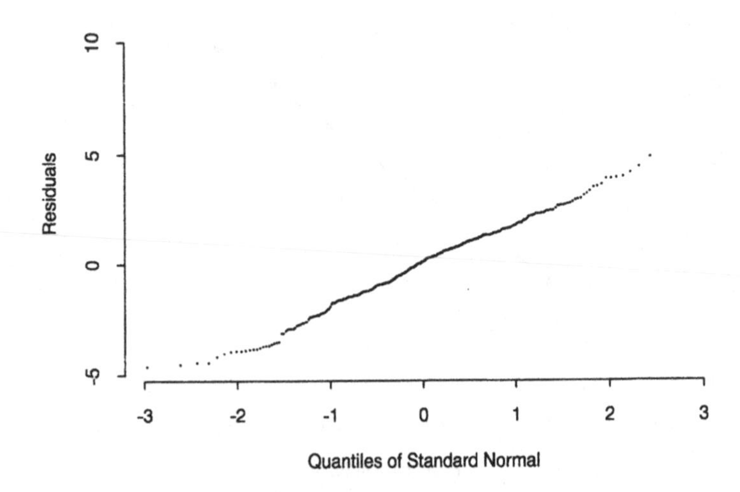

FIGURE 7.7. Normal-quantile plot of the residuals from the least squares fit of premolt size to postmolt size for 342 adult female Dungeness crabs (Hankin et al. [HDMI89]).

The root mean square or SD of the residuals is a measure of how informative the postmolt size is in determining premolt size (i.e., the size of a typical prediction error). In our example, the residual SD is 2 mm. This is much smaller than 11 mm, which is the SD of premolt size for all the crabs. The residual SD is also smaller than most of the standard deviations found in Table 7.3. This is in part due to the grouping together of the crabs into 5 mm bins, and using the center of the bin rather than the actual postmolt size to estimate a crab's premolt size.

The residual SD can be computed easily, as follows:

$$\sqrt{1-r^2}\,\mathrm{SD}(y) = 2\text{ mm}.$$

(We leave the derivation to the Exercises.) In our example, the residuals are roughly normally distributed (see Figure 7.7), so about 68% of the crabs' premolt sizes are within 2 mm of the regression line estimate of their size. This means that if we were to draw two lines parallel to the regression line, one 2 mm above the regression line and the other 2 mm below it, then roughly 68% of the points in the scatter plot would be expected to fall between these two lines. This is very useful in showing how the individual crabs may vary about the regression estimates.

In general, the residual SD is in the range from 0 to SD(y). It equals zero only when $r = \pm 1$; then there is an exact linear relationship between x and y, and knowing x allows us to predict y exactly. When $r \approx 0$, knowing x does not give us much information on what value y might take, so our prediction of y will have as much variability as if we did not know x. Notice that, in our example, an r of 0.98 corresponds to a reduction in variability to $\sqrt{1-r^2} = 0.20$ of the SD(y).

Although the correlation is extremely close to 1, the SD of the residuals is still 20% of the original SD. In fact,

$$\mathrm{SD}(y)^2 = \mathrm{SD}(\text{residuals})^2 + \mathrm{SD}(\hat{y})^2,$$

where the quantity $\mathrm{SD}(\hat{y})$ is the standard deviation of the fitted values. The $\mathrm{SD}(\hat{y})^2$ is that part of the variance of the observed data that can be attributed to variation along the regression line, which is $r^2\mathrm{SD}(y)^2$.

Bivariate Normal

The formula for the regression line follows from a special property of the joint normal distribution. If two standard normal random variables X and Y have correlation ρ, they are said to have a *bivariate normal density* with parameters $(0, 0, 1, 1, \rho)$, which represent the corresponding means, variances, and correlation of X and Y. We say the pair (X, Y) has a $\mathcal{N}_2(0, 0, 1, 1, \rho)$ distribution. The density function of (X, Y) is then

$$\frac{1}{2\pi\sqrt{1-\rho^2}} \exp\left\{-\frac{1}{2(1-\rho^2)}(x^2 - 2\rho xy + y^2)\right\},$$

and its contours are the values (x, y), where $x^2 - 2\rho xy + y^2$ is constant.

A special property of the bivariate normal is that

If (X, Y) have a $\mathcal{N}_2(0, 0, 1, 1, \rho)$ distribution, then X and $Y - \rho X$ are independent; X is $N(0, 1)$ and $Y - \rho X$ is $N(0, \sqrt{1-\rho^2})$.

From this property, we see that

$$Z = \frac{Y - \rho X}{\sqrt{1-\rho^2}}$$

has a standard normal distribution and is independent of X. Hence Y can be expressed as follows:

$$Y = \rho X + \sqrt{1-\rho^2}Z.$$

This means that given $X = x$, Y has a normal distribution with mean ρx and an SD of $\sqrt{1-\rho^2}$; namely,

$$Y = \rho x + \sqrt{1-\rho^2}Z.$$

We can apply this property to the more general case of bivariate normal random variables,

$$(X, Y) \sim \mathcal{N}_2(\mu_1, \mu_2, \sigma_1^2, \sigma_2^2, \rho).$$

By shifting and rescaling X and Y, we find that given $X = x$, the random variable Y has a normal distribution with mean

$$\mu_2 + \sigma_2\rho\frac{x - \mu_1}{\sigma_1}$$

and standard deviation $\sqrt{1-\rho^2}\sigma_2$.

Notice that if we estimate μ_1 by $\bar{x}$, μ_2 by $\bar{y}$, σ_1 by SD(x), σ_2 by SD(y), and ρ by r, then we arrive at $\hat{a}+\hat{b}x$ for an estimate of the conditional mean of Y given $X=x$ and $\sqrt{1-r^2}$SD(y) for its standard deviation. It can also be shown that these estimators of the parameters of the bivariate normal are maximum likelihood estimators.

Exercises

1. For crabs with premolt shells 140 mm wide, the regression line prediction of their postmolt size is ____mm, with an SD of ____mm. According to the normal approximation, we expect roughly 68% of the crabs to have a postmolt shell size within ____mm of this prediction. In fact, we find ____% of the crabs to be that close to the prediction. Table 7.4 contains summary statistics on the shell sizes for the crabs. The postmolt sizes of the subset of crabs with premolt shell size 140 mm are displayed below.
149.5 150.4 150.4 151.4 151.5 151.5 151.5 151.6 151.7 151.7 151.8 152.0
152.1 152.1 152.2 152.2 152.3 152.4 152.9 152.9 153.0 153.1 153.2 153.3
153.4 153.4 153.5 153.5 153.6 153.6 153.8 154.1 154.2 154.2 154.3 154.4
154.5 154.6 154.7 154.8 154.8 154.9 155.0 155.2 155.3 155.4 156.6 156.7

2. For a male Dungeness crab with a premolt carapace 150 mm wide, we use the regression line for female crabs to predict its postmolt carapace (Table 7.4):
$$\frac{150-129}{11}\times 0.98\times 10 + 144 = 163 \text{ mm}.$$
Comment on the appropriateness of this prediction.
3. The scatter plot in Figure 7.8 includes a number of juvenile crabs. Their premolt carapaces are under 100 mm. The regression line $-29+1.1\times$ postmolt does not include these juvenile crabs. Describe how you think the slope of the regression line will change when these crabs are included in the least squares procedure. Next compute the new regression line to confirm your expectations. With the extra crabs, the average premolt carapace width is now 126 mm with SD 16 mm, the average postmolt carapace width is 141 mm with SD 15 mm, and the correlation is 0.99.

TABLE 7.4. Summary statistics on carapace size for female Dungeness crabs (Hankin et al. [HDMI89]).

	Average	SD
Premolt size	129 mm	11 mm
Postmolt size	144 mm	10 mm

r=0.98

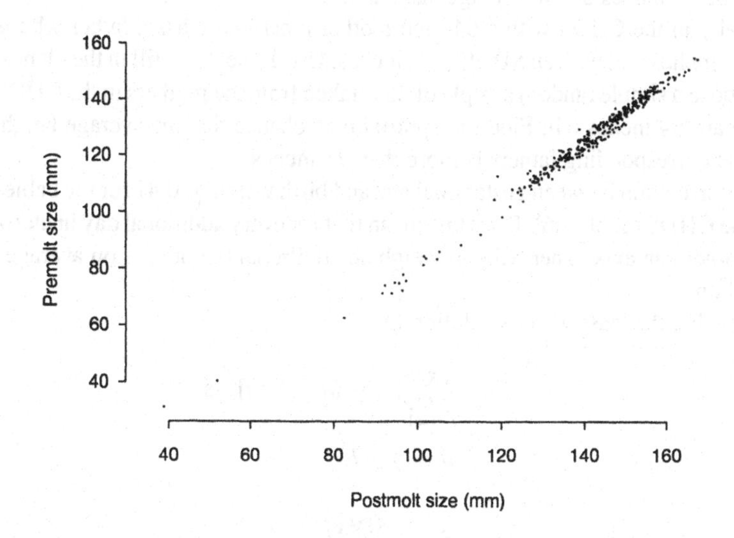

FIGURE 7.8. Scatter plot of premolt and postmolt carapace widths for 361 adult and juvenile female Dungeness crabs (Hankin et al. [HDMI89]).

TABLE 7.5. Summary statistics on heights of mothers and fathers in the CHDS (Yerushalmy [Yer64]).

	Average	SD
Mother's height	64 inches	2.5 inches
Father's height	70 inches	2.9 inches

r=0.34

TABLE 7.6. Summary statistics on gestational age and birth weight for babies in the CHDS (Yerushalmy [Yer64]).

	Average	SD
Birth weight	120 ounces	18 ounces
Gestational age	279 days	16 days

r=0.41

4. Use the summary statistics provided in Table 7.4 to find the regression line for predicting the increment in shell size from the postmolt shell size. The correlation between postmolt shell size and increment is −0.45. The increment in shell size is 14.7 mm on average, with SD 2.2 mm.

5. According to the summary statistics in Table 7.5, the mothers in the CHDS who are 69 inches tall tend to marry men 72 inches tall on average. Is it true

that if you take the fathers in the CHDS who are 72 inches tall, their wives will be 69 inches tall on average? Explain.

6. A baby in the CHDS with a 64-inch mother most likely has a father who is ____inches tall, with an SD of____inches. Use Table 7.5 to fill in the blanks.
7. Suppose a simple random sample of 25 is taken from the mothers in the CHDS who are 64 inches tall. Find the approximate chance that the average height of the corresponding fathers is more than 71 inches.
8. The correlation between gestational age and birth weight is 0.41 for the babies in the CHDS (Table 7.6). Does this mean that for every additional day in utero, a mother can expect her baby to weigh an additional 0.4 ounces on average? Explain.
9. Show that the least squares solution to

$$\sum_{i=1}^{n}[y_i - (a + bx_i)]^2 \text{ is}$$

$$\hat{a} = \bar{y} - \hat{b}\bar{x}$$

$$\hat{b} = r\frac{\text{SD}(y)}{\text{SD}(x)}.$$

10. Consider pairs (x_i, y_i), $i = 1, \ldots, n$. Show that

$$r(x, y) = r(ax + b, y),$$

where $r(x, y)$ denotes the correlation between x and y.

11. Show that if $y_i = ax_i + b$, then $r = +1$ if $a > 0$ and $r = -1$ if $a < 0$.
12. Show that $-1 \leq r \leq +1$.

    a. To simplify the argument, suppose $\bar{x} = \bar{y} = 0$ and $\text{SD}(x) = \text{SD}(y) = 1$. (This case is all that we need to prove, as shown in Exercise 10). Show that this implies

    $$r = \frac{1}{n}\sum x_i y_i \quad \text{and} \quad \text{SD}(x) = \frac{1}{n}\sum x_i^2.$$

    b. Use the following inequalities

    $$\sum(x_i - y_i)^2 \geq 0,$$
    $$\sum(x_i + y_i)^2 \geq 0,$$

    to complete the argument.

13. Show that the least squares line $y = \hat{a} + \hat{b}x$ passes through the point of averages $(\bar{x}, \bar{y})$.
14. Use the fact that $\hat{a}$ and $\hat{b}$ are the minimizers of

$$\sum[y_i - (a + bx_i)]^2$$

to show that the average of the residuals is 0.

15. Prove that the average of the fitted values, $\sum \hat{y}_i/n$, is $\bar{y}$.

16. Use Exercises 9 and 15 to show that

$$\mathrm{SD}(\hat{y}) = r\,\mathrm{SD}(y).$$

17. Show that the total sum of squares can be written as

$$\sum(y_i - \bar{y})^2 = \sum(\hat{y}_i - \bar{y})^2 + \sum(y_i - \hat{y}_i)^2.$$

Use this result and Exercise 16 to show that

$$\mathrm{SD(residuals)} = \sqrt{1 - r^2}\,\mathrm{SD}(y).$$

18. Consider the least squares estimates $\hat{a}$ and $\hat{b}$ based on the data (x_i, y_i), $i = 1, \ldots, n$. Suppose each x_i is rescaled as follows: $u_i = kx_i$. Find an expression in terms of $\hat{a}$ and $\hat{b}$ for the new coefficients obtained from the least squares fit of the line $y = c + du$ to the pairs (u_i, y_i), $i = 1, \ldots, n$. Also show that the fitted values remain the same (i.e., $\hat{a} + \hat{b}x_i = \hat{c} + \hat{d}u_i$).

19. Suppose the pairs $(X_1, Y_1), \ldots, (X_n, Y_n)$ are independent bivariate normal $\mathcal{N}_2(\mu_1, \mu_2, \sigma_1^2, \sigma_2^2, \rho)$ distributed random variables. Find the maximum likelihood estimates of μ_1 and μ_2 when σ_1^2, σ_2^2, and ρ are known.

Extensions

Growth Increment

The correlation between premolt and postmolt carapace size is 0.98 for the crabs collected in a premating embrace with premolt carapace size exceeding 100 mm. This correlation is very high because postmolt size is made up of premolt size plus a small growth increment. That is, when a group of crabs molt, the big crabs stay big and small crabs stay small, relatively speaking.

To see more formally how this works, let X and Z be normally distributed with means μ_x, μ_z, respectively, variances σ_x^2, σ_z^2, respectively, and correlation ρ. In our example, X represents premolt size and Z represents the growth increase from molting. For our data, the two sample means are 129 mm and 14.7 mm, the SDs are about 11 mm and 2.2 mm, respectively, and the sample correlation is -0.60. Consider the new random variable $Y = X + Z$. In our example, Y represents the postmolt size. The point is that when σ_x is considerably larger than σ_z, then Y is highly correlated with X, regardless of the correlation between X and Z. To see this, we find a lower bound for the correlation between X and Y that holds for all $\rho < 0$. First, we note that

$$\mathrm{Cov}(X, Y) = \sigma_x^2 + \rho\sigma_z\sigma_x.$$

$$\mathrm{SD}(Y) = \sqrt{\sigma_x^2 + \sigma_z^2 + 2\rho\sigma_z\sigma_x}.$$

Then plug these values for the covariance and SD into the correlation to find

$$\begin{aligned}\operatorname{corr}(X,Y) &= \frac{\sigma_x^2+\rho\sigma_z\sigma_x}{\sigma_x\sqrt{\sigma_x^2+\sigma_z^2+2\rho\sigma_z\sigma_x}} \\ &\geq \frac{\sigma_x^2-\sigma_z\sigma_x}{\sigma_x\sqrt{\sigma_x^2+\sigma_z^2}} \\ &= \frac{1-\lambda}{\sqrt{1+\lambda^2}},\end{aligned}$$

where $\lambda = \sigma_z/\sigma_x$.

In our example, from the two SDs alone, a lower bound on the correlation between premolt and postmolt size is 0.78. Adding a small increment to the premolt width gives a similar postmolt value. The increments have a small SD compared to the premolt SD. Thus pre- and postmolt sizes for an individual crab will be similar when compared to the variability of premolt values, and the correlation between pre- and postmolt values will be high. This implies that the regression equation found in the Theory section of this lab,

$$\text{premolt} = -29 + 1.1 \times \text{postmolt},$$

is not a very informative model for crab growth.

A more informative model should be based on the relationship between postmolt size and molt increment. In this section, we will explore several models for crab growth and use the crab data at hand to help determine which models do a good job of describing growth for this particular population of Dungeness crabs.

Growth Models

Crabs, and their growth patterns, have long been studied by biologists and statisticians. In 1894, Pearson examined the size frequency distribution of crabs; Thompson, in his 1917 book *On Growth and Form* ([Tho77]), compared the shape of the shell across species of crabs, and Huxley, in his 1932 book *Problems of Relative Growth* ([Hux93]), studied the changes over time in the relationship between the weight and carapace width of crabs.

One simple biological model for growth that is derived from the notion of cell duplication is the multiplicative growth model. It says that, for some size measurement y and time t,

$$\frac{\partial y}{\partial t} = ky;$$

that is, the rate of change in y is a multiple of y. For a multiplicative growth model, the relative growth is constant; namely,

$$\frac{1}{y}\frac{\partial y}{\partial t} = k.$$

A crab's molt increment is an approximation to its growth rate, and relative growth can be approximated by normalizing the molt increment by the premolt size. Notice that the relative increment,

$$\frac{(\text{post} - \text{pre})}{\text{pre}},$$

reports a crab's growth increment relative to its premolt size (i.e., it removes the size of the crab from its increase in size).

For our case, *constant relative growth* can be expressed as

$$\text{molt increment/premolt} = c.$$

More recently, empirical studies by Mauchline ([Mau77]) found that the growth of crabs as measured by weight followed a constant growth increment; that is,

$$\frac{\partial y}{\partial t} = k.$$

For our case, a *constant growth increment* is simply expressed as

$$\text{molt increment} = c.$$

Huxley studied the relationship between pairs of size measurements, such as width and weight. He showed that if two different size measurements, say x and y, both follow a multiplicative growth model, with constants k and l, respectively, then the measurements satisfy the *allometric* relationship

$$y = bx^{k/l}.$$

This can be shown via the relationship between their relative growths:

$$\frac{1}{y}\frac{\partial y}{\partial t} = \frac{k}{l}\frac{1}{x}\frac{\partial x}{\partial t}.$$

Thompson suggested that growth in weight y and length x should follow the relationship $y = bx^3$. The allometric relationship studied by Huxley is a generalization of this model.

More generally, a model for growth in terms of y and an allometric relationship between x and y can together provide a model for growth for x. For example, if we take Mauchline's constant increment model for crab weight (y) and an allometric relationship between carapace width (x) and weight, where $y = bx^c$, then

$$\frac{1}{x}\frac{\partial x}{\partial t} = \text{constant} \times \frac{1}{x^c}.$$

For molt increments, the *allometric relationship between weight and width and constant growth increment in weight* gives

$$\log(\text{molt increment/premolt}) = a - c \times \log(\text{premolt}),$$

where c is the exponent in the allometric relationship. What values of c make sense for this model?

Two other growth models that have been considered by biologists in their empirical study of crabs are also described here.

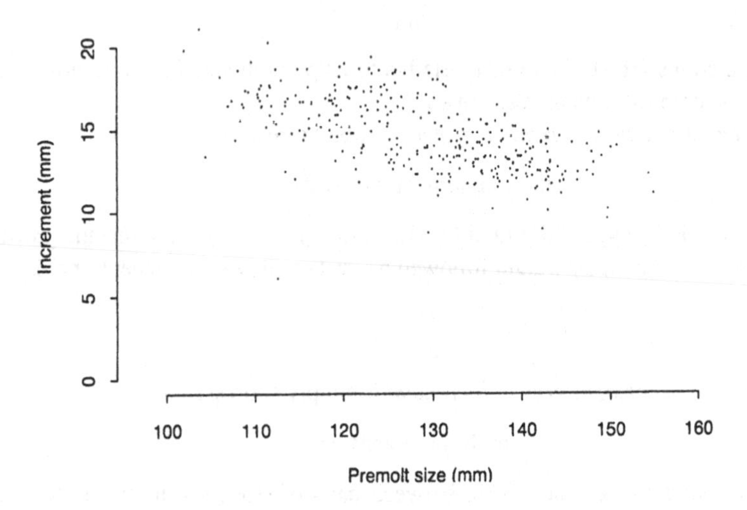

FIGURE 7.9. Scatter plot of molt increment against premolt size for adult female Dungeness crabs (Hankin et al. [HDMI89]).

Linear extension of constant relative growth:

$$\text{molt increment} = a + b \times \text{premolt}.$$

Log-linear extension of constant relative growth:

$$\log(\text{molt increment}/\text{premolt}) = a + b \times \text{premolt}.$$

Begin your investigation of the relationship between molt increment and premolt size by examining the scatter plot in Figure 7.9. The plot shows that larger crabs have smaller molt increments. This rules out some of the models described above. With which models are the data consistent? How would you choose among such models? Use the Dungeness crab data and take into account the biological explanations for growth in answering this question. Also keep in mind that it is reasonable to assume that the coefficients for the growth model may change for different stages of growth; that is, juvenile and adult crabs may have different growth rates.

Notes

Mike Mohr made the crab data for this lab available, and he assisted us with a description of the life cycle of the Dungeness crab. General information on the Dungeness crab can be found in Wickham's contribution to the Audubon Wildlife Report ([Wic88]).

Somerton ([Som80]) adapted the linear model for growth to a bent-line model in order to accommodate changes in growth from juvenile to adult stages. A discussion of the different models of growth can be found in Mauchline ([Mau77]). Wainwright and Armstrong ([WA93]) and Orensanz and Gallucci ([OG88]) contain comparisons of these models for other data.

References

[HDMI89] D.G. Hankin, N. Diamond, M.S. Mohr, and J. Ianelli. Growth and reproductive dynamics of adult female Dungeness crabs (*Cancer magister*) in northern California. *J. Conseil, Cons. Int. Explor. Mer*, **46**:94–108, 1989.

[Hux93] J.S. Huxley. *Problems of Relative Growth*. Johns Hopkins Press, Baltimore, MD, 1993. With introduction by F.B. Churchill and essay by R.E. Strauss.

[Mau77] J. Mauchline. Growth of shrimps, crabs, and lobsters—an assessment. *J. Cons. Int. Explor. Mer.*, **37**:162–169, 1977.

[MH89] M.S. Mohr and D.G. Hankin. Estimation of size-specific molting probabilities in adult decapod crustaceans based on postmolt indicator data. *Can. J. Fish. Aquat. Sci.*, **46**:1819–1830, 1989.

[OG88] J.M. Orensanz and V.F. Gallucci. Comparative study of postlarval life-history schedules in four sympatric species of *Cancer* (decapoda: Brachyura: Cancridae). *J. Crustacean Biol.*, **8**:187–220, 1988.

[Som80] D.A. Somerton. Fitting straight lines to Hiatt growth diagrams: a re-evaluation. *J. Conseil, Cons. Int. Explor. Mer*, **39**:15–19, 1980.

[Tho77] *On Growth and Form*. Cambridge University Press, 1977. An abridged edition edited by J.T. Bonner.

[WA93] T.C. Wainwright and D.A. Armstrong. Growth patterns in the Dungeness crab (*Cancer magister* Dana): Synthesis of data and comparison of models. *J. Crustacean Biol.*, **13**:36–50, 1993.

[Wic88] D.E. Wickham. *The Dungeness crab and the Alaska Red King crab*. Academic Press, San Diego, 1988, pp. 701–713.

[Yer64] J. Yerushalmy. Mother's cigarette smoking and survival of infant. *Am. J. Obstet. Gynecol.* **88**:505–518, 1964.

8

Calibrating a Snow Gauge

MONDAY, MAY 1, 1995 San Francisco Examiner[1]

Rain Prompts Small-Stream Flood Warning

By Larry D. Hatfield

A late-season cold front was moving off the Pacific Monday morning, bringing more heavy downpours to the rain-besotted coastal areas of Central and Northern California and more snow to the already towering snowpack in the Sierra. But the primary weather problem of the day was that the substantial rain and snow melt caused by balmy temperatures ahead of the new storm was boosting the runoff and increasing a flood threat in mountain streams.

Streams were running high in the foothills and higher elevations of Mariposa, Madera, Fresno and Tulare counties.

Bass Lake, near Oakhurst south of Yosemite National Park, was full and releases from the lake caused flood danger along Willow Creek in Madera County. The National Weather Service put a small-stream flood advisory into effect through Monday night and warned residents and travelers to be cautious there and on other streams flowing from the southern Sierra Nevada.

Rainfall totals from the latest extension of the seemingly endless parade of storms were heavy over most of the north part of the state. In the 24 hours ending at 5 a.m. Monday, 4.2 inches fell at Four Trees, in Plumas County northeast of Oroville. Three inches of rain fell there in just six hours ending at 5 p.m. Sunday.

Blue Canyon, at the mile-high level on Interstate 80 east of Auburn, had 2.48 inches in the 24-hour period ending at 5 a.m. and other heavy rain was reported in the Mount Shasta-Siskiyou region and in northwestern California.

In the Bay Area, 1.12 inches fell in San Rafael. San Francisco had 0.39 inches; Alameda 0.36; Oakland 0.32; and Redwood City 0.06.

Strong southerly winds gusted to 40 mph at Weed and Redding and a high wind warning was posted on the San Francisco Bay Bridge just before dawn.

[1]Reprinted by permission.

Introduction

The main source of water for Northern California comes from the Sierra Nevada mountains. To help monitor this water supply, the Forest Service of the United States Department of Agriculture (USDA) operates a gamma transmission snow gauge in the Central Sierra Nevada near Soda Springs, California. The gauge is used to determine a depth profile of snow density.

The snow gauge does not disturb the snow in the measurement process, which means the same snow-pack can be measured over and over again. With these replicate measurements on the same volume of snow, researchers can study snow-pack settlement over the course of the winter season and the dynamics of rain on snow. When rain falls on snow, the snow absorbs the water up to a certain point, after which flooding occurs. The denser the snow-pack, the less water the snow can absorb. Analysis of the snow-pack profile may help with monitoring the water supply and flood management.

The gauge does not directly measure snow density. The density reading is converted from a measurement of gamma ray emissions. Due to instrument wear and radioactive source decay, there may be changes over the seasons in the function used to convert the measured values into density readings. To adjust the conversion method, a calibration run is made each year at the beginning of the winter season. In this lab, you will develop a procedure to calibrate the snow gauge.

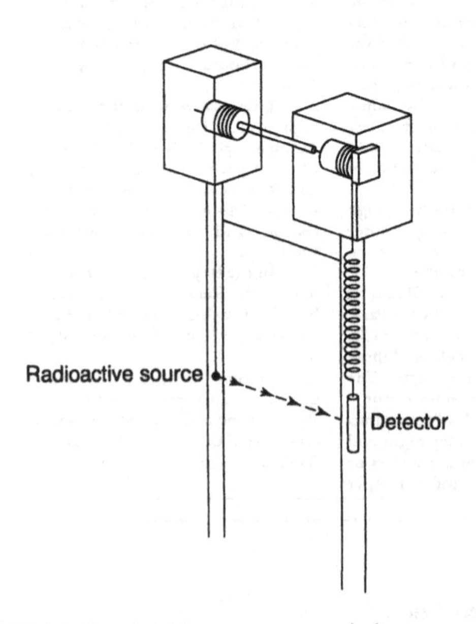

FIGURE 8.1. Sketch of the gamma transmission snow gauge.

TABLE 8.1. The measured gain for the middle 10 runs of 30 of the snow gauge, for each of 9 densities in grams per cubic centimeter of polyethylene blocks (USDA Forest Service).

Density	Gain									
0.686	17.6	17.3	16.9	16.2	17.1	18.5	18.7	17.4	18.6	16.8
0.604	24.8	25.9	26.3	24.8	24.8	27.6	28.5	30.5	28.4	27.7
0.508	39.4	37.6	38.1	37.7	36.3	38.7	39.4	38.8	39.2	40.3
0.412	60.0	58.3	59.6	59.1	56.3	55.0	52.9	54.1	56.9	56.0
0.318	87.0	92.7	90.5	85.8	87.5	88.3	91.6	88.2	88.6	84.7
0.223	128	130	131	129	127	129	132	133	134	133
0.148	199	204	199	207	200	200	205	202	199	199
0.080	298	298	297	288	296	293	301	299	298	293
0.001	423	421	422	428	436	427	426	428	427	429

The Data

The data are from a calibration run of the USDA Forest Service's snow gauge located in the Central Sierra Nevada mountain range near Soda Springs, California. The run consists of placing polyethylene blocks of known densities between the two poles of the snow gauge (Figure 8.1) and taking readings on the blocks. The polyethylene blocks are used to simulate snow.

For each block of polyethylene, 30 measurements were taken. Only the middle 10, in the order taken, are reported here. The measurements recorded by the gauge are an amplified version of the gamma photon count made by the detector. We call the gauge measurements the "gain."

The data available for investigation consist of 10 measurements for each of 9 densities in grams per cubic centimeter (g/cm^3) of polyethylene. The complete data appear in Table 8.1.

Background

Location

The snow gauge is a complex and expensive instrument. It is not feasible to establish a broad network of gauges in the watershed area in order to monitor the water supply. Instead, the gauge is primarily used as a research tool. The snow gauge has helped to study snow-pack settling, snow-melt runoff, avalanches, and rain-on-snow dynamics.

At one time, gauges were located on Mt. Baldy, Idaho, on Mt. Hood, Oregon, in the Red Mountain Pass, Colorado, on Mt. Alyeska, Alaska, and in the Central Sierra Nevada, California. The Central Sierra snow gauge provided the data to be analyzed in this lab. It is located in the center of a forest opening that is roughly 62 meters in diameter. The laboratory site is at 2099 meters elevation and is subject to

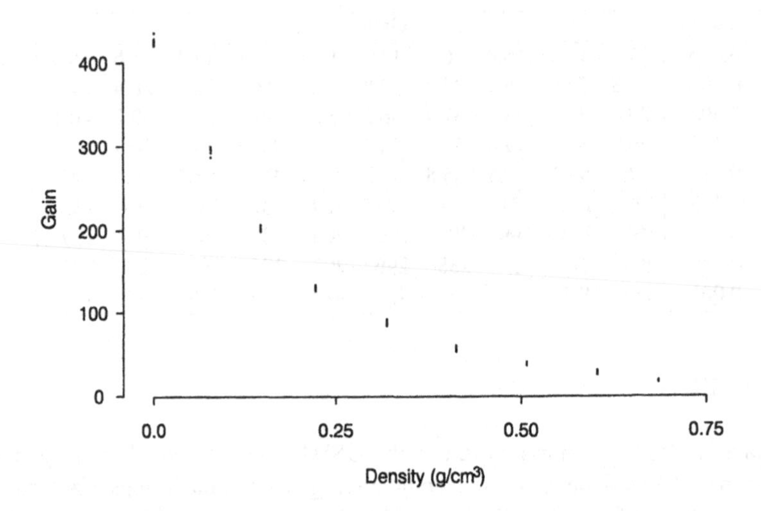

FIGURE 8.2. Plot of gain by density in g/cm^3 for ten blocks of polyethylene (USDA Forest Service).

major high-altitude storms which regularly deposit 5–20 centimeters of wet snow. The snow-pack reaches an average depth of 4 meters each winter.

The Gauge

The snow gauge consists of a cesium-137 radioactive source and an energy detector, mounted on separate vertical poles approximately 70 centimeters apart (Figure 8.1). A lift mechanism at the top of the poles raises and lowers the source and detector together. The radioactive source emits gamma photons, also called gamma rays, at 662 kilo-electron-volts (keV) in all directions. The detector contains a scintillation crystal which counts those photons passing through the 70-centimeter gap from the source to the detector crystal. The pulses generated by the photons that reach the detector crystal are transmitted by a cable to a preamplifier and then further amplified and transmitted via a buried coaxial cable to the lab. There the signal is stabilized, corrected for temperature drift, and converted to a measurement we have termed the "gain." It should be directly proportional to the emission rate.

The densities of the polyethylene blocks used in the calibration run range from 0.001 to 0.686 grams per cubic centimeter (g/cm^3). The snow-pack density is never actually as low as 0.001 or as high as 0.686. It typically ranges between 0.1 and 0.6 g/cm^3.

A Physical Model

The gamma rays that are emitted from the radioactive source are sent out in all directions. Those that are sent in the direction of the detector may be scattered or absorbed by the polyethylene molecules between the source and the detector. With denser polyethylene, fewer gamma rays will reach the detector. There are complex physical models for the relationship between the polyethylene density and the detector readings.

A simplified version of the model that may be workable for the calibration problem of interest is described here. A gamma ray on route to the detector passes a number of polyethylene molecules. The number of molecules depends on the density of the polyethylene. A molecule may either absorb the gamma photon, bounce it out of the path to the detector, or allow it to pass. If each molecule acts independently, then the chance that a gamma ray successfully arrives at the detector is p^m, where p is the chance, a single molecule will neither absorb nor bounce the gamma ray, and m is the number of molecules in a straight-line path from the source to the detector. This probability can be re-expressed as

$$e^{m \log p} = e^{bx},$$

where x, the density, is proportional to m, the number of molecules. A polyethylene block of high density can be roughly considered to be composed of the same molecules as a block that is less dense. Simply, there are more molecules in the same volume of material because the denser material has smaller air pores. This means that it is reasonable to expect the coefficient b in the equation above to remain constant for various densities of polyethylene.

The true physical model is much more complex, and in practice snow will be between the radioactive source and detector. However, it is expected that polyethylene is similar enough to snow (with respect to gamma ray transmission) to serve as its substitute and that the model described here is adequate for our purposes.

Investigations

The aim of this lab is to provide a simple procedure for converting gain into density when the gauge is in operation. Keep in mind that the experiment was conducted by varying density and measuring the response in gain, but when the gauge is ultimately in use, the snow-pack density is to be estimated from the measured gain.

- Use the data to fit gain, or a transformation of gain, to density. Try sketching the least squares line on a scatter plot. Do the residuals indicate any problems with the fit? If the densities of the polyethylene blocks are not reported exactly, how might this affect the fit? What if the blocks of polyethylene were not measured in random order?
- We are ultimately interested in answering questions such as: Given a gain reading of 38.6, what is the density of the snow-pack? or Given a gain reading

of 426.7, what is the density of the snow-pack? These two numeric values, 38.6 and 426.7, were chosen because they are the average gains for the 0.508 and 0.001 densities, respectively. Develop a procedure for adding bands around your least squares line that can be used to make interval estimates for the snow-pack density from gain measurements. Keep in mind how the data were collected: several measurements of gain were taken for polyethylene blocks of known density.

- To check how well your procedure works, omit the set of measurements corresponding to the block of density 0.508, apply your calibration procedure to the remaining data, and provide an interval estimate for the density of a block with an average reading of 38.6. Where does the actual density fall in the interval? Try this same test of your procedure for the set of measurements at the 0.001 density.
- Consider the log-polynomial model:

$$\log(\text{gain}) = a + b \times \text{density} + c \times \text{density}^2.$$

How well does it fit the data? Can you provide interval estimates here as well? Compare this model to the simple log-linear model in terms of its fit, predictive ability, physical model, and any other factors that you think are relevant.

Write a short instruction sheet for a lab technician on how the snow gauge should be calibrated for use in the winter months. Include an easy-to-use graph for determining snow density for a given gain. Accompany the instructions with a short technical appendix that explains the method you have chosen.

Theory

The Simple Linear Model

The simple linear model is that the expectation $\mathbb{E}(Y|x)$ of a random response Y at a known design point x satisfies the relation

$$\mathbb{E}(Y|x) = a + bx.$$

The *Gauss measurement model* supposes that measurement errors E have mean 0, constant variance (say σ^2), and are uncorrelated. A common practice is to express the response in the form

$$Y = a + bx + E,$$

with the understanding that the Es have the properties noted above. Notice that the Y is capitalized to denote that it is a random variable, and x is in lowercase to denote that it is fixed. Also, we use capitals when we explicitly represent Y as a random variable, and we use lowercase to represent an observed response, say y_1. Sometimes it is necessary to deviate from this convention when, for example, we take a function of the observations, such as the residual sum of squares $\sum(y_i - \hat{y}_i)^2$, and we also want to take the expectation of that quantity.

If we observe pairs $(x_1, y_1), \ldots (x_n, y_n)$, then the method of least squares can be used to estimate a and b. As in Chapter 7 on Dungeness crabs, the least squares estimates of a and b are

$$\begin{aligned} \hat{a} &= \frac{(\sum x_i^2)(\sum y_i) - (\sum x_i)(\sum x_i y_i)}{n \sum x_i^2 - (\sum x_i)^2}, \\ \hat{b} &= \frac{n \sum x_i y_i - (\sum x_i)(\sum y_i)}{n \sum x_i^2 - (\sum x_i)^2}. \end{aligned} \tag{8.1}$$

Note that $\hat{a}$ and $\hat{b}$ are linear functions of the responses y_i, so they are linear functions of the errors, even though we don't get to see them. It is left to the Exercises to show that $\hat{a}$ and $\hat{b}$ are unbiased and to find their variances and covariance under the Gauss measurement model. The residuals $y_i - (\hat{a} + \hat{b}x_i)$ are also unbiased, and we can think of the residuals as estimates of the errors. It is left to the Exercises to show that the residual sum of squares has expectation

$$\mathbb{E}(\sum [y_i - (\hat{a} + \hat{b}x_i)]^2) = (n-2)\sigma^2.$$

The residual sum of squares can thus provide an estimate of the variance in the Gauss measurement model.

Model Misfit

An alternative to the simple linear model is a polynomial model. For example, a quadratic model for the expectation $\mathbb{E}(Y|x)$ is

$$\mathbb{E}(Y|x) = c + dx + ex^2.$$

This model can also be expressed as a two-variable linear model if we rewrite x^2 as u;

$$\mathbb{E}(Y|x) = c + dx + eu. \tag{8.2}$$

The topic of fitting quadratics and other multivariable linear models is covered in Chapter 10.

Regardless of the model for the expected value of the response, we can fit a line to the data. For example, say that (8.2) is the true model for our data. If we observe pairs $(x_1, y_1), \ldots, (x_n, y_n)$, we can fit a line by the method of least squares to these observations: minimize, with respect to a and b, the sum of squares

$$\sum_{i=1}^{n} [y_i - (a + bx_i)]^2.$$

The solutions for $\hat{a}$ and $\hat{b}$ remain as shown in equation (8.1). However, the model has been misfitted. These sample coefficients $\hat{a}$ and $\hat{b}$ may be biased under the Gauss measurement model,

$$Y_i = c + dx_i + eu_i + E_i,$$

where the E_i are independent with mean 0 and variance σ^2. The expectation of $\hat{b}$ need not be d. It can be shown that

$$\mathbb{E}(\hat{b}) = d + e\left(n\sum x_i u_i - \sum x_i \sum u_i\right) / \left[n\sum x_i^2 - (\sum x_i)^2\right].$$

In the special case where the design is such that x is uncorrelated with u, we find

$$\mathbb{E}(\hat{b}) = d, \ \mathbb{E}(\hat{a}) = c.$$

Otherwise these fitted coefficients are biased. In any case, the residuals

$$r_i = y_i - \hat{a} - \hat{b}x_i$$

will be biased. For example, in this special case,

$$\mathbb{E}(r_i|x_i) = eu_i.$$

The residuals then include both measurement error from E_i and model misfit error from eu_i. If the root mean square (r.m.s.) of eu_i is not small in comparison to σ^2, the residual sum of squares does not provide a good estimate of σ^2. Residual plots of (x_i, r_i) help to indicate whether the model is misfitted. Residual plots are discussed below in greater detail. Note that even when the model is misfitted, the average of the residuals $\bar{r}$ is 0.

Transformations

Sometimes the relationship between the response and the design variable is not initially linear, but a transformation can put the relationship into a linear form. For example, the physical model for gain and density leads us to think that the simple linear model is inappropriate and it suggests the nonlinear relationship

$$\mathbb{E}(G) = ce^{bx},$$

where G denotes the gain and x denotes the density. By taking the natural logarithm of both sides of the equation, we find a linear relationship between $\log(\mathbb{E}(G))$ and x:

$$\log(\mathbb{E}(G)) = \log(c) + bx.$$

What happens to the errors? There are at least two approaches here. One is to continue with the Gauss measurement error model, in which case

$$G = ce^{bx} + U,$$

where the Us are measurement errors. Here the linearization obtained by taking logs of $\mathbb{E}(G)$ does not help:

$$\begin{aligned}\log(G) &= \log(ce^{bx} + U) \\ &= a + bx + V(x).\end{aligned}$$

In this case, we have a linear relationship between $\log(G)$ and x, but the variance of the errors changes with x.

We might use a multiplicative alternative to the Gauss measurement error model, which often describes data of this kind. In the multiplicative error model, random proportional error factors are nonnegative, have a common mean and variance, and are independent across measurements:

$$G = ce^{bx} W$$
$$\log(G) = \log(c) + bx + \log(W).$$

Now $\log(W)$ is like an error term in the Gauss model, except that its mean may not be zero (i.e., there may be bias).

Residual Plots

Graphical representations of the residuals are important diagnostic tools for detecting curvilinear relationships. If the simple linear model holds, then plots of the residuals against the explanatory variable —i.e., scatter plots of the pairs $(x_i, y_i - \hat{a} - \hat{b}x_i)$— should show a horizontal blur of points about the horizontal axis. Residuals can also be plotted against fitted values $\hat{y}_i$, another variable not used in the regression, or a function of x such as x^2.

Consider the residual plot (Figure 8.3) from the least squares fit of gain to density for the data in this lab. An obvious problem appears: the curvature in the residual plot indicates that it may be appropriate to include x^2 in the least squares fit or to transform G to $\log(G)$. The physical model for the relationship between gain and density suggests proceeding with the log transformation. In addition, the residual plots for the fit of log gain to density can help determine whether the errors are multiplicative or additive. If they are multiplicative, then the transformation should equalize the variance of the residuals. That is, for each value of x, the spread in the residuals should be roughly equal. However, any systematic error in the residuals indicates a problem other than whether the error may be multiplicative.

Figure 8.4 gives hypothetical examples of residual plots for different types of departures from the simple linear model. They show that:

- Patterns in the residuals may indicate a nonlinear relationship (top left).
- Unusually large residuals point to outliers that may have a large effect on the least squares line (bottom left).
- Funneling of the residuals may indicate that the variability about the least squares line may change with the values of x (top right).
- A linear trend in the residuals plotted against some variable not used in the regression may indicate that this variable should be included (bottom right).

Additionally, a normal-quantile plot of the residuals may indicate that they are non-normal, or more importantly that there is a long tail or skewness in the distribution of the residuals.

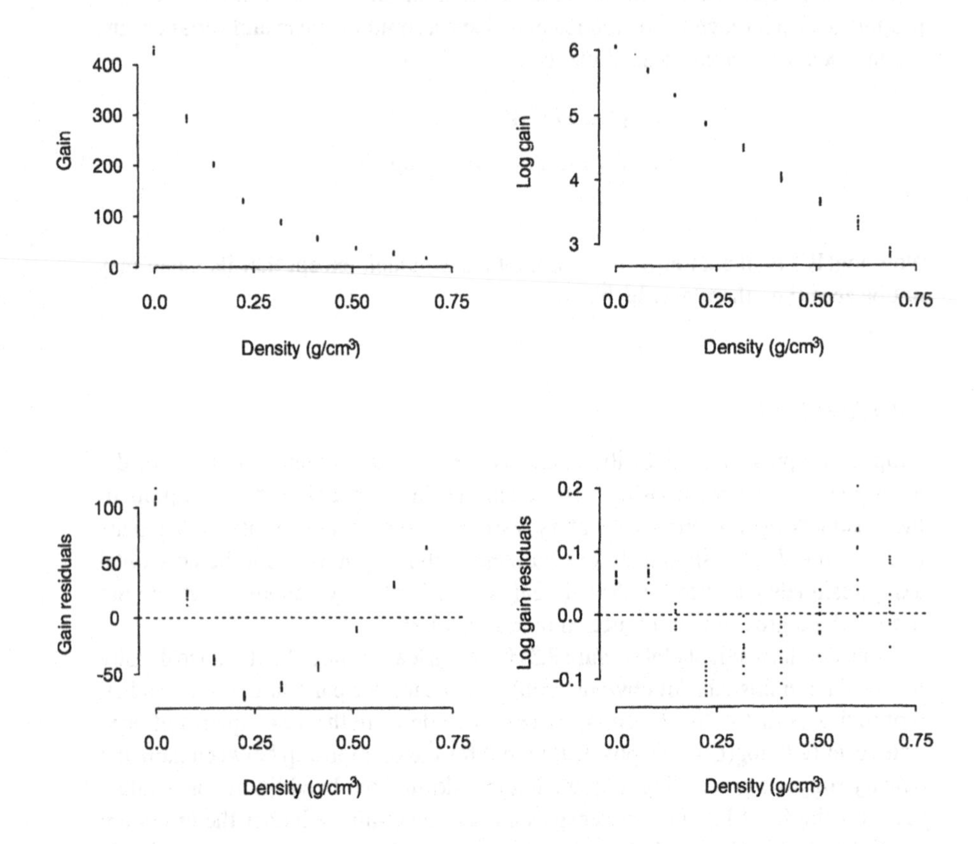

FIGURE 8.3. Observed data and residuals from fitting gain and log gain to density (USDA Forest Service).

Replicate Measurements

In this lab, the response is measured at 9 distinct values of the explanatory variable x, and for each x, 10 replicate measurements are taken. In general, suppose a random variable Y is measured k times at each one of m distinct values. Then if $x_1, \ldots, x_m$ are the m distinct values, the responses can be denoted by Y_{ij}, $i = 1, \ldots, m$ and $j = 1, \ldots, k$, where Y_{ij} is the jth measurement taken at x_i. That is, for our simple linear model,

$$Y_{ij} = a + bx_i + E_{ij}.$$

Note that the errors E_{ij} are assumed to be uncorrelated for all $i = 1, \ldots, m$ and all $j = 1, \ldots, k$.

The replicate measurements can provide an estimate of error variance σ^2 that does not rely on the model. If the model is incorrectly fitted as in the example dis-

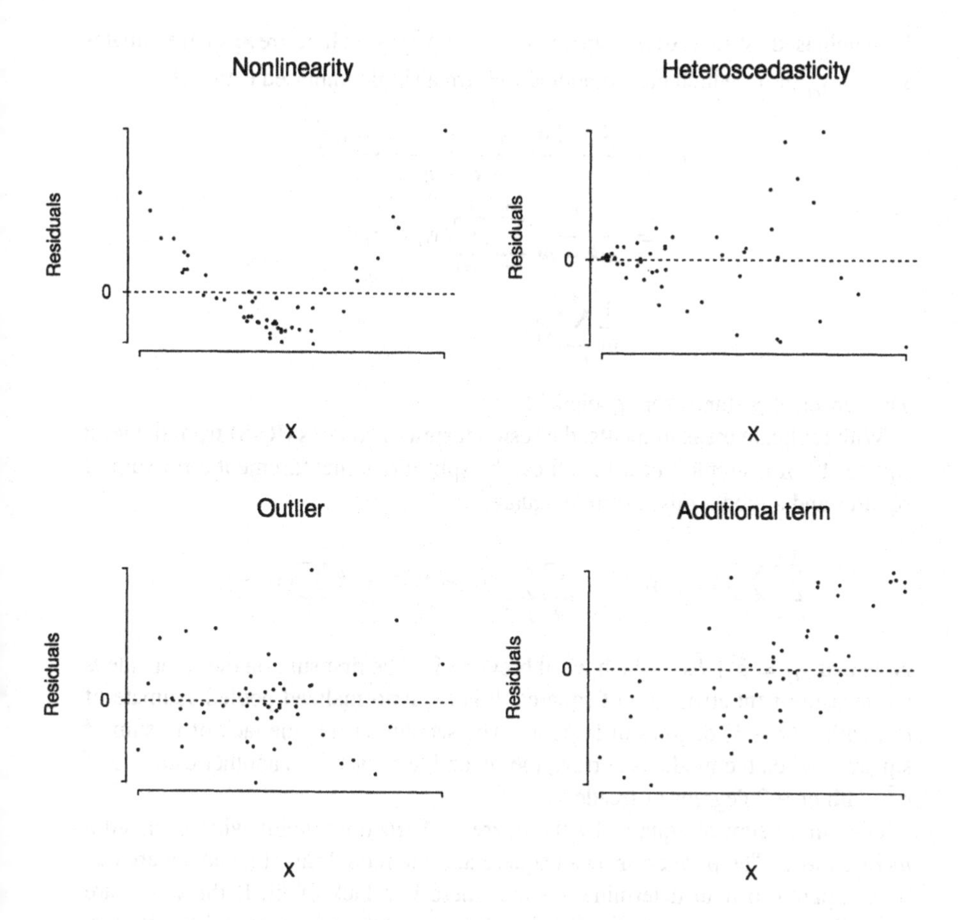

FIGURE 8.4. Sample residual plots indicating lack of fit.

cussed earlier, then the residuals include measurement error from the E_{ij} and model misfit. However, the replicate measurements allow us to estimate the measurement error separately from the model misfit.

To explain, suppose the model for the response is

$$Y_{ij} = c + dx_i + eu_i + E_{ij},$$

as discussed earlier, and the simple linear model is fitted by least squares. The replicates $Y_{11}, \ldots Y_{1k}$ are k uncorrelated random variables with mean $c + dx_i + eu_i$ and variance σ^2. Therefore

$$s_1^2 = \frac{1}{k-1}\sum_{j=1}^{k}(y_{1j} - \bar{y}_1)^2$$

is an unbiased estimate of σ^2, where $\bar{y}_1 = k^{-1}\sum_j y_{1j}$. There are m such estimates $s_1^2, \ldots, s_m^2$ of σ^2 which can be pooled to form a single unbiased estimate:

$$\begin{aligned} s_p^2 &= \frac{(k-1)s_1^2 + \cdots + (k-1)s_m^2}{mk-m} \\ &= \frac{1}{mk-m}\sum_{i=1}^{m}\sum_{j=1}^{k}(y_{ij}-\bar{y}_i)^2 \\ &= \frac{1}{m}\sum_{i=1}^{m}s_i^2. \end{aligned}$$

The subscript p stands for "pooled."

With replicate measurements, the residual sum of squares (RSS) from the least squares fit of a simple linear model can be split into a measurement error sum of squares and a model misfit sum of squares :

$$\sum_{i=1}^{m}\sum_{j=1}^{k}(y_{ij}-\hat{y}_i)^2 = \sum_{i=1}^{m}\sum_{j=1}^{k}(y_{ij}-\bar{y}_i)^2 + k\sum_{i=1}^{m}(\hat{y}_i-\bar{y}_i)^2.$$

Note that $\hat{y}_{ij} = \hat{a} + \hat{b}x_i$, which we abbreviate $\hat{y}_i$. The first sum on the right side is the measurement error sum of squares. It is an unnormalized pooled estimate of σ^2, with $m(k-1)$ degrees of freedom. The second term is the lack of fit sum of squares. When the model is correct, the second term provides another estimate of σ^2, with $m-2$ degrees of freedom.

Dividing a sum of squares by the degrees of freedom yields what is called a *mean square*. The pure error mean square and the model misfit mean square can be compared to help determine whether there is a lack of fit. If the errors are mutually independent and normally distributed, and there is no model misfit, then the ratio

$$\frac{k\sum_i(\hat{y}_i-\bar{y}_i)^2/(m-2)}{\sum_i\sum_j(y_{ij}-\bar{y}_i)^2/m(k-1)}$$

follows an F distribution with $m-2$ and $m(k-1)$ degrees of freedom. On the other hand, if there is model misfit, then the numerator should be larger than the denominator, with values of the ratio bigger than 3 to 5 indicative of misfit. We can use this ratio as a test of how well the data fit the model.

Confidence and Prediction Bands

We can predict the response y at any value x by $\hat{y} = \hat{a} + \hat{b}x$. An interval for this prediction can be based on the following variance for $k = 1$:

$$\mathrm{Var}(y-\hat{y}) = \mathrm{Var}(y-\hat{a}-\hat{b}x) = \sigma^2\left[1 + \frac{1}{m} + \frac{(x-\bar{x})^2}{\sum(x_i-\bar{x})^2}\right].$$

On the right side, the first term in square brackets is the contribution from the variation of y about the line; the second and third terms are the contributions from the uncertainty in the estimates of a and b, respectively.

An approximate 99% *prediction interval* for y, at x, is formed from the following two bounds:

$$(\hat{a} + \hat{b}x) + z_{0.995}\,\hat{\sigma}\left[1 + \frac{1}{m} + \frac{(x - \bar{x})^2}{\sum(x_i - \bar{x})^2}\right]^{1/2}$$

and

$$(\hat{a} + \hat{b}x) - z_{0.995}\,\hat{\sigma}\left[1 + \frac{1}{m} + \frac{(x - \bar{x})^2}{\sum(x_i - \bar{x})^2}\right]^{1/2},$$

where $z_{0.995}$ is the 0.995 quantile of the standard normal distribution, and $\hat{\sigma}$ is the standard deviation of the residuals. The prediction interval for y differs from a confidence interval for $a + bx$,

$$(\hat{a} + \hat{b}x) \pm z_{0.995}\,\hat{\sigma}\left[\frac{1}{m} + \frac{(x - \bar{x})^2}{\sum(x_i - \bar{x})^2}\right]^{1/2}.$$

The confidence interval is smaller because it does not include the variability of y about the line $a + bx$. It tells us about the accuracy of the estimated mean of y at x, whereas a prediction interval is for an observation, or the mean of a set of observations, taken at design point x.

Figure 8.5 displays these prediction intervals for all values of x over a range. These curves are called *prediction bands*. Notice that the size of an interval depends on how far x is from $\bar{x}$. That is, the interval is most narrow at $\bar{x}$ and gets wider the further x is from $\bar{x}$.

If y_0 is the average of r measurements, all taken for the same density x_0 and k replicate measurements are taken at each x_i, then the prediction variance becomes

$$\mathrm{Var}(y_0 - \hat{y}_0) = \sigma^2\left[\frac{1}{r} + \frac{1}{mk} + \frac{(x_o - \bar{x})^2}{k\sum(x_i - \bar{x})^2}\right].$$

As before, the first term in the variance is the contribution from the variation about the line, which is now σ^2/r because we have an average of r observations. The bands must be adjusted accordingly.

The bands shown here are pointwise bands. This means that they provide an interval for one future reading, or the average of a set of future readings.

Calibration

There is an important difference between the calibration run and how the gauge will be used in practice. In the calibration run, polyethylene blocks of known densities are placed in the gauge, and the gain is measured. At each density level, several measurements were taken and were not all the same. This variability is due to measurement error. In the future, the gain will be measured for an unknown snow density, and the gain measurement will be used to estimate snow density; that is,

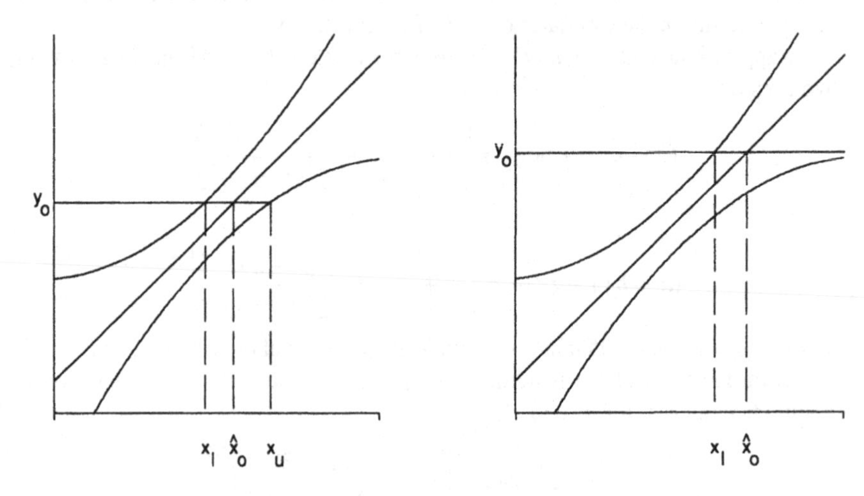

FIGURE 8.5. Prediction bands and inverse prediction estimates for x_0.

in the calibration run, we take the relationship between log gain and snow density to follow a standard linear model:

$$Y = a + bx + E,$$

where Y represents the log gain, x denotes the known polyethylene density, and the errors follow the Gauss measurement model. In the future, when the gauge is in operation, we will observe the new measurement y_0 for a particular snow-pack and use it to estimate the unknown density, say x_0.

One procedure for estimating the density first finds the least squares estimates $\hat{a}$ and $\hat{b}$ using the data collected from the calibration run. Then it estimates the density as follows:

$$\hat{x}_0 = \frac{y_0 - \hat{a}}{\hat{b}}.$$

This is called the *inverse estimator*.

Just as the least squares line can be inverted to provide an estimate for snow density, the bands in Figure 8.5 can be inverted to make interval estimates for snow density. In the left plot, read across from y_0, to the top curve, then read down to find the corresponding x-value, which we call x_l. This is the lower bound for the interval estimate. We similarly find the upper bound by reading across from y_0 to the lower curve and then down to the x-value, which we call x_u. The interval (x_l, x_u) is not symmetric about $\hat{x}_0$, and in some cases it may be a half-line, as in the example shown in the right plot in Figure 8.5.

Maximum Likelihood

In calibration problems, we have known design points, $x_1, \ldots x_m$, and it seems sensible to invert the bands about the regression line to provide an interval estimate of x_0. Here we show that under the Gauss measurement model and the additional assumption of normal errors, $\hat{x}_0$ is the maximum likelihood estimate of x_0, and the interval (x_l, x_u) is a confidence interval for x_0; that is, consider, for $k = 1$,

$$Y_i = a + bx_i + E_i \qquad i = 1, \ldots, m$$
$$Y_0 = a + bx_0 + E_0,$$

where we now suppose that $E_0, E_1, \ldots E_m$ are independent normal errors with mean 0 and variance σ^2. In our case, a, b, σ^2, and x_0 are unknown parameters.

Ignoring additive constants, the log-likelihood function for these four parameters is

$$l(\sigma, a, b, x_0)$$
$$= (m+1)\log(\sigma) - (2\sigma^2)^{-1}\left[\sum_{i=1}^{m}(y_i - a - bx_i)^2 + (y_0 - a - bx_0)^2\right],$$

and the maximum likelihood estimate $\hat{x}_0$ of x_0 satisfies

$$y_0 - \hat{a} - \hat{b}\hat{x}_0 = 0,$$

where $\hat{a}$ and $\hat{b}$ are the maximum likelihood estimates of a and b. The maximum likelihood estimates of a and b are simply the least squares estimates based on the m observations $y_1, \ldots y_m$.

Properties of these estimators can be used to show that

$$\frac{y_0 - \hat{a} - \hat{b}x_0}{\hat{\sigma}[1 + 1/n + (x_0 - \bar{x})^2/\sum(x_i - \bar{x})^2]^{1/2}} := W(x_0)$$

has an approximate normal distribution.

An approximate 95% confidence interval for x_0 is then the set of all x values that satisfy the inequalities

$$-z_{.975} \leq W(x) \leq z_{.975}.$$

From these inequalities, we see that (x_l, x_u) can be interpreted as a 95% confidence interval for x_0.

An Alternative

The inverse estimator of x_0 can be compared to an alternative procedure that treats the density as the response variable, even though the x_i are design points. That is, the alternative procedure uses the method of least squares to estimate c and d in the equation

$$x = c + dy.$$

Then for the log gain, y_0, the density is estimated as $\hat{c} + \hat{d}y_0$.

Halperin ([Hal70]) has shown that inverse estimation is indistinguishable from the alternative procedure when either (i) the calibration is intrinsically precise, meaning the slope b is steep relative to the error σ; or (ii) the design estimates the slope well (i.e., the x_i are widely dispersed). In practice, calibration procedures should not be run unless these conditions are present. Additionally, Halperin provided a simulation study that showed inverse estimation was preferable to the alternative under many other conditions.

Exercises

1. Show theoretically that the least squares estimates of a and b in the equation
$$\log(\text{gain}) = a + b \times \text{density}$$
are the same whether the sum of squares is minimized using the 9 average gains or all 90 gain measurements.
Show that the SDs of the residuals from the fit are different. Explain why this is the case.
2. Suppose the simple linear model holds, where
$$Y = a + bx + E,$$
and the errors are uncorrelated, with mean 0 and variance σ^2. Consider the least squares estimates $\hat{a}$ and $\hat{b}$ based on the pairs of observations $(x_1, y_1), \ldots (x_n, y_n)$. Derive the following expectations, variances, and covariance:
$$\mathbb{E}(\hat{a}) = a, \qquad \mathbb{E}(\hat{b}) = b,$$
$$\text{Var}(\hat{a}) = \frac{\sigma^2 \sum x_i^2}{n \sum x_i^2 - (\sum x_i)^2},$$
$$\text{Var}(\hat{b}) = \frac{n\sigma^2}{n \sum x_i^2 - (\sum x_i)^2},$$
$$\text{Cov}(\hat{a}, \hat{b}) = \frac{-\sigma^2 \sum x_i}{n \sum x_i^2 - (\sum x_i)^2}.$$
3. Use Exercise 2 to derive the variance of $\hat{a} + \hat{b}x$, and then provide an approximate 95% confidence interval for $\mathbb{E}(Y)$.
4. Use Exercise 3 to show that
$$\mathbb{E}\left(\sum_{i=1}^{n} [y_i - (\hat{a} + \hat{b}x_i)]^2\right) = (n-2)\sigma^2.$$

5. Explain why the confidence interval obtained in Exercise 3, for $\mathbb{E}(Y)$, is smaller than the prediction interval for y, using the prediction variance

$$\sigma^2\left[1 + \frac{1}{n} + \frac{(x-\bar{x})^2}{\sum(x_i-\bar{x})^2}\right].$$

6. Show that the cross product term in the residual sum of squares below must be 0.

$$\sum_{i=1}^{m}\sum_{j=1}^{k}(y_{ij}-\hat{y}_i)^2 = \sum_{i=1}^{m}\sum_{j=1}^{k}(y_{ij}-\bar{y}_i)^2 + k\sum_{i=1}^{m}(\hat{y}_i-\bar{y}_i)^2.$$

7. Show that both the pure error mean square and the model misfit mean square have expectation σ^2 under the assumption of the simple linear model.
8. Suppose there are k_i measurements $y_{ij}, j = 1, \ldots, k_i$ at each $x_i, i = 1, \ldots, m$. Provide a pooled estimate of σ^2.
9. Suppose n points, $x_1, \ldots, x_n$, are to be placed in $[-1, 1]$. Also suppose that both the simple linear model, $Y_i = a + bx_i + E_i$, and the Gauss measurement model hold. How should the x_i be chosen to minimize the variance of $\hat{b}$?
10. Suppose that $Y_i = a + bx_i + E_i$, where Y_i is the average of n_i replicates at x_i and the replicate errors are uncorrelated with common variance σ^2 (i.e., $\text{Var}(E_i) = \sigma^2/n_i$). Use transformations to re-express the linear model such that the errors have common variance, and find least squares estimates of a and b.
11. Suppose that the true model is

$$Y = c + dx + ex^2 + E,$$

where the Gauss measurement model holds; that is, errors are uncorrelated, with mean 0 and common variance σ^2. Suppose that the simple linear model is fit to the observations $(x_1, y_1), \ldots, (x_n, y_n)$; in other words, the quadratic term is ignored. Find the expectation of the residuals from this fit.
12. Use Exercise 4 to explain why W, which is defined in the subsection on maximum likelihood in the Theory section, has a t distribution with $n - 2$ degrees of freedom.

Notes

David Azuma of the USDA Forest Service Snow-Zone Hydrology Unit kindly provided the data for this lab, a detailed description of how the data were collected, and the references by Bergman ([Berg82]) and Kattelmann et. al. ([KMBBBH83]), which contain descriptions of the operation, location, and use of the cesium-137 snow gauge. Philip Price of the Lawrence Berkeley Laboratory generously helped us with a physical model for the relationship between gain and density.

Calibration and inverse regression is discussed in Draper and Smith ([DS81]) Chapter 1, Section 7, and in Carroll and Ruppert ([CR88]). Fisch and Strehlau

([FS93]) provide an accessible explanation of the maximum likelihood approach to inverse regression and confidence intervals. For more advance references, see Brown ([Bro93]) Chapter 2, Sections 3–7, and Seber ([Seb77]) Chapter 7, Section 2. For material on simultaneous confidence bands for calibration problems, see Lieberman et. al. ([LMH67]).

References

[Berg82] J.A. Bergman. Two standard precipitation gauges and the gamma transmission snow gauge: a statistical comparison. Research Note PSW-358, Pacific Southwest Forest and Range Experiment Station, Forest Service, U.S. Department of Agriculture, Berkeley, CA, 1982.

[Bro93] P.J. Brown. *Measurement, Regression, and Calibration*. Oxford University Press, Oxford, 1993.

[CR88] R. Carroll and D. Ruppert. *Transformation and Weighting in Regression*, Chapman and Hall, New York, 1988.

[DS81] N. Draper and H. Smith. *Applied Regression Analysis, 2nd Edition*. John Wiley & Sons, New York, 1981.

[FS93] R.D. Fisch and G.A. Strehlau. A simplified approach to calibration confidence sets. *Am. Stat.*, **47**:168–171, 1993.

[Hal70] M. Halperin. On inverse estimation in linear regression. *Technometrics* **12**:727–736, 1970.

[KMBBBH83] R.C. Kattelmann, B.J. McGurk, N.H. Berg, J.A. Bergman, J.A. Baldwin, and M.A. Hannaford. The isotope profiling snow gauge: twenty years of experience. Research Note 723-83, Pacific Southwest Forest and Range Experiment Station, Forest Service, U.S. Department of Agriculture, Berkeley, CA, 1983, 8p.

[LMH67] G.J. Lieberman, R.G. Miller, Jr., and M.A. Hamilton. Unlimited simultaneous discrimination intervals in regression. *Biometrika* **54**:133–145, 1967.

[Seb77] G.A.F. Seber. *Linear Regression Analysis*. John Wiley & Sons, New York, 1977.

9

Voting Behavior

The Voting Rights Act[1]

42 U.S.C. 1973

(a) No voting qualification or prerequisite to voting or standard, practice, or procedure shall be imposed or applied by any State or political subdivision in a manner which results in a denial or abridgment of the right of any citizen of the United States to vote on account of race or color, or in contravention of the guarantees set forth in section 1973b(f)(2) of this title as provided in subsection (b) of this section.

(b) A violation of subsection (a) of this section is established if, based on the totality of circumstances, it is shown that the political processes leading to nomination or election in the State or political subdivision are not equally open to participation by members of a class of citizens protected by subsection (a) of this section in that its members have less opportunity than other members of the electorate to participate in the political process and to elect representatives of their choice. The extent to which members of a protected class have been elected to office in the State or political subdivision is one circumstance which may be considered: Provided, that nothing in this section establishes a right to have members of a protected class elected in numbers equal to their proportion in the population.

[1]Reprinted by permission.

Introduction

In Stockton, California's November 4th 1986 election, the voters approved Measure C to discard the old system of district elections for City Council and Mayor, replacing it with a two tier system that included a district primary followed by a general city-wide election. After more than a year of controversy, six Stockton citizens filed a class-action lawsuit against the city. Their claim was that Measure C would dilute Hispanic and Black citizens' ability to elect their chosen representatives to the City Council. Hispanics make up about 20% of Stockton's population, and if the candidates preferred by Hispanics are often different from those preferred by Whites, then in a city-wide election it could prove very difficult for Hispanics to elect the candidate of their choice.

More than a year after the election to approve Measure C, in February, 1988, a United States District Court Judge blocked Stockton's municipal election that was to be held in June. According to the *Stockton Record* (Feb 20, 1998),

> U.S. District Court Judge Edward J. Garcia said an election can be held later if Measure C ultimately is ruled non-discriminatory. But he said the risk of harm to minorities is too great to allow an election under Measure C's ground rules before the measure can be tested in a trial.

The judge's decision was in response to the lawsuit filed by the six Stockton citizens. Although originally a class action suit, it was decertified before coming to trial. Instead, the plaintiffs made the claim that Measure C violated section 2 of the Voting Rights Act, which prohibits any practice which, even unintentionally, denies a citizen the right to vote because of race or color.

In voting-rights trials, statistical evidence is often given to demonstrate that a minority group generally prefers a candidate different from the one preferred by the majority group of voters. In this lab, you will have the opportunity to address the question of the validity of the statistical evidence commonly presented in these lawsuits.

Data

There are two sets of data available for this lab. The first set (Table 9.1) contains the election results from the 1988 Democratic presidential primary in Stockton, California. The election results are reported for each of the 130 electoral precincts in Stockton. In addition to the number of votes cast for presidential candidate Rev. Jesse Jackson and the total votes cast in the primary, census data are provided on the number of citizens in the precinct who are of voting age and on the proportions of these voting-age persons who are Hispanic and Black. The census information was derived from data for geographic regions that are larger than the precincts. To obtain precinct figures, the regional numbers were divided equally among the precincts in the region. This approximation leads to identical figures for precincts in the same region.

TABLE 9.1. Sample observations and data description for the 130 precinct results from the 1988 Democratic presidential primary results in Stockton, California.

Precinct	101	102	103	104	105	106	107
Jackson votes	120	192	201	129	60	113	122
Total votes	141	231	238	247	125	252	250
VAP	737	692	614	1349	955	1517	842
VAP Black	0.541	0.620	0.542	0.173	0.284	0.174	0.188
VAP Hispanic	0.381	0.236	0.381	0.476	0.439	0.469	0.412

Variable	Description
Precinct	Identification number for precinct.
Jackson votes	Votes cast in the primary for Jackson.
Total votes	Total votes cast in the primary.
Voting age population (VAP)	Number of citizens of voting age in precinct.
VAP Black	Proportion of voting-age population that is Black.
VAP Hispanic	Proportion of voting-age population that is Hispanic.

The second set of data is from an exit poll (Freedman et al. [FKSSE91]) conducted by the Field Research Corporation, a private consulting firm that conducts public opinion polls for state and federal agencies. The survey data were collected as voters left the polls. They were asked to anonymously complete a survey providing information on their race, income, education, and for whom they had cast their vote. See Table 9.2 for a description of these data. In the survey, the numbering scheme used to identify precincts does not match the precinct identification numbers used in the first data set.

The sampling method used for selecting the voters to be included in the exit poll is not described here. When analyzing the data from the exit poll, the sampled voters are treated as a population from a small city. These data are not used to compare against the city-wide election results, so the sampling procedure need not concern us.

Background

Election Rules in Stockton

Prior to Measure C, the city of Stockton was divided into nine districts, and one representative from each district sat on the City Council. Representatives were elected to the City Council via district elections. Candidates for a district's Council seat had to reside in the district and had to receive the most votes in the district.

TABLE 9.2. Sample observations and data description for the 1867 individuals in the 1988 exit poll (Freedman et al. [FKSSE91]).

Precinct	52	52	52	61	62	62	62	62	62
Candidate	3	1	4	3	3	1	1	1	3
Race	1	1	1	3	4	3	3	2	2
Income	5	0	4	0	3	3	1	3	1

Variable	Description
Precinct	Identification number for precinct.
Candidate	Vote cast for: 1 = Jackson; 2 = LaRouche; 3 = Dukakis; 4 = Gore.
Race	Voter's race: 1 = White; 2 = Hispanic; 3 = Black; 4 = Asian; 5 = other.
Income	Voter's income ($'000s): 1 = 00–10; 2 = 10–20; 3 = 20–30; 4 = 30–40; 5 = 40–50; 6 = 50–60; 7 = 60–70; 8 = 70+.

The mayor was elected in a city-wide election, where the mayoral candidates had to be City Council members.

Measure C changed the election of City Council members to a two-step process. First, each district held a primary election, where the top two vote-getters then ran in a general election. The general election was city-wide, with all voters in the city voting for one candidate in each district. Also under the new rules, the mayor was no longer required to be a sitting Council member, and the number of districts was reduced from nine to six.

Stockton Demographics

According to 1980 census data, 10% of the population of Stockton is Black; 22% is Hispanic; 59% is non-Hispanic White; and the remaining 9% are Asian or from other minority groups. The census data indicate that the Black and Hispanic communities are concentrated in south Stockton, while north Stockton is predominantly White.

At the time Measure C was adopted, three Blacks and one Hispanic sat on the City Council; two of the Blacks were elected from predominantly White districts. In 1990, the first election under the new rules was held; no Hispanic, one Black, and one Asian (the City's first Asian representative) were elected to the City Council.

Time Line of Badillo versus Stockton

11/86 Measure C approved by voters

12/87 Badillo, Fernandez, Villapando, Durham, Means, and Alston filed suit against the City of Stockton, California, stating that Measure C constitutes a violation of section 2 of the Voting Rights Act and that it violates the Fourteenth and Fifteenth Amendments of the United States Constitution.

> AMENDMENT XIV
> ... No State shall make or enforce any law which shall abridge the privileges or immunities of citizens of the United States; nor shall any State deprive any person of life, liberty, or property, without due process of law; nor deny to any person within its jurisdiction the equal protection of the laws. ...
> AMENDMENT XV
> The right of citizens of the United States to vote shall not be denied or abridged by the United States or by any State on account of race, color, or previous condition of servitude. ...

2/88 U.S. District Court Judge Garcia issued a preliminary injunction that blocked the June, 1988, Stockton primary election.

6/89 The U.S. District Court held a bench trial, where Judge Garcia dismissed the plaintiffs' charges against the city. According to 956 F.2d 884 (9th Cir. 1992)

> The district court held that the plaintiffs had failed to make out a valid section 2 claim. ... the plaintiffs had not presented sufficient evidence to establish that Blacks and Hispanics would vote together or separately as a politically cohesive group or that the White majority would vote as a block, overshadowing minority voting strength.

2/91 Durham appealed the decision.

2/92 Judges Schroeder, Canby, and Noonan of the U.S. Court of Appeals, Ninth Circuit, upheld the District Court's ruling, stating that the "city's change in election procedures increased the likelihood of minority voting strength dilution," but that it did not "mean that such dilution had actually occurred" and that the plaintiffs had failed to show that "minorities exhibited sufficient cohesion to cause election of their preferred candidates."

Vote Dilution and Polarization

Section 2 of the Voting Rights Act prohibits any practice that denies any citizen the right to vote on account of race. This includes any practice where members of a race "have less opportunity than other members of the electorate... to elect representatives of their choice."

The leading vote dilution case is Thornburg vs. Gingles, 478 US 30, which was decided in 1986. In that case, the Supreme Court ruled that plaintiffs must prove three things for a decision in their favor:

1. The minority group is sufficiently large and geographically compact to constitute a majority in a single-member district.
2. The minority group is politically cohesive.
3. The majority votes sufficiently as a block to enable it to usually defeat the minority preferred candidate.

To establish the first point, it only needs to be shown that a district can be drawn where at least 50% of the voting-age citizens belong to the minority group. The two other conditions are more difficult to establish. It must be shown that voting is polarized (i.e., the candidate generally preferred by the minority group is different from the candidate preferred by the majority group voters) and that the majority group's preferred candidate generally wins the election.

According to Klein and Freedman [KF93], who were consultants for the city of Stockton in the Badillo vs. Stockton case,

> Ecological regression has become the principle technique used by plaintiffs to demonstrate polarized voting.... Ecological regression was used in *Thornburg*, and the Supreme Court accepted the results. Since then, ecological regression has been used in voting-rights litigation in Arizona, California, Florida, Illinois, and Texas, among other states.

Ecological regression is described in the Theory section of this chapter. The basic premise is that the relationship between the votes cast in a precinct for a candidate and the ethnic make-up of a precinct can be used to determine individual voters' support for the candidate.

The 1988 Democratic Primary

Prior to the U.S. presidential election, which is held every four years, each political party selects a candidate to run in the election as the official party nominee. The Democratic party selects its nominee during a convention held in the summer before the November election. Each state sends delegates to the convention; these delegates determine the party's nominee for president. The state arm of the Democratic party sets the process for selecting its delegates for the convention. In California, part of that process includes a primary election, where adults who reside in the state and are registered with the party cast their vote in a secret ballot for the presidential candidate of their choice.

In 1988, Jesse Jackson, a Black minister, ran as a candidate for the Democratic nomination. Other candidates on the ballot included Michael Dukakis, Albert Gore, and Lyndon Larouche; all of these candidates were White. Dukakis was selected as the Democratic nominee, and in the presidential election, he lost to the Republican party nominee, George Bush.

Investigations

Consider the exit poll results from the 1988 Democratic primary in Stockton. If we treat the polled voters as if they constitute the entire voting-age population in some small town, say "Stocktette," then we have a special situation where we know how each eligible voter cast his or her vote in the election.

- To begin, summarize the election results for Stocktette by precinct. These are typical polling results, because a voter's ballot is secret.
- From this precinct-level summary, estimate the Hispanic support rate for Jackson, and attach an error to your estimate. The Hispanic support rate is defined as:

$$\frac{\text{Hispanic votes in Stocktette for Jackson}}{\text{Hispanic votes in the Stocktette primary}}.$$

  To help you make your estimate, assume that Hispanics support Jackson to the same extent regardless of where they live, as do non-Hispanics.
- Consider an alternative assumption that says voters in a precinct vote similarly regardless of race and that their support for Jackson is a function of income. More specifically, suppose the voters in a precinct vote in the same pattern, regardless of race, and the support for Jackson in a precinct is linearly related to the proportion of voters in the precinct with incomes exceeding $40,000. That is, the Hispanics who live in a precinct support Jackson to the same degree as all other ethnic groups in that precinct, and this support is determined by the relationship

$$\text{Proportion votes for Jackson} =$$
$$c + d \times \text{proportion voters earning over } \$40,000.$$

  For example, in a precinct where 20% of the voters have incomes exceeding $40,000, the support for Jackson is $c + 0.2d$ for all ethnic groups. Use this model to estimate the Hispanic support rate for Jackson in Stocktette.
- How do your two estimates compare to the truth and to each other? Is the true support rate within the error bounds you supplied for the first estimate? We can check our results against the truth only because we know how each individual voted in our hypothetical town. Typically, we do not know how individuals or groups of individuals, such as Hispanics, vote because the ballot is secret. Were the assumptions you used in making your estimates supported by the data on individual voters? Can these assumptions be checked if only precinct data are available?

Prepare a document to be submitted to the U.S. District Court Judge Garcia supporting or discouraging the use of the statistical methods you have investigated.

Theory

In the Stockton exit poll, among other things individual voters were asked to provide information on their education and income. These data are summarized in Table 9.3. The correlation between income and years of education for the voters in the sample is 0.45; voters with above average education levels tend to have above average incomes. Figure 9.1 shows a scatter plot of these data summarized to precincts. That is, each point in the plot represents a precinct, with the x-coordinate being the average number of years of education for voters in the precinct and the y-coordinate the average income. The correlation for the summarized data is 0.85, much higher than the correlation for individuals. But, it is individuals, not precincts, who earn incomes and attend school. Data reported for groups can be misleading because it is tempting to apply the correlation for groups to the individuals in the group. As in this case, there is typically a stronger correlation for groups than individuals. Correlation based on groups is called *ecological correlation*, and regression based on groups is called *ecological regression*. Here the ecological correlation is 0.85. (These calculations are done on binned data as in Table 9.3 because raw data is not available).

In this section, we will consider various models for the relationships between voter turnout and precinct population characteristics. We will also consider extending these relationships from precincts to individual voters. The data used in this section are from a 1982 Democratic primary for Auditor in Lee County, South Carolina (Loewen and Grofman [LG89]). There were two candidates running for

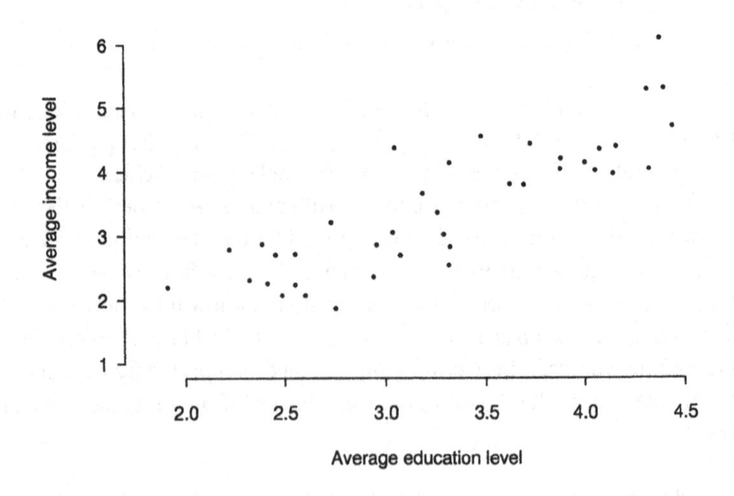

FIGURE 9.1. Average education and income levels in 39 precincts for voters surveyed in the exit poll from the 1988 Stockton Democratic primary.

TABLE 9.3. Table of voters by income and education for those surveyed in the exit poll from the 1988 Stockton Democratic primary (Freedman et. al. [FKSSE91]).

Education	Income ($'000s)								
(years)	<10	10–20	20–30	30–40	40–50	50–60	60–70	70+	Total
<12	96	60	48	15	7	3	3	5	237
12	88	87	89	87	49	26	9	9	444
13–15	70	103	117	113	75	62	28	25	603
16	9	22	33	32	31	27	20	26	200
17+	5	17	25	42	30	40	21	51	231
Total	268	289	312	299	192	158	81	116	1715

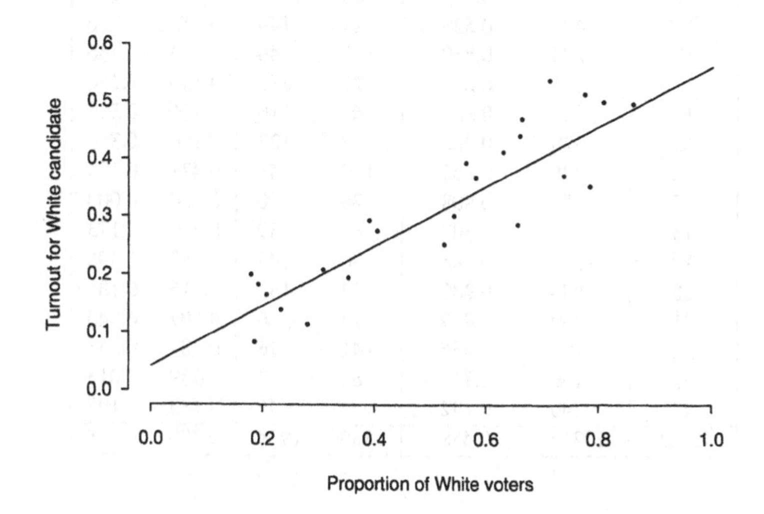

FIGURE 9.2. Scatter plot of the turnout for the White candidate against the proportion of voters in the precinct who are White, for the 1982 election results for Auditor in Lee County, South Carolina (Loewen and Grofman [LG89]). Also shown is the regression line for turnout in terms of proportion of voters who are White.

election: one White and the other Black. We will call the White candidate A and the Black candidate B. The data from this election are in Table 9.4.

Weighted Least Squares

Figure 9.2 is a scatter plot of the proportion of voters who turned out to vote for the White candidate against the proportion of voters in the precinct who are White.

TABLE 9.4. Table of 1982 election results for Auditor in Lee County, South Carolina (Loewen and Grofman [LG89]).

	Registered voters		Number of votes cast for		Proportion turnout for	
Precinct	Total number	Proportion White	*A*	*B*	*A*	*B*
1	567	0.783	200	43	0.353	0.076
2	179	0.659	79	1	0.441	0.006
3	659	0.404	182	146	0.276	0.222
4	651	0.562	257	72	0.395	0.111
5	829	0.581	306	170	0.369	0.205
6	965	0.389	284	159	0.294	0.165
7	226	0.774	116	21	0.513	0.093
8	418	0.541	126	29	0.301	0.069
9	456	0.178	91	144	0.200	0.316
10	172	0.738	64	30	0.372	0.174
11	472	0.523	119	149	0.252	0.316
12	278	0.630	115	46	0.414	0.166
13	871	0.232	121	218	0.140	0.250
14	353	0.351	69	116	0.196	0.329
15	320	0.206	53	127	0.166	0.397
16	276	0.663	130	48	0.471	0.174
17	52	0.808	26	0	0.500	0.000
18	225	0.307	47	39	0.209	0.173
19	362	0.185	30	143	0.083	0.395
20	844	0.280	97	153	0.115	0.181
21	149	0.859	74	6	0.497	0.040
22	492	0.655	141	16	0.287	0.033
23	156	0.712	84	2	0.539	0.013
24	240	0.192	44	46	0.183	0.192
Total	10212	0.458	2855	1924	0.280	0.188

Turnout for a candidate in a precinct is computed as follows:

$$\frac{\text{number of votes cast for candidate in precinct}}{\text{number of persons of voting age in precinct}}.$$

It appears that precincts with more White voters tend to have a higher turnout for the White candidate and that this relationship is approximately linear. In the next section, we show how this linearity can be used to estimate the support among Whites and Blacks for each candidate. Here we concentrate on fitting a line to the data.

The precincts are numbered 1 through 24 in Table 9.4. For precinct i, $i = 1, \ldots, 24$, let x_i represent the proportion of the voting-age persons who are White, and let y_i be the proportion of the voting-age persons who turned out and voted for the White candidate; these proportions appear in the third and sixth columns of Table 9.4, respectively. We can use the method of *weighted least squares* to fit a line to the data by minimizing the following sum of squares with respect to a

and b;

$$\sum_{i=1}^{24} w_i[y_i - (a + bx_i)]^2,$$

where $w_i = n_i/n$, the n_i is the number of registered voters in the ith precinct, and $n = \sum n_i$. The w_i in the least squares formula above are weights. The weights reflect the variation in the proportions y_i; larger precincts play a greater role than smaller precincts in determining the least squares coefficients. To further explain, consider each eligible voter as having a propensity, or chance, p to choose to turn out and vote for a candidate, independent of the other voters. Then the number of persons who actually do turn out and vote for a candidate has expectation $n_i p$ and variance $n_i p(1-p)$, and the proportion of voters who turn out for the candidate has expected value p and variance $[p(1-p)/n_i]$. Therefore, weighting each term in the sum of squares by w_i gives the terms equal variance.

The least squares solutions to the above sum of squares is:

$$\hat{a}_w = \bar{y}_w - \hat{b}_w \bar{x}_w,$$
$$\hat{b}_w = \frac{\sum w_i(y_i - \bar{y}_w)(x_i - \bar{x}_w)}{\sum w_i(x_i - \bar{x}_w)^2},$$

where

$$\bar{x}_w = \sum w_i x_i,$$
$$\bar{y}_w = \sum w_i y_i.$$

The subscript w reminds us that we use weights in the sum of squares. See the Exercises for other examples of weighted regression and for the derivation of these quantities.

For the White candidate in the Lee County primary, the least squares line is

$$0.04 + 0.52 \times x_i, \qquad (9.1)$$

and for the Black candidate (see Figure 9.3), the least squares line is

$$0.36 - 0.37 \times x_i. \qquad (9.2)$$

In ecological regression, the least squares lines in equations (9.1) and (9.2) are used to estimate support rates for candidates.

Support Rates for Candidates

The White support rate for candidate A is defined to be:

$$\frac{\text{number of Whites who voted for candidate } A}{\text{number of Whites who voted}}.$$

Because the ballot is secret, both the numerator and denominator are unknown, and we use the least squares lines in equations (9.1) and (9.2) to provide estimates of these quantities. To do this, a key assumption is first made:

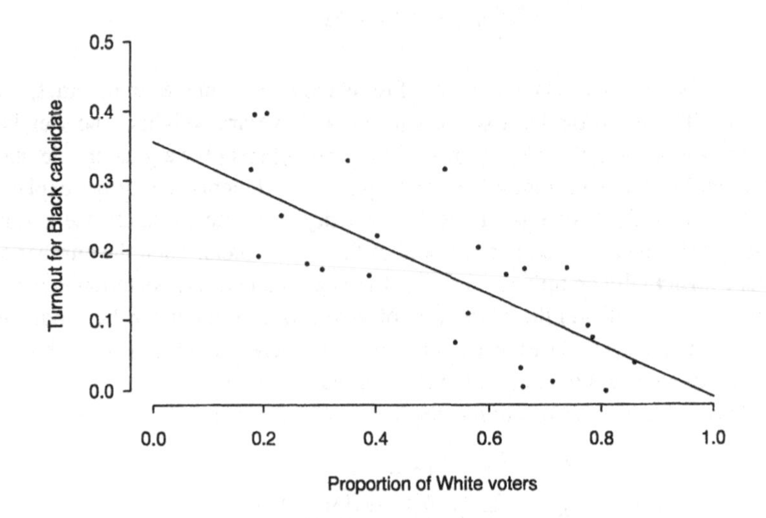

FIGURE 9.3. Scatter plot of the turnout for the Black candidate against the proportion of voters in the precinct who are White for the 1982 election results for Auditor in Lee County, South Carolina (Loewen and Grofman [LG89]). Also shown is the regression line for turnout in terms of proportion of voters who are White.

ETHNICITY ASSUMPTION: *Whites have the same propensity to turn out and vote for a particular candidate, regardless of the precinct they vote in. Blacks have the same propensity, but possibly different from the Whites' propensity, to turn out and vote for the particular candidate, regardless of precinct.*

How is this assumption used to determine a candidate's support rates? First we estimate the turnout for the White candidate; we take $x = 1$ and then $x = 0$ in equation (9.1) to estimate the turnout for candidate A in an all-White precinct and in an all-Black precinct, respectively:

$$0.04 + 0.52 \times 1 = 0.56,$$
$$0.04 + 0.52 \times 0 = 0.04.$$

According to the Ethnicity Assumption, the turnout for a candidate among White voters is the same across precincts, so the turnout for a candidate in a 100% White precinct is an estimate of the White turnout for the candidate. Similarly, to estimate turnout for the White candidate among Black voters, we can use the regression line prediction of turnout in a precinct with 0% Whites.

Using equation (9.2), we estimate the turnout for candidate B to be -1% among White registered voters and 36% among Black registered voters. Notice that the regression line gives an estimate for the White turnout for candidate B that is impossible because it is negative. The negative estimate arises because we are using the regression line at an x-value that is far from the observed proportion of Whites in any precinct.

Finally, to estimate the White support for candidate A, we use the ratio

$$\frac{0.56}{0.56 - 0.01} = 1.02,$$

The numerator is the estimated proportion of White voters who turned out and voted for A, and the denominator is the estimated proportion of White voters who turned out to vote. Again, we see an impossible estimate; the White support for A is larger than 1 due to the negative estimate for the turnout rate of Whites for candidate B. Nonetheless, voting definitely appears to be polarized because the Black support for the White candidate is estimated as

$$\frac{0.04}{0.04 + 0.36} = 0.11.$$

Standard errors for these support rates are discussed later in this section.

In estimating the support for a candidate among a subgroup of voters, we have used the line fitted to precincts, which we have just shown can lead to improbable results. The Ethnicity Assumption justifies the use of precinct regression in estimating White and Black support for a candidate. Unfortunately, neither assumption can be fully checked because they require data on individual voters, which we do not have because the ballot is secret.

Alternatives to the Ethnicity Assumption

An alternative to the Ethnicity Assumption on voting patterns is the following:

> NEIGHBORHOOD ASSUMPTION: *All voters in a precinct have the same propensity to turn out and vote for a particular candidate, regardless of a voter's race.*

The Neighborhood Assumption leads to an alternative method for estimating support for a candidate. According to this assumption, all eligible voters in a precinct support a candidate to the same extent, so the support for a candidate among White registered voters in a precinct can be estimated by the support for the candidate among all voters in the precinct. The overall support among eligible White voters for a candidate can be estimated by

$$\frac{\sum m_i x_i}{\sum (m_i + l_i) x_i}, \tag{9.3}$$

where m_i is the number of votes cast for A, and l_i is the number of votes cast for B (columns 4 and 5 in Table 9.4). To compute the numerator, the proportion of Whites in a precinct is applied to the number of votes a candidate received in the precinct. The denominator is an estimate of the total votes cast for either candidate by White voters. According to the Neighborhood Assumption, the support for the White candidate among Whites is estimated at 67%, and the support for this candidate among Blacks is estimated at 53%.

These estimates are very different from those calculated under the Ethnicity Assumption. The Neighborhood Assumption is in some sense opposite to the Ethnicity Assumption. One assumption is that the propensity to vote is determined by location, not race. The other assumption states that a voter's actions are determined by his or her race, not location. We don't expect either assumption to be fully correct but perhaps a reasonable model of voting behavior. However, both require precinct characteristics to apply to individual voters, and the question as to which assumption is most appropriate cannot be checked with the data at hand. Data on individual registered voters are needed, such as data from an exit poll.

A third assumption that we will also consider here says:

INCOME ASSUMPTION: *All voters in a precinct have the same propensity to turn out and vote for a particular candidate, and this propensity is a linear function of the average income in a precinct.*

Under this assumption, to estimate the White support for a candidate, we would use the formula in (9.3), except that we would substitute $n_i(\hat{c} + \hat{d}z_i)$ for m_i, where $\hat{c}$ and $\hat{d}$ are the least squares estimates of c and d obtained from minimizing

$$\sum w_i[y_i - (c + dz_i)]^2,$$

and where z_i is a summary measure of income for voters in precinct i. Similarly, for l_i we substitute the least squares estimate of the support for the Black candidate. We see that the Income Assumption is a Neighborhood-type Assumption.

Adding more ethnic groups or terms, such as income, in the regression typically improves the situation, but this addition neglects the assumption of linearity, and how these extra variables are involved in group relations remains unknown.

Standard Errors

In Chapter 7 on Dungeness crabs, the two variables of interest were the premolt and postmolt size of the crab shell, and they jointly exhibited natural variability. Here, we tend to think of the ethnic make-up of a precinct as a predetermined quantity and that the voter turnout exhibits natural variability. As in Chapter 8 on calibrating a snow gauge, a simple linear model for turnout would be that the expected turnout $\mathbb{E}(Y_i|x_i)$ is a linear function of x_i, the given proportion of eligible voters in the precinct;

$$\mathbb{E}(Y_i|x_i) = a + bx_i.$$

We place additional assumptions on the random Y_i—that they are independent with variance σ^2/n_i and hence

$$\mathbb{E}(Y_i|x_i) = a + bx_i,$$
$$\text{Var}(Y_i|x_i) = \sigma^2/n_i.$$

With this model, the weighted least squares solutions $\hat{a}_w$ and $\hat{b}_w$ are random, and their expectations and variances can be used to compute the standard error of the estimate:

$$\hat{y} = \hat{a}_w + \hat{b}_w \times 1,$$

$$\mathrm{SE}(\hat{y}) = \sigma\sqrt{\frac{1}{n} + \frac{(1-\bar{x}_w)^2}{\sum n_i(x_i - \bar{x}_w)^2}}.$$

Using an estimate of σ, the standard error above can provide confidence intervals for $\hat{y}$ at $x = 1$.

Exercises

1. Consider observations $(x_1, y_1), \ldots, (x_n, y_n)$. Show that if these observations are split into two groups, and the group means are $(\bar{x}_A, \bar{y}_A)$ and $(\bar{x}_B, \bar{y}_B)$, then the ecological correlation of the groups is ± 1 regardless of the correlation of the original observations.
2. Consider the hypothetical town with 10 precints, each with 100 voters, shown in Table 9.5. For simplicity, suppose all voters turn out to vote. Also suppose that 20% of the White voters, 40% of the Hispanic voters, and 80% of the Black voters support canidate A, regardless of the precinct in which they reside (i.e., candidate A receives 32 votes in precinct #1 and 50 votes in precinct #10).

   a. Use the Ethnicity Assumption, the number of votes candidate A receives in each precinct, and the percentage of White voters in each precinct to estimate the White support for candidate A (i.e., lump Hispanic and Black voters together as non-White).
   b. Use the Neighborhood Assumption and these same numbers to estimate the White support for candidate A.

3. As in Exercise 2, consider the hypothetical town shown in Table 9.5. Suppose the Neighborhood Assumption holds and all voters in a precinct have the same

TABLE 9.5. Table of voter ethnicity by precinct for a hypothetical town with 10 precincts.

	Registered voters		
Precinct	White	Black	Hispanic
1	80	20	0
2	80	15	5
3	70	20	10
4	60	30	10
5	60	35	5
6	40	40	20
7	40	35	25
8	20	50	30
9	20	30	50
10	10	30	60
Total	480	305	215

propensity to support candidate A. Also suppose that this propensity is 0.8, 0.7, 0.7, 0.6, 0.6, 0.5, 0.4, 0.2, 0.2, and 0.2 for precincts #1–10, respectively. For example, this means that candidate A receives 80 votes in precinct #1 and 20 in #10. As in Exercise 2, estimate the support for candidate A using only the number of votes candidate A receives in each precinct and the percentage of White voters in each precinct under the Ethnicity Assumption and again under the Neighborhood Assumption.

4. Assume the Neighborhood model. Then in precinct #1 in Lee County (Table 9.4, we would estimate that candidate A received ________ votes and candidate B received ________ votes from the White voters in the precinct.

5. Assume the Neighborhood model. Estimate the support for Jesse Jackson among White voters for those Stockton precincts reported in Table 9.1.

6. For the precincts in Lee County (Table 9.4), find the unweighted least squares line that fits turnout for candidate B to the proportion of White voters. Use this least squares line to estimate the support for the candidate among White and among Black voters. Compare your estimate to the one obtained by minimizing the weighted sum of squares in equation (9.2).

7. For pairs $(x_1, y_1), \ldots, (x_n, y_n)$ and weights w_i, find the a and b that minimize the following weighted sum of squares:

$$\sum w_i[y_i - (a + bx_i)]^2.$$

8. Suppose that in the model

$$Y_i = a + bx_i + E_i$$

the errors E_i have mean 0 and are uncorrelated, but $\mathrm{Var}(E_i) = \sigma^2/w_i$. The model may be transformed as follows:

$$\sqrt{w_i}Y_i = \sqrt{w_i}a + b\sqrt{w_i}x_i + \sqrt{w_i}E_i,$$

or equivalently,

$$Z_i = au_i + bv_i + D_i,$$

where $Z_i = \sqrt{w_i}Y_i$, $u_i = \sqrt{w_i}$, $v_i = \sqrt{w_i}x_i$, and $D_i = \sqrt{w_i}E_i$. Show that this new model satisfies the Gauss measurement model (Chapter 8). Also show that minimizing the sum of squares

$$\sum (z_i - au_i - bv_i)^2$$

is equivalent to minimizing the weighted sum of squares

$$\sum w_i(y_i - a - bx_i)^2.$$

9. Use the results from Exercise 8 to find the expected values, variances, and covariances of $\hat{a}_w$ and $\hat{b}_w$.

10. Use the result of Exercise 9 to show that

$$\text{SE}(\hat{y}) = \sigma\sqrt{\frac{1}{n} + \frac{(1-\bar{x}_w)^2}{\sum n_i(x_i - \bar{x}_w)^2}},$$

where $\hat{y} = \hat{a}_w + \hat{b}_w \times 1$.

Notes

The data for this lab were made available by David Freedman of the University of California, Berkeley, and Stephen Klein of the Rand Corporation. In addition, David Freedman provided answers to our many questions on the Stockton election.

References

[FKSSE91] D.A. Freedman, S.P. Klein, J. Sacks, C.A. Smyth, and C.G. Everett. Ecological regression and voting rights. *Eval. Rev.*, **15**:673–711, 1991.

[KSF91] S.P. Klein, J. Sacks, and D.A. Freedman. Ecological regression versus the secret ballot. *Jurimetrics*, **31**:393–413, 1991.

[KF93] S.P. Klein and D.A. Freedman. Ecological regression in voting rights cases. *Chance*, **6**:38–43, 1993.

[LG89] J. Loewen and B. Grofman. Recent developments in methods used in vote dilution litigation. *Urban Lawyer*, **21**:589–604, 1989.

10

Maternal Smoking and Infant Health (continued)

Introduction

In Chapter 1, the first lab on Maternal Smoking and Infant Health, it was found that the babies born to women who smoked during their pregnancy tended to weigh less than those born to women who did not smoke. But smokers may differ from nonsmokers in some essential ways that may affect the birth weight of the baby, whether or not the mother smoked. The 1989 Surgeon General's Report addresses this problem:

> ... cigarette smoking seems to be a more significant determinant of birth weight than the mother's prepregnancy height, weight, parity, payment status, or history of previous pregnancy outcome, or the infant's sex. The reduction in birthweight associated with maternal tobacco use seems to be a direct effect of smoking on fetal growth.
> Mothers who smoke also have increased rates of premature delivery. The newborns are also smaller at every gestational age.

In this lab, you will use data from the Child Health and Development Studies (CHDS) to investigate the assertions in the Surgeon General's Report. The CHDS provides a comprehensive study of all babies born between 1960 and 1967 at the Kaiser Foundation Hospital in Oakland, California (Yerushalmy [Yer64]). Many of the factors discussed in the Surgeon General's Report are reported for the families in the CHDS data set.

TABLE 10.1. Sample observations and data description for the 1236 babies in the Child Health and Development Studies subset.

Birth weight	120	113	128	123	108	136	138
Gestation	284	282	279	NA	282	286	244
Parity	1	0	1	0	1	0	0
Age	27	33	28	36	23	25	33
Height	62	64	64	69	67	62	62
Weight	100	135	115	190	125	93	178
Smoking status	0	0	1	1	1	1	0

Variable	Description
Birth weight	Baby's weight at birth, to the nearest ounce.
Gestation	Duration of the pregnancy in days, calculated from the first day of the last normal menstrual period.
Parity	Indicator for whether the baby is the first born (1) or not (0).
Age	Mother's age at the time of conception, in years.
Height	Height of the mother, in inches.
Weight	Mother's prepregnancy weight, in pounds.
Smoking status	Indicator for whether the mother smokes (1) or not (0).

The Data

As described in Chapter 1, the data for this lab are a subset of a much larger study: the Child Health and Development Studies (CHDS). The full data set includes all pregnancies that occurred between 1960 and 1967 among women in the Kaiser Foundation Health Plan in Oakland, California. The data here are from one year of the study; it includes all 1236 male single births where the baby lived at least 28 days. The Kaiser Health Plan is a prepaid medical care program; the members represent a broadly based group that is typical of an employed population. Approximately 15,000 families participated in the study. Some of the information collected on all single male births in one year of the study is provided here. The variables that are available for analysis are described in Table 10.1. Summary statistics on these variables appear in Table 10.2.

The mothers in the CHDS were interviewed early in their pregnancy, and a variety of medical, genetic, and environmental information was ascertained on both parents, including height, weight, age, smoking habits, and education. The initial interview was thorough and careful. The information on smoking was obtained before the birth of the baby and as such was not influenced by the outcome of the pregnancy. At birth, the baby's length, weight, and head circumference were recorded. Yerushalmy ([Yer64]) describes the initial interview process:

> The interviewers are competent; their relationship with the women is excellent; and the interview is obtained in a relaxed and unhurried fashion. The

TABLE 10.2. Summary statistics on the 1236 babies and their mothers from the Child Health and Development Studies subset.

Variable	Number	Average	SD	units
Birth weight	1236	120	18	ounces
Gestation	1223	279	16	days
Parity	1236	0.25	0.44	proportion
Age	1234	27	6	years
Height	1214	64	2.5	inches
Weight	1200	129	21	pounds
Smoking status	1236	0.39	0.49	proportion

greatest advantage, however, lies in the fact that the information is obtained prospectively, and therefore cannot be biased by knowledge of the outcome.

Background

Smokers versus Nonsmokers

Several studies of birth weight and smoking show that smokers differ from nonsmokers in many ways. In the CHDS, Yerushalmy ([Yer71]) found that the nonsmokers were more likely to use contraceptive methods and to plan the baby, and less likely to drink coffee and hard liquor (Table 10.3).

A very large database of all births in the state of Missouri between 1979 and 1983 (the smoking status of the mother was recorded on the birth certificate in Missouri) has been analyzed by several researchers (Malloy et al. [MKLS88], Kleinman et al. [KPMLS88]). They too found many important factors by which the pregnant women who smoked differed from those who did not smoke. On average, they were less well educated, younger, and less likely to be married than their nonsmoking counterparts (Table 1.6 in Chapter 1). These dissimilarities could contribute to the difference found in birth weight between the babies born to smokers and those born to nonsmokers.

A 1987 study of pregnant women in Northern California (Alameda County Low Birth Weight Study Group [Ala90]) excluded those women from their study

TABLE 10.3. Characteristics of mother according to smoking status for the families in the Child Health and Development Studies (Yerushalmy [Yer71]).

Characteristic	Percent of smokers	Percent of nonsmokers
Not using contraceptive	33	25
Planned baby	36	43
Drink 7+ cups of coffee a day	22	7
Drink 7+ glasses of whiskey a week	5	2

who used cocaine, heroin, or methadone. These women were removed from the analysis because prior studies indicated that the use of these substances is strongly associated with cigarette smoking and low-birth-weight. In the California study, it was found that smokers had a higher risk than nonsmokers of giving birth to low-birth-weight and preterm babies, even when these comparisons were made for women with similar alcohol use, prepregnancy weight, age, socio-economic status, prenatal care, and number of children.

Association versus Causation

Any study involving the smoking habits of people must necessarily be observational because people choose to smoke or not. This means that it is difficult to determine whether differences in birth weight are due to smoking or to some other factor by which the smokers and nonsmokers differ. In statistical jargon, observational studies are not controlled experiments. The investigator has no control over the assignment of subjects to the treatment (smoking), so the effect of smoking on the response (birth weight) can be confused, or *confounded*, with the effect due to other factors, such as alcohol consumption.

This same problem arises in studies of lung cancer and smoking, yet the Surgeon General concluded that smoking causes lung cancer and placed a warning on the side of cigarette packages that reads:

Cigarette smoking has been determined to cause lung cancer.

This conclusion is based on a large amount of evidence collected on lung cancer and smoking. The evidence is from various sources: large prospective studies that follow smokers and nonsmokers over years to observe their medical history and cause of death; large retrospective studies that compare the smoking habits of lung cancer patients with those of matched controls; twin studies that compare twins where one smokes and the other does not smoke; and controlled animal experiments that show smoking causes tumors in laboratory animals. There is also a plausible physiological explanation for how smoking can directly influence lung cancer, and this explanation is missing for other possible explanatory factors. Taking all of this evidence together, the Surgeon General determined that smoking does cause lung cancer, and this opinion is commonly accepted among the health professionals and researchers today.

There are far fewer studies on the relationship between smoking and birth weight and on the relationship between smoking and infant death. The Surgeon General's warning about maternal smoking and low birth weight that appears on the side panel of cigarette packages reads:

Smoking by pregnant women may result in fetal injury, premature birth, and low birthweight.

This warning is not as strongly worded as the Surgeon General's warning about lung cancer, which may be a reflection of the amount of evidence collected on the problem.

Investigations

The excerpt from the Surgeon General's Report contains three assertions.

- Mothers who smoke have increased rates of premature delivery.
- The newborns of smokers are smaller at every gestational age.
- Smoking seems to be a more significant determinant of birth weight than the mother's prepregnancy height and weight, parity, payment status, or history of previous pregnancy outcomes, or the infant's sex.

You are to investigate each of these statements.

Premature Delivery

- Do the mothers in the CHDS who smoked during pregnancy have higher rates of premature delivery than those who did not smoke? Consider collapsing the values for gestational age into a few manageable subgroups for making the comparison. Examine the rates directly and graphically.

Size at Every Gestational Age

The Surgeon General's Report states that the newborns of smokers are *smaller* at every gestational age. Babies could be smaller in length, weight, head circumference, or in another common measure of size, length^2/weight. Only birth weight is available for this lab, although Table 1.5 in Chapter 1 contains measurements on body length for the newborns in the CHDS.

- For different gestational ages, graphically compare the birth weights of babies born to smokers to those born to nonsmokers. In Chapter 1 on infant health, it was found that, overall, the average weight of a baby born to a smoker was 9 ounces lower than the average weight of a baby born to a nonsmoker. Is smoking associated with a 9 ounce reduction in birth weight at every gestational age? If not, can you describe how the difference changes with gestational age? Is there a simple relationship between birth weight and gestational age? Does it hold for both smokers and nonsmokers?

Determinants of Birth Weight

Address the claim of the Surgeon General that "smoking seems to be a more significant determinant of birth weight than the mother's prepregnancy height and weight, parity, payment status, or history of previous pregnancy outcomes, or the infant's sex." All of the CHDS families have the same payment status, all babies included in this lab are the same sex, and information on the history of previous pregnancies is not available. Therefore you will need to restrict your analysis to the mother's height, prepregnancy weight, and parity.

- When we compared the babies of smokers to those of nonsmokers, we found a difference of 9 ounces in their average birth weights. How does this difference compare to the difference between the birth weight of babies who are firstborn and those who are not? What about the birth weights of babies born to mothers who are above the median in height or weight and babies born to mothers who are below the median in height or weight?
- Recall that we also found that the spread and shape of the birth-weight distributions for the babies born to smokers and nonsmokers were quite similar. How do the birth-weight distributions compare for the groupings of the babies by parity, mother's height, or weight?
- Set aside the information as to whether they smoke or not, and divide the mothers into many groups where within each group the mothers are as similar as possible. For example, take into account a mother's height, weight, and parity when placing her in a group. Although the mothers within a group are similar in many respects, they may differ according to whether the mother smokes or not and by other characteristics as well. *Within* each group, compare the birth weights of babies born to smokers and nonsmokers. Do you see any patterns *across* groups?
- Another way to compare the effect of a mother's smoking on birth weight to the effect of her height on birth weight is to see if the variation in birth weight is related to these factors and, if so, in what way. First, consider the standard deviation of birth weight for all babies. Determine how much it is reduced when we take into consideration whether the mother smokes or not. To do this, for babies born to smokers, find the root mean square (r.m.s.) deviation in birth weight. Compare this to the r.m.s. deviation in birth weight for babies born to nonsmokers. If they are roughly the same, then they can be pooled to get a single r.m.s. deviation of each baby's birth weight from the average birth weight for its group.
- We can also see how the deviation in birth weight varies with mother's height. The babies can be split into groups, where each group consists of those babies whose mothers have the same height to the nearest inch. For each group, the remaining variation in birth weight can be computed using the r.m.s. deviation in birth weight about the group average. If the deviations do not vary much across groups, then they can be combined to form a single r.m.s. deviation in birth weight given mother's height. This single measurement of variation in birth weight can then be compared to the variation remaining in birth weight given the mother's smoking status. These types of comparisons are useful in determining the relative importance of smoking versus other factors that may affect birth weight.

Write up your findings to be included in a new handbook on pregnancy. The handbook is directed at the college-educated population of pregnant women and their families to help them make informed decisions about their pregnancy. Your findings are to appear in a section on smoking during pregnancy. Report your findings in "plain English" so the intended audience can understand your view of the problem and your conclusions. Include an appendix with supporting statistical analysis.

Theory

In Chapter 7 on Dungeness crabs, we saw that the natural variability in the size of crab shells before and after molting was well described by the regression line. In this chapter, it is the variability in size of babies and their mothers that are of interest. This section generalizes the regression line with one variable that was introduced in Chapter 7. It covers regression with indicator variables; techniques for comparing regression lines that use different explanatory variables; a geometric interpretation of least squares; and, in the Extensions section, two-variable regression.

Recall that the method of least squares involves finding the values of the intercept a and the slope b that minimize the sum of squared deviations

$$\sum_{i=1}^{n}[y_i - (a + bx_i)]^2$$

for the n points $(x_1, y_1), \ldots (x_n, y_n)$. Throughout this section, we will take y_i to represent the birth weight of the ith baby in the CHDS data set, and x_i to denote the length of gestation for the baby.

Indicator Variables

Separate regression lines for the babies born to smokers and those born to nonsmokers can fit birth weight to gestational age. By splitting the babies and their mothers into two groups according to whether or not the mother smokes, we can minimize the sum of squares, with respect to a_s, b_s, a_n, and b_n, separately for each group,

$$\sum_{\text{(smokers)}} [y_i - (a_s + b_s x_i)]^2 \quad \text{and} \quad \sum_{\text{(nonsmokers)}} [y_i - (a_n + b_n x_i)]^2,$$

to obtain two regression lines, one for smokers and one for nonsmokers. If these two regression lines are roughly parallel over the typical range for gestational age, then we can proceed to fit these two lines simultaneously such that they have exactly the same slope. When the two lines have a common slope ($b_s = b_n$), then we can obtain a more accurate estimate of birth weight by using the data from both groups to fit two parallel lines. We use an *indicator variable* to accomplish this task.

Indicator variables take on only two possible values: 0 and 1. An indicator for smoking status would be 1 for a mother who smoked during her pregnancy and 0 for a mother who did not. It "indicates" the smoking status of the mother. Whether the baby is firstborn or not can also be represented via an indicator variable.

To make this definition more precise, for the ith baby in the CHDS, $i = 1, \ldots, n$, let $z_i = 1$ if the baby's mother smoked during her pregnancy and $z_i = 0$ if she did not. Then we can find the values of a_p, b_p, and c_p that minimize the following sum of squared deviations:

$$\sum_{i=1}^{n}[y_i - (a_p + b_p x_i + c_p z_i)]^2.$$

The subscript p reminds us that the data from the two groups are pooled. Notice that this sum of squares is equivalent to the following sum of squares:

$$\sum_{\text{(smokers)}} [y_i - (a_p + c_p + b_p x_i)]^2 + \sum_{\text{(nonsmokers)}} [y_i - (a_p + b_p x_i)]^2.$$

The minimization fits two parallel lines to the data, one for smokers and one for nonsmokers. If the minimizing values for a_p, b_p, and c_p are $\hat{a}_p$, $\hat{b}_p$, and $\hat{c}_p$, then $\hat{a}_p + \hat{c}_p + \hat{b}_p x$ is the regression line for the mothers who smoke and $\hat{a}_p + \hat{b}_p x$ is the regression line for those who do not smoke. It follows that $\hat{c}_p$ is the average difference in birth weight between the two groups of babies, after we adjust (equivalently control) for gestational period. This qualification states that we calculate the difference using a linear regression that includes gestation period. By fitting two parallel lines, we are saying that this difference is roughly the same for each gestation period. In other words, whether the babies are born at 37 weeks or 40 weeks, the average difference in weight between those born to smokers and those born to nonsmokers is roughly $\hat{c}_p$.

Residual plots that include information as to whether the baby belongs to the smoking group or nonsmoking group can be helpful in determining whether the fitted parallel lines model is appropriate.

Comparing Regressions

The Surgeon General uses the term "a more significant determinant" when comparing the relationship between smoking status and birth weight to the relationship between mother's height and birth weight. There are a variety of ways to interpret this comparative statement. We present one here that uses two regression lines — one that fits birth weight to a smoking indicator and one that fits birth weight to mother's height.

Recall that y_i represents the birth weight of the ith baby and that z_i is an indicator for the smoking status of the mother. Now let v_i represent the ith mother's height. We write the regression line for expressing birth weight in terms of mother's height as $\hat{d} + \hat{e}v$.

To fit birth weight to the smoking indicator alone, we would minimize, with respect to f and g, the sum of squares

$$\sum_{i=1}^{n}[y_i - (f + g z_i)]^2 = \sum_{\text{nonsmokers}} (y_i - f)^2 + \sum_{\text{smokers}} [y_i - (f + g)]^2.$$

It is easy to show by completing the square that $\hat{f}$ is the average birth weight for the nonsmokers, $\hat{f}+\hat{g}$ is the average birth weight for the smokers, and the regression "line" is $\hat{f}+\hat{g}z$.

To determine whether smoking or mother's height is a more significant determinant of birth weight, we can compare the *root mean square (r.m.s.) deviations* of the residuals for each of the fits. First, the r.m.s. of the deviations of birth weight from their average is 18.4 ounces; this is simply another name for the SD of birth weight. The r.m.s. deviation of the residuals from fitting birth weight to mother's height is 18.0 ounces, and for fitting birth weight to the smoking indicator it is 17.8 ounces. The residuals from the regression for the smoking indicator have a smaller r.m.s. than those from the regression for mother's height. An alternative description is that the smoking indicator reduces the r.m.s. in birth weight by 3% $(1-17.8/18.4)$. Mother's height reduces it by 2%. Both reductions are quite small; there is a lot of variability in birth weight that remains "unexplained" by either the smoking indicator or the mother's height. However, this percentage is still meaningful, as millions of babies are born annually to smokers in the U.S.

We now give a few words of caution about the interpretation of the reduction in standard deviation. Although we have used the conventional terminology "explained" variation, it should not be taken to imply a causal relationship. It means only that the deviations of birth weight from their average are reduced when babies are compared to those in a group more like themselves (i.e., those born to smokers, or those born to mothers of a particular height). In addition, the variation "explained" by one of two or more variables could, in some instances, also be explained by another of the variables. This is not the case for the pair of smoking and mother's height because whether the mother smokes or not is uncorrelated with her height. To explain, note that the variance in birth weight is 337 ounces^2. Smoking reduces this variance by 20 ounces^2 (i.e., the residuals from the least squares fit of birth weight to smoking status have a variance of 317 ounces^2). Mother's height reduces the variance in birth weight by 13 ounces^2. Usually this reduction in variance is expressed as a percentage. In this case, the reduction in variance explained by smoking is 6%, and the reduction explained by mother's height is 4%. When we fit birth weight to both smoking status and mother's height (i.e., when we fit two parallel lines one for smokers and one for nonsmokers), the variability in birth weight is reduced by 10% from 337 ounces^2 to 303 ounces^2. This additive relationship holds because the sample correlation between mother's height and the smoking indicator is .01, which is essentially 0; that is, the explanatory variables smoking and height are nearly uncorrelated. When the two variables are correlated, then both can explain some of the same variability in the response. This is described in more detail in the Extensions section.

Geometry of Least Squares

The geometric interpretation of least squares is intuitively appealing and easily leads to the generalization of simple regression to two-variable and multiple regression.

To begin, we consider a simpler problem where we have only the observations $y_1, \ldots, y_n$. Suppose our goal is to find k that minimizes

$$\sum_{i=1}^{n}(y_i - k)^2.$$

This problem can be re-expressed in terms of vectors by representing the n observations as an $n \times 1$ column vector $(y_1, \ldots, y_n)$, which we call $\mathbf{y}$. Then the sum of squares above can be rewritten as

$$|\mathbf{y} - k\mathbf{1}|^2,$$

where $\mathbf{1}$ is the $n \times 1$ column vector of 1's—namely, $(1, \ldots, 1)$—and where $| \ |^2$ is the squared Euclidean length of the vector,

$$|\mathbf{y}|^2 = \sum_{i=1}^{n} y_i^2.$$

The problem can now be recognized as one of finding the closest point to $\mathbf{y}$ in the linear span of $\mathbf{1}$. The linear span of the vector $\mathbf{1}$ consists of all those vectors of the form $k\mathbf{1}$ for some constant k. Figure 10.1 shows a picture of the problem.

The closest point to $\mathbf{y}$ in the linear span of $\mathbf{1}$ is $\bar{y}\mathbf{1}$, where as usual $\bar{y} = (y_1 + \cdots y_n)/n$. To show that this is indeed the case, note that $\bar{y}\mathbf{1}$ is in the linear span of $\mathbf{1}$. Also note that $\mathbf{y} - \bar{y}\mathbf{1}$ is orthogonal to $\mathbf{1}$ because the dot product of the two vectors is 0, as shown below:

$$\begin{aligned}(\mathbf{y} - \bar{y}\mathbf{1}) \cdot \mathbf{1} &= \mathbf{y} \cdot \mathbf{1} - \bar{y}\mathbf{1} \cdot \mathbf{1} \\ &= n\bar{y} - n\bar{y} \\ &= 0.\end{aligned}$$

Pythagoras' formula says that for any k,

$$\begin{aligned}|\mathbf{y} - k\mathbf{1}|^2 &= |\mathbf{y} - \bar{y}\mathbf{1} + (\bar{y} - k)\mathbf{1}|^2 \\ &= |\mathbf{y} - \bar{y}\mathbf{1}|^2 + |(\bar{y} - k)\mathbf{1}|^2.\end{aligned}$$

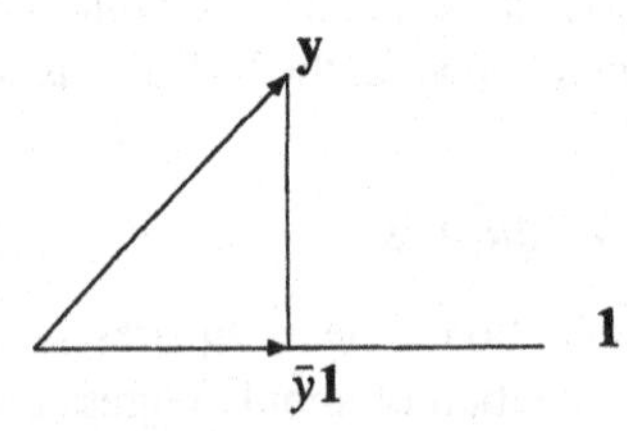

FIGURE 10.1. Projection of the vector $\mathbf{y}$ onto the linear span of $\mathbf{1}$.

The cross product disappears because $\mathbf{y} - \bar{y}\mathbf{1}$ is orthogonal to $\mathbf{1}$. Hence $|\mathbf{y} - k\mathbf{1}|^2$ is minimized for $k = \bar{y}$. (Notice that the squared length of $\mathbf{y} - \bar{y}\mathbf{1}$ is n times the variance of $\{y_1, \ldots, y_n\}$.)

In general, the closest point to a vector $\mathbf{y}$ in the linear span of a vector $\mathbf{v}$ is the orthogonal projection

$$P_{\mathbf{v}}\mathbf{y} = \frac{\mathbf{y} \cdot \mathbf{v}}{|\mathbf{v}|^2}\mathbf{v},$$

where $\mathbf{y} \cdot \mathbf{v}$ is the dot product $\sum_i y_i v_i$.

Now we can consider the original problem of minimizing, with respect to a and b, the sum of squares

$$\sum [y_i - (a + bx_i)]^2,$$

which we can re-express as

$$|\mathbf{y} - (a\mathbf{1} + b\mathbf{x})|^2.$$

Figure 10.2 shows the solution to be the projection of $\mathbf{y}$ onto the linear span of the two vectors $\mathbf{1}$ and $\mathbf{x}$. In general, the orthogonal projection of $\mathbf{y}$ onto a linear space $\mathcal{L}$ is that unique vector $\mathbf{p}$ in $\mathcal{L}$ such that the residual $\mathbf{y} - \mathbf{p}$ is orthogonal to $\mathcal{L}$.

To determine the projection onto $\mathbf{1}$ and $\mathbf{x}$, it is simpler to consider the equivalent problem of projecting $\mathbf{y}$ onto the linear span of the two orthogonal vectors $\mathbf{1}$ and $\mathbf{x} - \bar{x}\mathbf{1}$. The linear span of $\mathbf{1}$ and $\mathbf{x}$ is also the linear span of $\mathbf{1}$ and $\mathbf{x} - \bar{x}\mathbf{1}$.

The projection $P_{1,\mathbf{x}}\mathbf{y}$ is the same as the projection $P_{1,\mathbf{x}-\bar{x}1}\mathbf{y}$, which, due to orthogonality, is equivalent to the sum of the two projections,

$$P_{1,\mathbf{x}}\mathbf{y} = P_{1}\mathbf{y} + P_{\mathbf{x}-\bar{x}1}\mathbf{y} = \bar{y}\mathbf{1} + \frac{\mathbf{y} \cdot (\mathbf{x} - \bar{x}\mathbf{1})}{|\mathbf{x} - \bar{x}\mathbf{1}|^2}(\mathbf{x} - \bar{x}\mathbf{1}).$$

This projection is the vector of fitted values for the original regression problem. To see this, rearrange the terms to find that the ith element of the projected vector is

$$\bar{y} - b_{yx\cdot 1}\bar{x} + b_{yx\cdot 1}x_i,$$

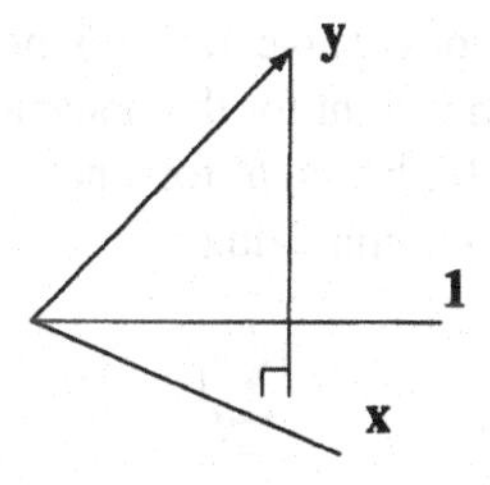

FIGURE 10.2. Projection of the vector **y** onto the linear span of **1** and **x**.

where $b_{yx\cdot 1}$ is the coefficient of $\mathbf{x}-\bar{x}\mathbf{1}$. This coefficient is identical to the coefficient $\hat{b}$ obtained from minimizing the sum of squared deviations

$$\sum[y_i - (a + bx_i)]^2$$

in Chapter 7.

This geometric representation of least squares can be used to establish many of the properties of the fitted values and residuals, where the fitted value and residual vectors are denoted

$$\hat{\mathbf{y}} = P_{1,x}\mathbf{y},$$
$$\mathbf{r} = \mathbf{y} - \hat{\mathbf{y}}.$$

Of particular interest is the fact that $\bar{y}\mathbf{1}$, $b_{yx\cdot 1}(\mathbf{x} - \bar{x}\mathbf{1})$, and $\mathbf{y} - \bar{y}\mathbf{1} - b_{yx\cdot 1}(\mathbf{x} - \bar{x}\mathbf{1})$ are pairwise orthogonal. This fact makes many of the following statements self-evident.

- $\bar{r} = 0.$
- $\hat{\mathbf{y}} \cdot \mathbf{r} = 0.$
- $\mathbf{x} \cdot \mathbf{r} = 0.$
- $|\mathbf{y} - \bar{y}\mathbf{1}|^2 = |\mathbf{r}|^2 + |\hat{\mathbf{y}} - \bar{y}\mathbf{1}|^2.$

The proofs of these relationships are left to the Exercises.

Exercises

1. Consider the relationship between low birth weight, parity, and smoking. Construct an artificial example where smokers have a higher rate of low-birth-weight babies than the nonsmokers, but, when parity is also considered, the relationship inverts. In other words, when examined within parity status, smokers in comparison to nonsmokers have a lower rate of low-birth-weight babies. This oddity is called Simpson's Paradox.
2. Consider the minimization with respect to f and g of the expression

$$\sum_{i=1}^{n}[y_i - (f + gz_i)]^2,$$

where z_i is the indicator variable for smoking and y_i is birth weight. Show that $\hat{f}$ is the average birth weight for the nonsmokers and that $\hat{g}$ is the difference between the average birth weight for smokers and nonsmokers.
3. Find the solution for b minimizing

$$\sum_{i=1}^{n}(y_i - bx_i)^2.$$

4. Prove that $P_{\mathbf{v}}\mathbf{y}$ is the closest point to $\mathbf{y}$ in the linear span of $\mathbf{v}$.

5. Show that the expression for $\hat{b}$ obtained by minimizing
$$\sum[y_i - (a + bx_i)]^2$$
with respect to a and b,
$$\hat{b} = r\frac{\text{SD}(y)}{\text{SD}(x)},$$
is the same as
$$\frac{\mathbf{y} \cdot (\mathbf{x} - \bar{x}\mathbf{1})}{|\mathbf{x} - \bar{x}\mathbf{1}|^2}.$$
6. Prove that
$$b_{(y\cdot1)(x\cdot1)} = b_{yx\cdot1},$$
where $b_{(y\cdot1)(x\cdot1)}$ is the coefficient obtained by projecting $\mathbf{y} - \bar{y}\mathbf{1}$ onto the linear span of $\mathbf{x} - \bar{x}\mathbf{1}$.
7. Show that the following vectors are pairwise orthogonal.
   a. $\mathbf{1}$.
   b. $\mathbf{x} - \bar{x}\mathbf{1}$.
   c. $\mathbf{y} - \bar{y}\mathbf{1} - b_{yx\cdot1}\mathbf{x}$.
8. Use the results from Exercise 7 to establish the following properties of the residual and fitted value vectors:
   a. $\bar{r} = 0$.
   b. $\hat{\mathbf{y}} \cdot \mathbf{r} = 0$.
   c. $\mathbf{x} \cdot \mathbf{r} = 0$.
   d. $|\mathbf{y} - \bar{y}\mathbf{1}|^2 = |\mathbf{r}|^2 + |\hat{\mathbf{y}} - \bar{y}\mathbf{1}|^2$.

Extensions

In this section, we consider two-variable linear regression. We already encountered a special case of two-variable linear regression when we included the indicator variable for smoking status in the regression that fitted two parallel lines for smokers and nonsmokers. Now we examine the more general case. As an example, we use the regression of birth weight on mother's height and prepregnancy weight.

By examining a scatter plot, it is easy to see if the relationship between birth weight and mother's height is roughly linear. It is not as easy to see if the relationship between three variables is roughly linear. To assess whether two-variable linear regression is appropriate, we can divide the babies and their mothers into groups according to the mother's weight, so that each group consists of mothers who are roughly the same weight. All plots of birth weight on mother's height for each group of babies should show linear relationships, and the lines should have similar slopes. Also, for each group we can fit the simple (i.e., one variable) linear regression of baby's weight on mother's height. This would give us a series of regression lines, one for each weight group. If these regression lines are roughly parallel, then two-variable linear regression can be used to fit a common slope.

In two-variable linear regression, we minimize, with respect to a, b, and c, the sum of squares

$$\sum_{i=1}^{n}[y_i - (a + bv_i + cw_i)]^2,$$

where y_i is the baby's birth weight, v_i is the mother's height, and w_i is the mother's weight for the ith baby in the CHDS. We have performed this minimization for the CHDS babies and found that $\hat{a} = 35$ ounces, $\hat{b} = 1.2$ ounces/inch, and $\hat{c} = 0.07$ ounces/pound. Specifically, the two-variable linear regression is

$$\hat{y}_i = 35 + 1.2v_i + 0.07w_i.$$

The coefficient for mother's height is different from the coefficient in the simple regression of birth weight on mother's height. There we found that

$$\hat{y}_i = 27 + 1.4v_i.$$

The two coefficients for mother's height differ because the 1.4 arises from fitting mother's height alone to birth weight and the 1.2 is from fitting mother's height in the presence of mother's weight. To interpret the coefficient 1.2, we need to keep in mind that both the height and weight variables were fitted in the regression. This means that among mothers with a given fixed weight, 1.2 ounces is the average increase in birth weight for each additional inch of height. We see that there is less of a height effect on birth weight when we know the mother's weight.

We compare the variance (mean square) of the residuals from the two-variable fit to the variances from each of the one-variable fits. Birth weight is reduced from 337 to 322 ounces^2 in the two-variable linear regression that incorporates both mother's height and weight. For the one-variable linear regressions, the reduction in mean square of the residuals for mother's height is 13 ounces^2, and the reduction for mother's weight is 8 ounces^2. Notice that mother's height reduces the variation in birth weight more than mother's weight. This is because the correlation between mother's height and birth weight is higher than the correlation between mother's weight and birth weight. Also notice that the reduction is not additive as it was for the two-variable model that included mother's height and the smoking indicator. This is because mother's height and mother's weight are correlated. See Table 10.4 for the correlations.

TABLE 10.4. Correlations among birth weight, mother's height, and mother's weight for 1197 babies from the CHDS.

	Birth weight	Mother's height	Mother's weight
Birth weight	1.00	0.20	0.16
Mother's height	0.20	1.00	0.44
Mother's weight	0.16	0.44	1.00

Geometry

The geometric picture of two-variable linear regression is useful in understanding the interpretation of the coefficients. As before, write $\mathbf{y}$, $\mathbf{v}$, and $\mathbf{w}$ for the $n \times 1$ column vectors, $(y_1, \ldots, y_n)$, $(v_1, \ldots, v_n)$, and $(w_1, \ldots, w_n)$, respectively. Also let $\mathbf{1}$ denote the $n \times 1$ column vector of 1's. The two-variable least squares minimization described here is equivalent to the minimization, with respect to a, b, and c, of the squared length

$$|\mathbf{y} - (a + b\mathbf{v} + c\mathbf{w})|^2.$$

By analogy with the geometric solution of the one-variable linear regression, the two-variable solution is the projection of $\mathbf{y}$ onto the space spanned by the vectors $\mathbf{1}$, $\mathbf{v}$, and $\mathbf{w}$. This space is equivalent to the space spanned by $\mathbf{1}$, $\mathbf{v} - \bar{v}\mathbf{1}$, and $\mathbf{w} - \bar{w}\mathbf{1} - b_{wv\cdot 1}(\mathbf{v} - \bar{v}\mathbf{1})$. It is also equivalent to the space spanned by $\mathbf{1}$, $\mathbf{w} - \bar{w}\mathbf{1}$, and $\mathbf{v} - \bar{v}\mathbf{1} - b_{vw\cdot 1}(\mathbf{w} - \bar{w}\mathbf{1})$. See the Exercises for a proof of this result. This implies that

$$\begin{aligned} P_{1,v,w}\mathbf{y} &= \bar{y}\mathbf{1} + b_{yv\cdot 1}(\mathbf{v} - \bar{v}\mathbf{1}) + b_{yw\cdot 1,v}[\mathbf{w} - \bar{w}\mathbf{1} - b_{wv\cdot 1}(\mathbf{v} - \bar{v}\mathbf{1})] \\ &= \bar{y}\mathbf{1} + b_{yw\cdot 1}(\mathbf{w} - \bar{w}\mathbf{1}) + b_{yv\cdot 1,w}[\mathbf{v} - \bar{v}\mathbf{1} - b_{vw\cdot 1}(\mathbf{w} - \bar{w}\mathbf{1})], \end{aligned}$$

where

$$b_{yw\cdot 1,v} = \frac{\mathbf{y} \cdot [\mathbf{w} - \bar{w}\mathbf{1} - b_{wv\cdot 1}(\mathbf{v} - \bar{v}\mathbf{1})]}{|\mathbf{w} - \bar{w}\mathbf{1} - b_{wv\cdot 1}(\mathbf{v} - \bar{v}\mathbf{1})|^2}$$

and

$$b_{yv\cdot 1,w} = \frac{\mathbf{y} \cdot [\mathbf{v} - \bar{v}\mathbf{1} - b_{vw\cdot 1}(\mathbf{w} - \bar{w}\mathbf{1})]}{|\mathbf{v} - \bar{v}\mathbf{1} - b_{vw\cdot 1}(\mathbf{w} - \bar{w}\mathbf{1})|^2}.$$

Collect terms to find that

$$\begin{aligned} &(\bar{y} - b_{yv\cdot 1}\bar{v} + b_{yw\cdot 1,v}b_{wv\cdot 1}\bar{v} - b_{yw\cdot 1,v}\bar{w})\mathbf{1} \\ &\quad + (b_{yv\cdot 1} - b_{yw\cdot 1,v}b_{wv\cdot 1})\mathbf{v} + b_{yw\cdot 1,v}\mathbf{w} \\ &= (\bar{y} - b_{yw\cdot 1}\bar{w} + b_{yv\cdot 1,w}b_{vw\cdot 1}\bar{w} - b_{yv\cdot 1,w}\bar{v})\mathbf{1} \\ &\quad + b_{yv\cdot 1,w}\mathbf{v} + (b_{yw\cdot 1} - b_{yv\cdot 1,w}b_{vw\cdot 1})\mathbf{w}. \end{aligned}$$

Since the coefficients of $\mathbf{1}$, $\mathbf{v}$, and $\mathbf{w}$ must be equal, we find that

$$b_{yv\cdot 1,w} = b_{yv\cdot 1} - b_{yw\cdot 1,v}b_{wv\cdot 1}$$

and

$$b_{yw\cdot 1,v} = b_{yw\cdot 1} - b_{yv\cdot 1,w}b_{vw\cdot 1}.$$

The two-variable linear regression equation can now be more compactly expressed as

$$(\bar{y} - b_{yv\cdot 1,w}\bar{v} - b_{yw\cdot 1,v}\bar{w})\mathbf{1} + b_{yv\cdot 1,w}\mathbf{v} + b_{yw\cdot 1,v}\mathbf{w}.$$

Here we see that the coefficients of $\mathbf{v}$ and $\mathbf{w}$ from the two-variable linear regression depend on the presence of the other variable. In other words, the coefficient for

v results from the projection of **y** onto that part of **v** that is orthogonal to **1** and **w**. If **v** and **w** are orthogonal, then the coefficients in the two-variable model are the same as those in the one-variable model; that is, $b_{yv\cdot 1,w} = b_{yv\cdot 1}$ and $b_{yw\cdot 1,v} = b_{yw\cdot 1}$.

Plots of residuals from the fit of **y** to **1** and **v** against the residuals from the fit of **w** to **1** and **v** can help determine whether the slope $b_{yw\cdot 1,v}$ is roughly 0 or not. This is because $b_{yw\cdot 1,v} = b_{(y\cdot 1,v)(w\cdot 1,v)}$ (see the Exercises).

The two-variable linear regression can be easily generalized to multiple linear regression using this geometric picture. But, linearly fitting multiple variables should be done carefully. There is usually not enough data even for the two-variable regression to see whether the simple linear regression involving one of the variables has the same slope for all values of the second variable. For three or more variables, we need a huge data set. For example, to determine whether a three-variable regression of birth weight on gestation and mother's height and weight is supported by the data would involve producing plots of birth weight and gestation for groups of babies with mothers of the same height and weight. Often the linear relationship between variables is assumed to hold and no checking of the assumptions is made. At other times, the multiple linear regression is simply regarded as the best linear fit to the data.

Extension Exercises

1. Polynomial regression is a special case of multiple linear regression. Consider fitting a polynomial,

$$y = bx + cx^2,$$

   to the points (x_i, y_i), $i = 1, \ldots n$. Express the polynomial as a two-variable linear equation with variables x and $u = x^2$, and find the least squares estimates of the coefficients.

2. Suppose you try to minimize the following sum of squares with respect to a, b, c, and d:

$$\sum_{i=1}^{n} [y_i - (a + bx_i + cu_i + dv_i)]^2,$$

   where $v_i = x_i + 2u_i + 3$. Explain geometrically why it would be problematic to estimate the coefficients at the unique minimum.

3. Prove

$$P_{1,\mathbf{x}}\mathbf{y} = P_{1}\mathbf{y} + P_{\mathbf{x}-\bar{x}1}\mathbf{y}.$$

4. Consider the minimization with respect to a, b, c, and d of the following sum of squares:

$$\sum_{i=1}^{n} [y_i - (a + bx_i + cz_i + dx_iz_i)]^2,$$

where x_i is mother's height, z_i is an indicator variable for smoking, and y_i is birth weight. Show that the minimization above yields the same result as fitting birth weight to mother's height separately for smokers and nonsmokers.

5. Show that the projection onto the sum of three orthogonal one-dimensional subspaces is the sum of the projections onto each of the subspaces.

6. Show that

$$b_{yw\cdot 1,v} = b_{(y\cdot 1,v)(w\cdot 1,v)}.$$

Notes

The Surgeon General's Report and other information on smoking and reproduction can be found in *Reducing the health consequences of smoking: 25 years of progress.* A report of the Surgeon General ([DHHS89]) and in *Vital and Health Statistics* ([NCHS88]).

The data sources are discussed in the Notes section of Chapter 1, the first part of Maternal Smoking and Infant Health.

References

[Ala90] Alameda County Low Birth Weight Study Group. Cigarette smoking and the risk of low birth weight: A comparison in black and white women. *Epidemiology,* **1**:201–205, 1990.

[NCHS88] National Center for Health Statistics. Health promotion and disease prevention: United States, 1985. *Vital and Health Statistics,* Series 10, No. 163, Public Health Service. DHHS Publication No. (PHS) 88–1591, U.S. Department of Health and Human Services, Washington, D.C., February 1988.

[DHHS89] United States Department of Health and Human Services. *Reducing the health consequences of smoking: 25 years of progress.* A report of the Surgeon General. DHHS publication No. (CDC) 89-8411. U.S. Department of Health and Human Services, Public Health Service, Office on Smoking and Health, Washington, D.C., 1989.

[KPMLS88] J.C. Kleinman, M.B. Pierre, J.H. Madans, G.H. Land, and W.F. Schramm. The effects of maternal smoking on fetal and infant mortality. *Am. J. Epidemiol.,* **127**:274–282, 1988.

[MKLS88] M. Malloy, J. Kleinman, G. Land, and W. Schramm. The association of maternal smoking with age and cause of infant death. *Am. J. Epidemiol.,* **128**:46–55, 1988.

[Yer64] J. Yerushalmy. Mother's cigarette smoking and survival of infant. *Am. J. Obstet. & Gynecol.,* **88**:505–518, 1964.

[Yer71] J. Yerushalmy. The relationship of parents' cigarette smoking to outcome of pregnancy—implications as to the problem of inferring causation from observed associations. *Am. J. Epidemiol.,* **93**:443–456, 1971.

11

A Mouse Model for Down Syndrome

[1]

THURSDAY, FEBRUARY 9, 1995 San Francisco Examiner

Scientists Create Mouse with Alzheimer's

By Sally Lehrman

Bay Area scientists have created a mouse that exhibits the same brain abnormalities as humanswith Alzheimer's, offering an important new tool for understanding the disease and developing a treatment.

"This mouse is a real breakthrough ... providing the first real hope we can progress toward meaningful drugs," said John Groom, president and chief executive of Athena Neurosciences, the South San Francisco company that led the collaborative project.

Alzheimer's is one of the top killers of Americans, with at least 100,000 people dying of the disease each year. The condition is marked by a progressive loss of memory, dementia, and eventually death.

Researchers have found it difficult to pinpoint the cause of Alzheimer's, how the brain changes as it loses memory, and how to go about fixing it. The lack of an animal that developed anything like the disease has made it especially hard to study.

Mice with rodent versions of cystic fibrosis, obesity and other conditions have become valuable tools for scientists to learn about these diseases and the ways drugs can interfere. Scientistshave tried for many years to engineer an Alzheimer's mouse through genetic manipulation, but never have been able to achieve changes in the brain that mimic the disease.

In the Thursday issue of Nature magazine, Athena, its partners Eli Lilly and Co., and their collaborators describe a mouse that develops remarkably similar characteristics to humans with Alzheimer's. The animal develops plaque in its brain, abnormal nerve fibers that surround the plaquelike a web, inflammatory cells and deterioration of the connections between its nerve cells.

The problems arise in two key areas of the brain related to spatial memory and associative learning. Later this year, the Athena scientists intend to test the animals in a complicated maze and see whether they have memory difficulties just like humans.

[1]Reprinted by permission.

Introduction

Down Syndrome is a congenital syndrome that occurs when a child inherits an extra chromosome 21 from his or her parents. The syndrome is associated with some degree of physical and mental retardation and is one of the most common congenital syndromes. One-quarter of a million people in the U.S. have Down Syndrome.

In the 1980s, it was discovered that only the genes at the bottom of chromosome 21 cause the syndrome, and scientists today are working to further identify and isolate the genes on the chromosome responsible for the disorder. To do this, they genetically alter the DNA of lab mice by adding to it cloned human DNA that comes from a small part of chromosome 21. If the transgenically altered mice exhibit symptoms of Down Syndrome, then that fragment of DNA contains the genes responsible for the syndrome.

Although much is known about the physical and mental retardation associated with Down Syndrome, very little is known about the equivalent syndrome in mice. To determine if the mice with extra DNA fragments exhibit the syndrome, they are typically put through tests for learning disabilities and intelligence. These tests rely on the mice being able to react to visual cues.

Unfortunately, over 500 of the lab mice were born blind and as a result could not partake in the usual tests to diagnose the syndrome. For these mice, only their body weight was measured. While it is known that people with Down Syndrome have a tendency for obesity, it is not known whether obesity is one of the features of the syndrome in mice. Nonetheless, it is hoped that comparisons of the weights of these blind mice will offer additional evidence in the quest to further narrow the region of chromosome 21 that is identified with the syndrome.

Data

The Human Genome Center at the Lawrence Berkeley Laboratory constructed a panel of transgenic mice, each containing one of four fragments of cloned human chromosome 21 (Smith et al. [SZZCR95], [SSS97], Smith and Rubin [SR97]). The mice in the panel were created by artificially introducing cloned human DNA into fertilized eggs. These first-generation transgenic mice were bred with nontransgenic mice in order to increase the pool of genetically altered mice. A first-generation mouse may pass on its extra piece of DNA to some of its off-spring. Those second-generation mice that inherit the cloned DNA are mated with nontransgenic mice, and so on, in order to continue increasing the numbers of transgenic mice available for experimentation. The data provided here are from over 500 descendants of the panel of mice. All mice that have descended from the same first-generation transgenic mouse are said to belong to a family line.

Four different cloned fragments of DNA were used to create the transgenic mice. They are called 141G6, 152F7, 230E8, and 285E6. Some of these DNA

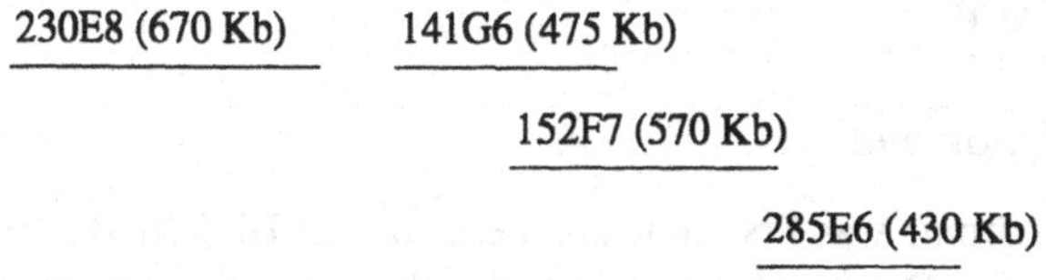

230E8 (670 Kb) 141G6 (475 Kb)

152F7 (570 Kb)

285E6 (430 Kb)

FIGURE 11.1. Map of the four fragments of DNA from chromosome 21 in kilobases (Smith et al.[SZZCR95]).

TABLE 11.1. Sample observations and data description for 532 offspring of the panel of transgenic mice.

DNA	C	C	C	C	A	A	A	A	A
Line	50	50	50	50	4	4	28	28	28
Transgenic	1	0	0	1	1	1	0	1	0
Sex	1	1	0	0	1	1	1	1	1
Age	113	113	112	112	119	119	115	115	115
Weight	31.6	30.1	23.1	26.3	31.2	28.4	28.1	30.1	29.1
Cage	1	1	5	5	7	7	9	9	10

Variable	Description
DNA	Fragment of chromosome 21 integrated in parent mouse (A=141G6; B=152F7; C=230E8; D=285E6).
Line	Family line.
Transgenic	Whether the mouse contains the extra DNA (1) or not (0).
Sex	Sex of mouse (1=male; 0=female).
Age	Age of mouse (in days) at time of weighing.
Weight	Weight of mouse in grams, to the nearest tenth of a gram.
Cage	Number of the cage in which the mouse lived.

fragments overlap: fragment 141G6 overlaps with 152F7, and 152F7 overlaps 285E6 (Figure 11.1). With the exception of a small region between 230E8 and 141G6, the four pieces of DNA completely cover the part of chromosome 21 that is known to contain the Down Syndrome genes.

The data available to us are for blind family lines, a subset of the family lines in the study. The information available on the mice are sex, age, weight, whether they are transgenic or not, family line, the extra DNA fragment in the family line, and cage (Table 11.1). Mice of the same sex and litter are housed in the same cage. However, cages may contain multiple litters if the number of same-sex mice in a litter is small.

Background

Down Syndrome

Down Syndrome was first clinically described in 1866 by the English physician John Langdon Down, who described the characteristic features of the syndrome but was unable to determine its cause. In 1932, de Waardenburg suggested that the syndrome resulted from a chromosomal abnormality. This was not confirmed until 1959, when Lejeune and Jacobs independently discovered that the syndrome was caused by an extra chromosome 21.

A syndrome is a collection of symptoms that occur together to indicate the presence of the condition. As many as 120 features have been ascribed to Down Syndrome, but most people with this syndrome have only 6 or 7 of these symptoms. All have some degree of physical and mental retardation.

The typical Down Syndrome individual's eyes slant slightly upward and there is a fold of skin that runs vertically from the inner corner of the eye to the bridge of the nose. The face is noticeably rounded from the front and appears flattened from the side. The mouth cavity is slightly smaller and the tongue is slightly larger than usual, which results in a habit of the individual putting out his or her tongue. Also, the hands are broad with short fingers and a single crease in the palm.

Body weight and length at birth are usually below that expected, and, in adulthood, those with Down Syndrome are generally short and have a tendency for obesity.

Aside from these obvious physical traits, one-third of those with Down Syndrome suffer from heart problems. Six different types of heart conditions are known to afflict those with the syndrome, including a hole between the right and left side of the heart, narrow blood vessels to the lungs, and malformed valves.

Finally, the Down Syndrome child learns at a steady pace, but in comparison to a child without the syndrome the rate is much slower. The gap in abilities widens as the child grows. However, the Down Syndrome child can continue to learn new skills in adolescence and well into adulthood.

Trisomy 21

Down Syndrome is congenital, meaning it is present at birth, and it is due to the abnormal development of the fetus. All cells in the body originate from the single cell that is formed from the fusion of sperm and egg. In the center of each cell is a nucleus that contains the genetic material inherited from the parents. There are approximately 100,000 genes in the cell nucleus, and they are arranged into chromosomes: 23 from the mother and 23 from the father. These cells are called diploid cells.

The only cells that do not have two sets of chromosomes are the sperm and egg cells. They each contain a single set of 23 chromosomes. In the formation of eggs and sperm, a diploid cell replicates its DNA and then divides into four new cells, which are called haploid cells. Each haploid cell contains one set of

23 chromosomes. These new cells develop into the eggs and sperm, although in women only one in four of the haploid cells will mature into an egg. Sometimes, the cell division does not proceed correctly, and a haploid cell will have two chromosomes of one type. This mix-up causes some genetic disorders. When the egg and sperm join, there will be three copies of the chromosome instead of two. This is called a trisomy.

Ninety-five percent of all Down Syndrome cases occur when an extra 21st chromosome appears in all cells (i.e., a trisomy 21). The extra 21st chromosome nearly always comes from the egg, and the eggs of older women are more likely to have an extra chromosome. Although women over 35 have only 5–8% of the total number of pregnancies, their babies account for 20% of Down Syndrome births. This is because the odds of having an affected child for a woman aged 40 to 44 is about 1 in 40, whereas at age 20 it is only 1 in 2300 (Hotchner [Hot90]). In the United States, Down Syndrome occurs on average at a rate of 1 per 800 births, and the national population of individuals with Down Syndrome is estimated to be 250,000 (Selikowitz [Sel90]).

The Transgenic Mouse

A transgenic mouse is a mouse that has had its DNA altered by humans. This is accomplished by removing a newly fertilized egg from a female mouse and injecting it with a fragment of DNA. In the first 12 hours after fertilization, there are still two nuclei present in the egg, one from the egg and one from the sperm. During this time, the egg is placed under a dissecting microscope, and a micro-injection needle pierces the egg and injects the extra DNA into the sperm nucleus. Then the new DNA integrates with the DNA in the sperm. The egg is transplanted into a female, and after 12 hours have elapsed, the egg nucleus and the altered sperm nucleus fuse to form the mouse embryo.

Theoretically, the mouse that develops from this genetically altered DNA should display the characteristics of the inserted DNA. However, expression of the inserted DNA does not always occur; in fact, only about 1 injection in 100 is successful. To determine if a mouse is in fact transgenic, the tip of its tail is removed and examined to see if the nuclei of the cells have the extra fragment of DNA.

For experimenters to create a mouse model that will exhibit characteristics of human Down Syndrome, four or five separate lines of mice were created for each DNA fragment. (See the Data section for a description of a family line.) To determine whether a mouse with a particular fragment of DNA exhibits Down Syndrome, the mouse is given tests for learning disabilities, fearfulness, and low intelligence. Only male mice are tested because female mice ovulate frequently (every few days), and it is not known what effect ovulation may have on their performance on these tests. The tests are visual, so the blind mice could not perform them. Instead, their weights were recorded, and we will look for differences in weight as signifying expression of human Down Syndrome.

Laboratory Conditions

The life expectancy of laboratory mice is 12 to 18 months, and they are infertile after 1 year. The health of these mice is extremely fragile because they are from fully inbred strains (i.e., because they are genetically identical). Precautions are taken to prevent the spread of disease among mice. For example, an air filter is placed over the top of each cage, and an air vent (or hood) is used when the scientist works directly with a mouse. These cautionary measures are designed to keep disease-causing bacteria out of the environment.

The mice live in 15×30 cm plastic cages stacked on shelves in a small windowless room. Typically five mice occupy one cage. Males and females are housed separately to prevent unwanted breeding. Mice of the same sex and litter share a cage, and if the litter is small, two litters may share the same cage. The mice have a constant supply of food and water but no means of exercise. The cages are cleaned once a week by a lab assistant, who is the only person to handle the mice other than the experimenter.

Investigations

Is there a difference between the weights of transgenic and nontransgenic mice? If so, for which fragment or fragments of the DNA are there differences?

To answer these questions, make graphical and numerical comparisons of groups of mice. Try to make comparisons between groups that are as similar as possible. For example, male mice on average tend to weigh about 5 grams more than female mice. Comparing groups of mixed-sex mice may reveal very little because a small transgenic difference could be hidden or canceled out by the difference between sexes.

- Begin by graphically comparing the weights of mice of the same sex. How do the weights of mice with the extra 141G6 fragment of DNA compare to mice with no extra copy of DNA? What about the mice with extra copies of the other fragments of DNA?
- Investigate the variability in weight. Is it roughly the same across different groups of mice? If so, then the pooled standard deviation for all nontransgenic mice would provide a better estimate of the variability in weight.
- Formalize your comparison of the means of these different groups of mice using a linear model with indicator variables for the sex and DNA fragment effects. Consider an additive model in sex and DNA fragment. Be sure to consider the possibility of interactions. In multi-way classifications, it is advisable to check the cell counts for all the classifications. In other words, check to see that there are plenty of mice in each classification: male-141G6, female-141G6, etc. If there are too few data to justify an additive model, the model may be misleading, and fitting separate models for subgroups of mice would be preferable.

- Do the residuals indicate whether mice of the same sex and DNA fragment have roughly the same weight, or do age and litter make a difference? Because all mice from the same litter and the same sex are kept in one cage together, we can use cage number as a surrogate for litter. Indeed, controlling for cage controls for age, sex, and DNA fragment, although some cages contain mice from more than one litter. Do the transgenic mice in a cage tend to be above average in weight in comparison to the nontransgenic mice from the same cage?
- If you found differences between transgenic and nontransgenic mice, assess the size of the difference. That is, are the differences you found big enough to indicate that a particular region of DNA contains the Down Syndrome genes?

The scientists at the lab want to know if the data on the weight of the mice support their findings from the visual tests. They need assistance in making sense out of the mass of data collected on weight. Be complete in writing up your findings; include results that may not be conclusive as well as those that seem to be definitive.

Theory

Fitting Means

Throughout this section, we will take y_i to represent the weight of the ith mouse in the experiment. We saw in Chapter 10 on Maternal Smoking and Infant Health that the mean $\bar{y}$ is the least squares solution when we minimize

$$\sum_{i=1}^{n}(y_i - c)^2$$

with respect to c. We also saw that if our plan is to fit two means to the data, say one for female mice and one for male mice, then this can be accomplished with an indicator variable e_M, where $e_{M,i} = 1$ if the ith mouse is male and 0 if female. The least squares solution, minimizing with respect to a and b, of the sum of squares

$$\sum_{i=1}^{n}[y_i - (a + be_{M,i})]^2$$

is $\hat{a} = \bar{y}_F$ and $\hat{b} = \bar{y}_M - \bar{y}_F$, where

$$\bar{y}_F = \frac{1}{n_F}\sum_{(F)} y_i,$$

$$\bar{y}_M = \frac{1}{n_M}\sum_{(M)} y_i,$$

n_F and n_M are the number of females and number of males, respectively, and the summation is over the female and male mice, respectively. The predicted values from the fit are one of two values:

$$\hat{a} = \bar{y}_F,$$
$$\hat{a} + \hat{b} = \bar{y}_M,$$

depending on the sex of the mouse.

We can generalize this technique to fit means to more than two groups, such as which extra fragment of DNA a mouse has. Here there are five categories, one for each of the four possible DNA fragments and one for the absence of any trisomy. Note that this is a combination of the DNA and transgenic variables from Table 11.1. Let e_A, e_B, e_C, e_D indicate which extra fragment of DNA is present in the mouse. That is, $e_{A,i}$ is 1 if the ith mouse carries the extra DNA fragment denoted 141G6, and 0 if not. For a mouse with a trisomy, only one of these indicators is 1 and the others are 0, and if a mouse has no extra DNA, then all four indicators are 0. Consider the least squares solution, minimizing with respect to b_0, b_A, b_B, b_C, and b_D, of the following sum of squares:

$$\begin{aligned}
&\sum_{i=1}^{n}[y_i - (b_0 + b_A e_{A,i} + b_B e_{B,i} + b_C e_{C,i} + b_D e_{D,i})]^2 \\
&\quad = \sum_{\text{(no trisomy)}} (y_i - b_0)^2 + \sum_{(A)}(y_i - b_0 - b_A)^2 + \sum_{(B)}(y_i - b_0 - b_B)^2 \\
&\qquad + \sum_{(C)}(y_i - b_0 - b_C)^2 + \sum_{(D)}(y_i - b_0 - b_D)^2.
\end{aligned}$$

From the equation on the right side, we see that each sum is over a different set of mice and so can be minimized separately. The predicted values are $\bar{y}_0$, $\bar{y}_A$, $\bar{y}_B$, $\bar{y}_C$, and $\bar{y}_D$, the mean weights of the nontransgenic mice and the mice with an extra copy of the 141G6, 152F7, 230E8, and 285E6 fragments, respectively. This implies that

$$\begin{aligned}
&\hat{b}_0 = \bar{y}_0, \\
&\hat{b}_A = \bar{y}_A - \bar{y}_0, \qquad \hat{b}_B = \bar{y}_B - \bar{y}_0, \\
&\hat{b}_C = \bar{y}_C - \bar{y}_0, \qquad \hat{b}_D = \bar{y}_D - \bar{y}_0.
\end{aligned}$$

When multiple groups are to be compared, graphical comparisons using box-and-whisker plots may be useful. In Figure 11.2, one box-and-whisker plot is made for each group, and box-and-whisker plots for the groups are plotted on the same graph.

A Model for Means

In the previous section, we showed how means could be fitted to groups of observations using least squares. In the DNA example, the model for the weights of the

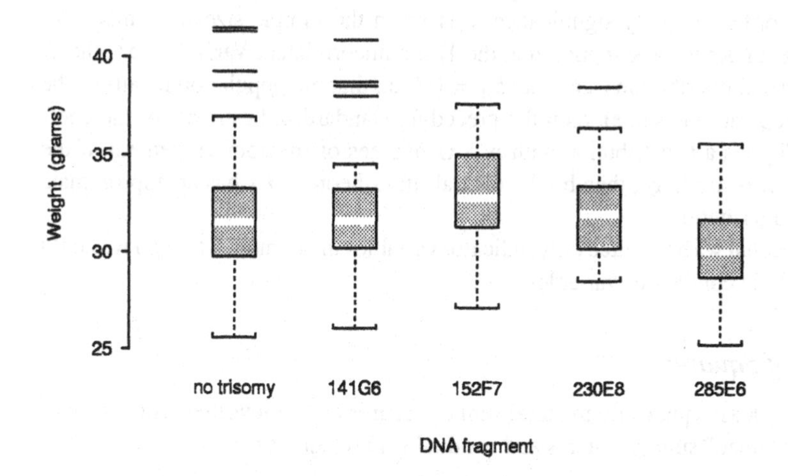

FIGURE 11.2. Comparative box-and-whisker plots of weight, in grams, for male mice grouped by presence and type of DNA fragment (Smith et al. [SZZCR95]).

mice is simply that

$$\mathbb{E}(Y_i) = \begin{cases} b_0 & \text{if no trisomy present} \\ b_0 + b_A & \text{if 141G6 fragment present} \\ b_0 + b_B & \text{if 152F7 fragment present} \\ b_0 + b_C & \text{if 230E8 fragment present} \\ b_0 + b_D & \text{if 285E6 fragment present.} \end{cases}$$

This simple model, where the mean of a continuous valued response Y_i is determined according to the group or category to which the ith unit belongs, is called a *one-way classification.* An alternative parameterization of the model is presented in a later section. One advantage to this parameterization is that b_A represents the difference between the nontransgenic mean and the mean for those mice with the extra 141G6 fragment. If the mean weights are the same, then $b_A = 0$, and the estimated coefficient may be used to test this hypothesis: could $\hat{b}_A$ be this far from zero by chance. To do this, we standardize the estimate:

$$\hat{b}_A / \widehat{SE}(\hat{b}_A).$$

Estimation of the standard error of $\hat{b}_A$ is discussed in the Exercises. To determine the distribution of this statistic, we place the following typical assumptions on the Y_i:

NORMAL MEANS MODEL. *The Y_i are uncorrelated, Var(Y_i)= σ^2, and the means $\bar{Y}_A, \bar{Y}_B, \ldots$ have approximate normal distributions.*

Traditionally the additional assumption is made that the Y_i have normal distributions in order to justify significance tests when the sample sizes are small. For example, under the assumption that the Y_i are uncorrelated, $\text{Var}(Y_i) = \sigma^2$, the Y_i have normal distributions, and that $b_A = 0$ (i.e., that the population means of the two groups are the same), then the preceding standardized estimate of the coefficient $\hat{b}_A$ has a t-distribution with $n - G$ degrees of freedom. In general, if the sample sizes are large, then by the central limit theorem, $\hat{b}_A$ has an approximate normal distribution.

Notice that we have fitted only indicator variables in our model, though we could also include continuous variables.

Sums of Squares

As in any least squares fit, the total sum of squares can be rewritten as the sum of the "explained" sum of squares and residual sum of squares,

$$\sum_{i=1}^{n}(y_i - \bar{y})^2 = \sum_{i=1}^{n}(\hat{y}_i - \bar{y})^2 + \sum_{i=1}^{n}(y_i - \hat{y}_i)^2.$$

But with a one-way classification, these sums of squares have simple forms because the fitted values are exactly the group means:

$$\sum_{i=1}^{n}(\hat{y}_i - \bar{y})^2 = \sum_{g=1}^{G} n_g(\bar{y}_g - \bar{y})^2,$$

$$\sum_{i=1}^{n}(y_i - \hat{y}_i)^2 = \sum_{g=1}^{G}\sum_{(g)}(y_i - \bar{y}_g)^2,$$

where

$$\bar{y}_g = \frac{1}{n_g}\sum_{(g)} y_i$$

is the average of the n_g subjects in the gth group.

These sums of squares are traditionally laid out in an *analysis of variance (ANOVA) table* as shown in Table 11.2. In an analysis of variance table, the explained, or group, sum of squares is typically called the *between group* sum of squares, and the residual sum of squares is called the *within group* sum of squares. These two sums of squares can be used to simultaneously compare the means from all groups. In other words, under the hypothesis that $\mathbb{E}(Y_1) = \cdots = \mathbb{E}(Y_G)$ and the Y_i are uncorrelated, $\text{Var}(Y_i) = \sigma^2$, and the Y_i have normal distributions, the ratio

$$\frac{\sum(\hat{y}_i - \bar{y})^2/(G-1)}{\sum(y_i - \hat{y}_i)^2/(n-G)}$$

TABLE 11.2. Analysis of Variance Table for a one-way classification.

Source	df	Sum of squares	Mean square	F statistic
Group	$G-1$	$\sum_{g=1}^{G} n_g(\bar{y}_g-\bar{y})^2$	$\frac{\sum n_g(\bar{y}_g-\bar{y})^2}{G-1}$	$\frac{\sum n_g(\bar{y}_g-\bar{y})^2/(G-1)}{\sum(y_i-\hat{y}_i)^2/(n-G)}$
Residual	$n-G$	$\sum_{i=1}^{n}(y_i-\hat{y}_i)^2$	$\frac{\sum(y_i-\hat{y}_i)^2}{n-G}$	
Total	$n-1$	$\sum_{i=1}^{n}(y_i-\bar{y})^2$		

has an F distribution with $G-1$ and $n-G$ degrees of freedom. Here the degrees of freedom for the numerator are the number of groups less 1, and for the denominator they are the number of observations minus the number of groups. This ratio (the F statistic) is robust against moderate departures from normality. It is also robust against unequal error variances, and in the absence of normality, the F statistic can still act as a measure of variation between groups relative to that within groups.

An Alternative Parameterization

We have seen that the model for the means of observations from G groups can be parameterized as $\mathbb{E}(Y_i) = b_0 + b_g$ when the ith subject is in the gth group, $g = 1, \ldots, G-1$, and $\mathbb{E}(Y_i) = b_0$ for the Gth group. An alternative parameterization often used in analysis of variance is

$$\mathbb{E}(Y_i) = \mu + \alpha_g,$$

when the ith subject is in the gth group, $g = 1, \ldots, G$. This representation of the model for group means is overparameterized because there are $G+1$ parameters and only G groups.

We can still find a unique minimum, with respect to $\mu, \alpha_1, \alpha_2, \ldots, \alpha_G$, of the sum of squares

$$\sum_{i=1}^{n}[y_i - (\mu + \alpha_1 e_{1,i} + \alpha_2 e_{2,i} + \ldots \alpha_G e_{G,i})]^2,$$

where $e_{g,i}$ is 1 if the ith subject belongs to group g and is 0 otherwise. Differentiating with respect to these parameters and setting the expressions to zero, we get a set of $G+1$ equations:

$$\sum_{i=1}^{n}[y_i - (\hat{\mu} + \hat{\alpha}_1 e_{1,i} + \hat{\alpha}_2 e_{2,i} + \ldots + \hat{\alpha}_G e_{G,i})] = 0;$$

$$\sum_{(g)}[y_i - (\hat{\mu} + \hat{\alpha}_g e_{g,i})] = 0, \quad \text{for } g = 1, \ldots, G.$$

These equations determine the unique minimum of the sum of squares, and from this it is easy to check that $\bar{y}_g = \hat{\mu} + \hat{\alpha}_g$ for $g = 1, \ldots, G$. However, there are many combinations of estimates that achieve this unique minimum sum of squares (see Exercise 7). A unique set of estimates can be singled out by placing a linear constraint on the parameters $(\mu, \alpha_1, \alpha_2, \ldots, \alpha_G)$ or, more simply, by taking $\hat{\mu} = \bar{y}$. This is permissible as soon as we know there is at least one solution of the above $G + 1$ equations. It then follows that

$$\hat{\alpha}_g = \bar{y}_g - \bar{y}, \quad g = 1, \ldots, G.$$

Regardless of the parameterization, the fitted values remain the same. They are the G group means: $\bar{y}_1, \ldots, \bar{y}_G$. That is, if mouse i is in the gth group, $g = 1, \ldots, G - 1$,

$$\begin{aligned} \hat{y}_i &= \hat{\mu} + \hat{\alpha}_g \\ &= \hat{b}_0 + \hat{b}_g \\ &= \bar{y}_g, \end{aligned}$$

and if mouse i is in the Gth group, then $\hat{y}_i = \bar{y}_G$ as well. We would obtain the same set of solutions $(\hat{\mu}, \hat{\alpha}_1, \ldots, \hat{\alpha}_G)$ if we minimized our sum of squares subject to the constraint

$$n_1\alpha_1 + n_2\alpha_2 + \ldots + n_G\alpha_G = 0,$$

where n_g is the number of mice in group g. Note that our estimates $\hat{\alpha}_g = \bar{y}_g - \bar{y}$ satisfy this linear constraint. In the special case when there are an equal number of responses in each group, the constraint has a particularly simple form: $\sum \alpha_g = 0$. This special case is called a *balanced one-way classification*.

Multi-way Classifications

Mice may be classified in many different ways, according to sex (male or female), DNA fragment (141G6, 152F7, 230E8, 285E6), whether they are transgenic (yes, no), family line, and cage. We may want to consider more than one classification simultaneously, or we may want to stratify and look at groups of mice separately.

Here is an example where we consider two classifiers and stratify by a third. Restrict attention to all those mice that have descended from a transgenically altered mouse with an extra 141G6 fragment of DNA. In other words, stratify according to the type of DNA fragment in the family line. These mice may or may not be transgenic, and they also may be male or female. The sex of the mouse and whether it is transgenic are the two classifiers that we will consider here for the 141G6 subgroup.

We can fit means for each of the 2×2 classifications of the mice: male-transgenic, male-nontransgenic, female-transgenic, and female-nontransgenic. For the 141G6 subgroup, denoted by A,

$$\sum_{(A)}[y_i - (a_0 + a_{MT}e_{MT,i} + a_{MN}e_{MN,i} + a_{FT}e_{FT,i})]^2,$$

where the variables e_{MT}, e_{MN}, and e_{FT} are indicators for the male-transgenic, male-nontransgenic, and female-transgenic classifications, respectively, and the summation is over those mice in a 141G6 family. Then

$$\hat{a}_0 = \bar{y}_{FN},$$
$$\hat{a}_{MT} = \bar{y}_{MT} - \bar{y}_{FN},$$
$$\hat{a}_{MN} = \bar{y}_{MN} - \bar{y}_{FN},$$
$$\hat{a}_{FT} = \bar{y}_{FT} - \bar{y}_{FN}.$$

That is, the least squares solution fits four means to each of the four groups. We call this the *full model.*

An alternative, simpler model is the *additive model* that minimizes, with respect to d_0, d_M, and d_T, the sum of squares

$$\sum_{(A)} [y_i - (d_0 + d_M e_{M,i} + d_T e_{T,i})]^2,$$

where e_M indicates the sex of the mouse, and e_T indicates whether the mouse is transgenic or not. With this model, only three parameters are fitted; this implies that the expected weights of the four groups are:

d_0 for female nontransgenic mice,

$d_0 + d_T$ for female transgenic mice,

$d_0 + d_M$ for male nontransgenic mice, and

$d_0 + d_M + d_T$ for male transgenic mice.

The model is called additive because the mean levels for female and male nontransgenic mice differ by d_M, and the mean levels for female and male transgenic mice differ by d_M as well. That is, d_M, the sex *effect*, is the same regardless of whether the mice are transgenic or not. With the additive model, there is one fewer parameter to fit than in the previous full model. An additional indicator variable may be added to the additive model, $e_{MT} = e_M \times e_T$, to obtain the full model. Including this additional variable allows the difference in mean levels between the male and female transgenic mice to be different from that for the nontransgenic mice. It is called an *interaction* term because it allows the sex effect to "interact" with the transgenic effect, so differences between the male and female means depend on whether the mice are transgenic. As in the one-way classification, the standardized estimate for the coefficient of the interaction term has an approximate normal distribution provided the group means are additive and the normal means model holds. Additionally, for small samples, if the Y_i have normal distributions, then the coefficient has a t distribution with $n - 4$ degrees of freedom. Therefore, if the standardized estimate is large compared to the normal (or t) distribution, this suggests that the additive assumption is not true in this case.

Plots and F tests provide other ways to check the additivity assumption. Figure 11.3 shows plots of mean weights for the mice in the 141G6 family lines. The means for female transgenic and female nontransgenic mice are connected with

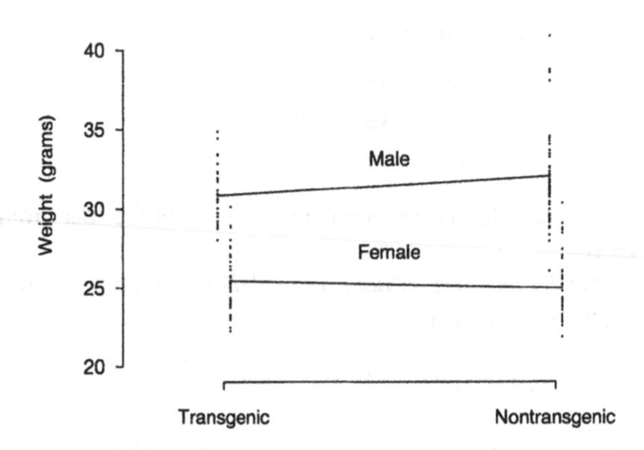

FIGURE 11.3. Plot of weight, in grams, for male and female mice from 141G6 family lines, according to whether they are transgenic or not (Smith et al. [SZZCR95]). The lines illustrate the difference in mean weight for transgenic and nontransgenic mice by sex.

a line, as are the means for the male transgenic and nontransgenic mice. Additivity would be indicated by parallel line segments. However, here we see that the difference in mean weights of male and female mice is larger for nontransgenic mice. In particular, male transgenic mice weigh more than male nontransgenic mice, whereas the female transgenic mice weigh less than the female nontransgenic mice. Though the lines are close to being parallel, this relationship suggests that an additive model is not suitable for these data. An F test can be used to test this statement.

We can also consider different parameterizations for multi-way classifications. In the two-way classification, with G levels of the first classifier and H levels of the second, we parameterize the additive model as follows:

$$\mathbb{E}(Y_i) = \tau + \gamma_g + \delta_h,$$

where the ith subject belongs to the gth category of the first classifier and the hth category of the second classifier. To fit this model, we minimize the following sums of squares with respect to $\tau, \gamma_1, \ldots, \gamma_G, \delta_1, \ldots, \delta_H$, where $f_{h,i}$ is 1 if the ith unit belongs to the hth category of the second classifier, and 0 otherwise:

$$\sum_{i=1}^{n} \left[y_i - \left(\tau + \sum_{g=1}^{G} \gamma_g e_{g,i} + \sum_{h=1}^{H} \delta_h f_{h,i} \right) \right]^2 .$$

As with the one-way classification, this linear model is overparameterized, and again we can single out a unique set of parameter estimates achieving the minimum sum of squares in different ways. In the balanced case, where there are K mice, say, in each of the $G \times H$ classifications, two convenient linear constraints which single out a unique solution are

$$\sum \gamma_g = \sum \delta_h = 0.$$

Assuming that our two-way classification is balanced, these constraints lead to the following least-squares estimates:

$$\hat{\tau} = \bar{y},$$
$$\hat{\gamma}_g = \bar{y}_{g.} - \bar{y},$$
$$\hat{\delta}_h = \bar{y}_{.h} - \bar{y},$$

where $\bar{y}_{g.}$ is the average response for the HK mice in group g, and $\bar{y}_{.h}$ is the average response for the GK mice in group h. Note that K of the mice in group g of the first classifier belong to group h of the second. (It is easy to check that $\sum \hat{\gamma}_g = \sum \hat{\delta}_h = 0$.) The advantage of this parameterization is that the sums of squares have particularly simple forms. However, if group sizes are unequal, then, in general, there are no simple formulas for the unique solution. See the Exercises for an alternative way of subscripting the y_i.

Two-way Classification

For the balanced two-way classification, the explained sum of squares for the additive model can be written

$$\sum_{i=1}^{n} (\hat{y}_i - \bar{y})^2 = SS_1 + SS_2,$$

where

$$SS_1 = HK \sum_g \hat{\gamma}_g^2 = HK \sum_g (\bar{y}_{g.} - \bar{y})^2,$$
$$SS_2 = GK \sum_h \hat{\delta}_h^2 = GK \sum_h (\bar{y}_{.h} - \bar{y})^2,$$

and the residual sum of squares can be written as

$$\sum_{i=1}^{n} (y_i - \hat{y}_i)^2 = SS_{1\times 2} + WSS,$$

where

$$SS_{1\times 2} = K \sum_g \sum_h [\bar{y}_{gh} - (\hat{\mu} + \hat{\gamma}_g + \hat{\delta}_h)]^2$$
$$= K \sum_g \sum_h [\bar{y}_{gh} - \bar{y}_{g.} - \bar{y}_{.h} + \bar{y}]^2$$

TABLE 11.3. Analysis of variance table for the balanced two-way classification.

Source	df	SS	MS	F
Group 1	$G-1$	SS_1	$\frac{SS_1}{G-1}$	$\frac{SS_1/(G-1)}{WSS/GH(K-1)}$
Group 2	$H-1$	SS_2	$\frac{SS_2}{H-1}$	$\frac{SS_2/(H-1)}{WSS/GH(K-1)}$
Interaction	$(G-1)(H-1)$	$SS_{1\times 2}$	$\frac{SS_{1\times 2}}{(G-1)(H-1)}$	$\frac{SS_{1\times 2}/(G-1)(H-1)}{WSS/GH(K-1)}$
Residual	$GH(K-1)$	WSS	$\frac{WSS}{GH(K-1)}$	
Total	$GHK-1$	$\sum(y_i-\bar{y})^2$		

and

$$WSS = \sum_g \sum_h \sum_{(gh)} (y_i - \bar{y}_{gh})^2.$$

Here $\bar{y}_{gh}$ is the average of the K mice belonging to both group g of the first classifier and group h of the second classifier, and the inner sum in WSS is over the K observations in this group.

The respective degrees of freedom for SS_1, SS_2, $SS_{1\times 2}$, and WSS are: $G-1$, $H-1$, $(G-1)(H-1)$, and $GH(K-1)$. Each of the ratios of mean squares in the last column of Table 11.3 can be used for model checking. The first ratio tests whether $\alpha_g = 0$, $g = 1, \ldots, G$, and the second can be used in a similar test that $\delta_h = 0$, $h = 1, \ldots, H$. The third ratio can be used to check the additivity assumption. It compares the reduction in residual sum of squares of the additive model when we include the interaction term to the residual sum of squares for the full model:

$$\frac{[\sum(y_i - \bar{y}_{g.} - \bar{y}_{.h} + \bar{y})^2 - \sum(y_i - \bar{y}_{gh})^2]/(G-1)(H-1)}{\sum(y_i - \bar{y}_{gh})^2/(n - GH)}.$$

These test statistics each have an F distribution, given the typical assumptions on Y_i. For example, the test statistic for the interaction sum of squares has an F distribution with $(G-1)(H-1)$ and $GH(K-1)$ degrees of freedom.

Exercises

1. Table 11.4 contains the averages and SDs for the weights of male and female mice from family lines with DNA fragment 141G6. Use these summary statistics to construct an ANOVA table for the 141G6 mice with sex as the only classifier.

2. Consider the summary statistics in Table 11.4 for male and female mice from the 141G6 families. Use these statistics to estimate the coefficients in the linear

TABLE 11.4. Summary statistics for the weights (grams) of male and female mice in 141G6 family lines (including both transgenic and nontransgenic mice).

	Number	Average	SD
Male	94	31.70	2.62
Female	83	25.23	2.00

model

$$\mathbb{E}(Y_i) = a + be_{M,i}.$$

Construct a test of the hypothesis that b is 0. *Hint*: Recall that $\hat{b} = \bar{y}_M - \bar{y}_F$, and use a pooled estimate of σ^2 (page 172) in your estimate of

$$\mathrm{SE}(\hat{\mathrm{b}}) = \sigma\sqrt{\frac{1}{\mathrm{n_M}} + \frac{1}{\mathrm{n_F}}}.$$

3. Show that the square of the test statistic in Exercise 2 is the same as the F statistic in the ANOVA table derived in Exercise 1. Also show that, in general, a t test that tests for a difference in two means and uses a pooled estimate of the SD is equivalent to the F test in a one-way ANOVA table.
4. Consider the one-way classification based on DNA fragment,

$$\mathbb{E}(Y_i) = b_0 + b_A e_{A,i} + b_B e_{B,i} + b_C e_{C,i} + b_D e_{D,i},$$

where, for example, $e_{A,i} = 1$ if the ith mouse is transgenic and belongs to one of the 141G6 families, and otherwise it is 0.

a. Suppose we assume that the Y_i's are uncorrelated with variance σ^2. Show that

$$\mathrm{SE}(\hat{b}_A) = \sigma\sqrt{\frac{1}{n_0} + \frac{1}{n_A}}.$$

b. Show that

$$\frac{1}{n-G}\Big[\sum_{\text{(no trisomy)}} (y_i - \bar{y}_0)^2 + \sum_{(A)} (y_i - \bar{y}_A)^2 + \sum_{(B)} (y_i - \bar{y}_B)^2 + \sum_{(C)} (y_i - \bar{y}_C)^2 + \sum_{(D)} (y_i - \bar{y}_D)^2\Big]$$

is an unbiased estimate of σ^2.

5. Consider the coefficients in Exercise 4. To conduct four hypothesis tests that each of the coefficients is 0 at the $\alpha = 0.05$ level, we set the cutoff according to $\alpha/4 = 0.0125$ because we want the overall level of all four tests to be at most 0.05. In other words, we want the probability of a Type I error to be at most 0.05 simultaneously for all four tests. Prove that this is the case; that is,

$$\mathbb{P}(\text{ any test rejects} \mid \text{all hypotheses correct}) < 0.05.$$

6. Consider the following simple model for the weights of the mice,

$$\mathbb{E}(Y_i) = a + bx_i,$$

where for the ith mouse

$$x_i = \begin{cases} -1/n_F & \text{if it is female} \\ +1/n_M & \text{if it is male,} \end{cases}$$

n_F is the number of female mice and n_M is the number of male mice. Show that the least squares predicted values are $\bar{y}_F$ and $\bar{y}_M$, according to whether the mouse is male or female. Use this fact to derive the least squares estimates of the coefficients:

$$\hat{a} = \bar{y},$$
$$\hat{b} = (\bar{y}_M - \bar{y}_F)/\left(\frac{1}{n_M} + \frac{1}{n_F}\right).$$

7. Show that for a one-way classification with G groups, if $(\hat{\mu}, \hat{\alpha}_1, \ldots, \hat{\alpha}_G)$ minimizes

$$\sum_{i=1}^{n} [y_i - (\mu + \alpha_1 e_{1,i} + \ldots + \alpha_G e_{G,i})]^2,$$

then so does $(\hat{\mu} - t, \hat{\alpha}_1 + t, \ldots, \hat{\alpha}_G + t)$ for any real t.
8. Multiple subscripts are often used in analysis of variance models. For example, suppose there are n observations, and each observation is classified into one of G groups. We can denote the observation by Y_{gi} using two subscripts, the first denoting the group g to which the unit belongs, $g = 1, \ldots, G$, and the second the index of the unit within the group, $i = 1, \ldots, n_g$, where n_g is the size of group g. We can then fit the linear model

$$\mathbb{E}(Y_{gi}) = c_g, \quad g = 1, \ldots, G,$$

by minimizing

$$\sum_{g=1}^{G} \sum_{i=1}^{n_g} (y_{gi} - c_g)^2$$

with respect to $c_1, \ldots, c_G$. Show that the minimizing values are $\hat{c}_g = \bar{y}_g$, where, as before,

$$\bar{y}_g = \frac{1}{n_g} \sum_{i=1}^{n_g} y_{gi}.$$

9. Consider the alternative parameterization for the one-way classification,

$$\mathbb{E}(Y_{gi}) = \mu + \alpha_g,$$

where Y_{gi} is defined in Exercise 8. Find the set of equations that are satisfied by all solutions in the minimization of

$$\sum_{g=1}^{G}\sum_{i=1}^{n_g}(y_{gi} - \mu - \alpha_g)^2$$

with respect to $\mu, \alpha_1, \ldots, \alpha_G$. Show that, if we set

$$\hat{\mu} = \bar{y} = \frac{1}{n}\sum_g\sum_i y_{gi},$$

the solutions are uniquely given by $\hat{\alpha}_g = \bar{y}_g - \bar{y}$.

10. For the one-way classification with G groups, show that the total sum of squares can be written as

$$\sum_{g=1}^{G}\sum_{i=1}^{n_g}(y_{gi} - \bar{y})^2 = \sum_{g=1}^{G} n_g(\bar{y}_g - \bar{y})^2 + \sum_{g=1}^{G}\sum_{i=1}^{n_g}(y_{gi} - \bar{y}_g)^2.$$

Be sure to show that the cross product is 0.

11. Consider a one-way classification with G groups and n_g observations in each group, $g = 1, \ldots, G$. Under the Gauss measurement model, show that the within-group sum of squares has expectation

$$\sum_{g=1}^{G}(n_g - 1)\sigma^2.$$

Also, under the additional assumption that the group means are all equal, show that the between-group sum of squares has expectation $(G-1)\sigma^2$. Hint: Adapt the result on page 38 that gives $\mathbb{E}(s^2)$ to this situation.

12. Table 11.5 is an ANOVA table for the one-way classification by transgenic for the 74 male mice in the 285E6 family line. Many of the cells in the table have been left blank. Fill in the blank cells using only the other entries in the table and 74, the number of mice. Determine whether the F test is statistically significant.

13. Consider the following model for a 2×2 classification, including a factor for sex, for transgenic, and for their interaction for mice in the families with the 141G6 fragment:

$$\mathbb{E}(Y_i) = d_0 + d_M e_{M,i} + d_T e_{T,i} + d_{MT,i} e_{M,i} \times e_{T,i}.$$

TABLE 11.5. ANOVA table for the one-way classification of weight, in grams, by transgenic for 74 male mice in the 285E6 lines.

Source	df	SS	MS	F
Transgenic	1	19.4	___	2.4
Residual	___	___	___	
Total	___	___		

Use the fact that the predicted values are the group means to determine the least squares estimates for the coefficients:

$$\hat{d}_0 = \bar{y}_{FN},$$
$$\hat{d}_T = \bar{y}_{FT} - \bar{y}_{FN},$$
$$\hat{d}_M = \bar{y}_{MN} - \bar{y}_{FN},$$
$$\hat{d}_{MT} = \bar{y}_{MT} - \bar{y}_{FT} + \bar{y}_{FN} - \bar{y}_{MN}.$$

14. Show that under the assumption that the weights are uncorrelated with finite variance σ^2, the estimate $\hat{d}_{MT}$ derived in Exercise 13 has the following variance:

$$\sigma^2\left(\frac{1}{n_{MT}} + \frac{1}{n_{MN}} + \frac{1}{n_{FN}} + \frac{1}{n_{FT}}\right).$$

15. In Exercise 13, we saw that the interaction coefficient was estimated by

$$\hat{d}_{MT} = \bar{y}_{MT} - \bar{y}_{FT} + \bar{y}_{FN} - \bar{y}_{MN}.$$

a. Use the summary statistics in Table 11.6 to find a numeric value for this estimate.

b. Estimate the variance of $\hat{d}_{MT}$.

c. Test the hypothesis of no interaction.

16. In the two-way classification, three subscripts can be used to denote the response of the ith unit in both the gth group of the first classifier and the hth group of the second classifier,

$$\mathbb{E}(Y_{ghi}) = \theta_{gh},$$

for $g = 1, \ldots, G, h = 1, \ldots, H$, and $i = 1, \ldots, n_{gh}$. Consider minimizing

$$\sum_{g=1}^{G}\sum_{h=1}^{H}\sum_{i=1}^{n_{gh}}(y_{ghi} - \theta_{gh})^2,$$

with respect to θ_{gh}, for $g = 1, \ldots, G$ and $h = 1, \ldots, H$. Show that the least squares solution is $\hat{\theta}_{gh} = \bar{y}_{gh}$, where

$$\bar{y}_{gh} = \frac{1}{n_{gh}}\sum_{i=1}^{n_{gh}} y_{ghi}.$$

TABLE 11.6. Summary statistics of weight (grams) for mice in the 141G6 family lines.

Group	Number	Average	SD
Female nontransgenic	43	25.5	1.9
Female transgenic	40	25.0	2.1
Male nontransgenic	30	30.9	1.8
Male transgenic	64	32.1	2.1

17. In the linear model

$$\mathbb{E}(Y_{ghi}) = \tau + \gamma_g + \delta_h,$$

for the balanced two-way classification with $n_{gh} = K$ for all g and h, we saw that

$$\hat{\tau} = \bar{y},$$
$$\hat{\gamma}_g = \bar{y}_{g.} - \bar{y}$$
$$\hat{\delta}_h = \bar{y}_{.h} - \bar{y}$$

a. Show that the model sum of squares can be written as

$$HK\sum_{g=1}^{G}\hat{\gamma}_g^2 + GK\sum_{h=1}^{H}\hat{\delta}_h^2.$$

Be sure to show that the cross product term is 0.

b. Show that the residual sum of squares can be written as

$$K\sum_{g=1}^{G}\sum_{h=1}^{H}(\bar{y}_{gh} - \hat{\tau} - \hat{\gamma}_g - \hat{\delta}_h)^2 + \sum_{g=1}^{G}\sum_{h=1}^{H}\sum_{i=1}^{K}(y_{ghi} - \bar{y}_{gh})^2.$$

Again, be sure to show that the cross product term is 0.

c. Use part (b) to prove that the residual sum of squares for the additive model is always at least as large as the residual sum of squares for the full model with every cell having its own mean.

18. Consider the geometric approach to least squares (see the Extensions section of Chapter 10), where $\mathbf{y}$ represents the $n \times 1$ vector of weights y_i, $\mathbf{1}$ is a vector of 1s, and $\mathbf{e}_M$ is the vector that indicates the sex of the mouse (i.e., $\mathbf{e}_{M,i} = 1$ if the ith mouse is male and 0 otherwise). Then the sum of squares can be expressed as

$$|\mathbf{y} - (b_0\mathbf{1} + b_M\mathbf{e}_M)|^2.$$

a. Show that space spanned by $\mathbf{1}$ and $\mathbf{e}_M$ is the same as the space spanned by $\mathbf{e}_M$ and $\mathbf{e}_F$, where the vector $\mathbf{e}_F$ is an indicator for the female mice.

b. Explain why the projection of $\mathbf{y}$ onto $\mathbf{e}_M$ is $\bar{y}_M\mathbf{e}_M$.

c. Show that $\mathbf{e}_M$ and $\mathbf{e}_F$ are orthogonal.

d. Use the orthogonality to determine the projection of $\mathbf{y}$ onto this space and to find $\hat{b}_0$ and $\hat{b}_M$.

19. Use the geometric approach to explain why the following sum of squares is overparameterized:

$$\left|\mathbf{y} - \left(\mu\mathbf{1} + \sum_{g=1}^{G}\alpha_g\mathbf{e}_g\right)\right|^2,$$

where the $n \times 1$ vectors $\mathbf{e}_g$ are indicators for the groups. Further, use the constraint that $\sum n_g \alpha_g = 0$ to re-express the sum of squares above as

$$\left|(\mathbf{y} - \bar{y}\mathbf{1}) - \left(\sum_{g=1}^{G} \alpha_g \mathbf{e}_g\right)\right|^2$$

and therefore show that, under this constraint, $\hat{\alpha}_g = \bar{y}_g - \bar{y}$.

20. Apply the geometric approach to least squares to Exercise 6. That is, first show that $\mathbf{1}$ is orthogonal to $\mathbf{x}$. Then use this fact to derive $\hat{a}$ and $\hat{b}$ as the coefficients of the projections of $\mathbf{y}$ onto the vectors $\mathbf{1}$ and $\mathbf{x}$, respectively.

Notes

Much of the background information on the creation of transgenic mice and on the care and testing of the mice was obtained from Desmond Smith, a scientist at the Human Genome Center, Lawrence Berkeley National Laboratory. Smith provided a guided tour of the laboratory and was very helpful in answering our many questions on the subject. The data were made available by Edward Rubin, senior scientist on the project. A complete and nontechnical introduction to Down Syndrome can be found in *Down Syndrome: The Facts* ([Sel90]), and current information on the syndrome can be found on the web site http://www.nas.com/downsyn/. Cheryl Gay assisted in the preparation of the background material for this lab.

References

[Hot90] T. Hotchner. *Pregnancy & Childbirth.* Avon Books, New York, 1990.

[Sel90] M. Selikowitz. *Down Syndrome: The Facts.* Oxford University Press, New York, 1990.

[SZZCR95] D.J. Smith, Y. Zhu, J. Zhang, J. Cheng and E. Rubin. Construction of a panel of transgenic mice containing a contiguous 2-Mb set of YAC/P1 clones from human chromosome 21q22.2. *Genomics*, **27**:425–434, 1995.

[SSS97] D.J. Smith, M.E. Stevens, S.P. Sudanagunta, R.T. Bronson, M. Makhinson, A.N. Watabe, T.J. O'Dell, J. Fung, H.G. Weier, J. Cheng, and E. Rubin. Functional screening of 2 Mb of human chromosome 21q22.2 in transgenic mice implicates minibrain in learning defects associated with Down syndrome. *Nat. Genet.*, **16**:28–36, 1997.

[SR97] D.J. Smith and E. Rubin. Functional screening and complex traits: human 21q22.2 sequences affecting learning in mice. *Hum. Mol. Genet.*, **6**:1729–1733, 1997.

12

Helicopter Design

[1]

---

[1]DILBERT © United Features Syndicate. Reprinted by permission.

Introduction

Figure 12.1 contains a diagram for making a helicopter from a single sheet of 8.5 × 5.5 inch paper and a paper clip. Can you build a better one—one that takes *longer* to reach the ground when dropped from a height of eight feet? To develop this new, superior helicopter, at most 25 sheets of paper may be used and at most two hundred flight times may be recorded.

Unlike the other labs, this one does not have a real-world question with a set of data collected that addresses the question. Instead, it is an artificial setup in which you, the investigator, design and conduct an experiment. The analysis of the experimental results is only one part of the statistical work involved in the experimentation process. In this lab, you will encounter many of the problems faced in the design and implementation of an experiment, including finding out how accurately flight time can be measured, figuring out how to decide which helicopter to build first, and deciding how to proceed in the experimentation as you analyze your experimental results.

Data

The data for this lab have not been previously collected. As part of your lab work, you will collect it in a notebook for analysis.

The information collected will be the wing widths and lengths of model helicopters, and their flight times. The wing dimensions will be determined by you, the investigator. They are called the design variables; they are the variables over which you have control. Their units of measurement are to be determined by the investigator.

The response is the time it takes the helicopter to reach the ground from a fixed height. To record these times, you will need a stop watch that measures time to the nearest hundredth of a second.

Along with the helicopter measurements and flight times, you may find it useful to annotate your numeric records. For example, a helicopter may hit a wall during flight, or a helicopter may have very erratic flights, sometimes sailing smoothly and other times plummeting wildly to the ground. Record all flight times and keep track of any additional information that may prove useful in your search for the best helicopter.

Background

All helicopters are restricted to have the general form shown in Figure 12.1. The wing length and the wing width of the helicopter are the only two design variables to be investigated in the search for the best helicopter. The body width and length are to remain constant for all helicopters. A single, 32 mm, metal paper clip is

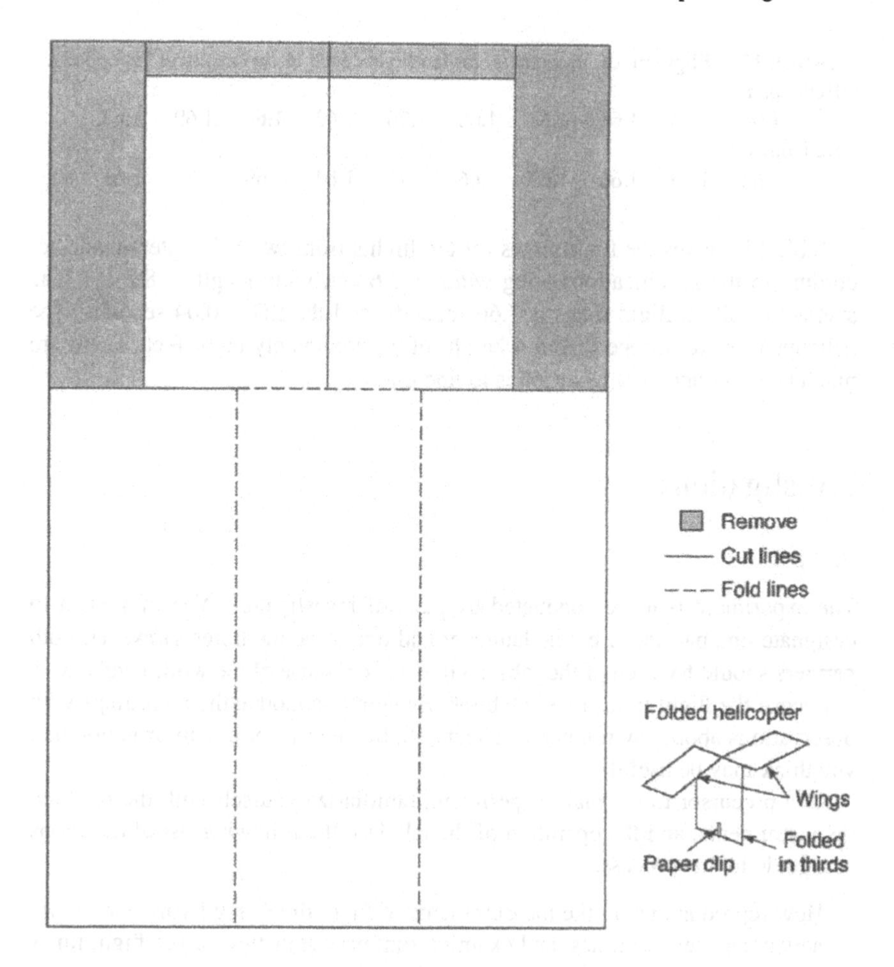

FIGURE 12.1. Directions for making a paper helicopter.

to be attached to the bottom of the helicopter, as in Figure 12.1. Please use only authorized materials in the construction of the helicopters. No additional folds, creases, or perforations are allowed.

Here are some comments on helicopter building that previous students have made in their notebooks. They may save you time in getting started.

> "Rich creased the wings too much and the helicopters dropped like a rock, turned upside down, turned sideways, etc."
>
> "Andy proposes to use an index card to make a template for folding the base into thirds."
>
> "After practicing, we decided to switch jobs. It worked better with Yee timing and John dropping. 3 – 2 – 1 – GO."

TABLE 12.1. Flight times, in seconds, for ten flights each of two identical helicopters.

Helicopter										
Helicopter #1	1.64	1.74	1.68	1.62	1.68	1.70	1.62	1.66	1.69	1.62
Helicopter #2	1.62	1.65	1.66	1.63	1.66	1.71	1.64	1.69	1.59	1.61

Table 12.1 gives the flight times for ten flights from two helicopters made according to the specifications wing width = 4.6 cm, wing length = 8.2 cm. The average of all 20 flight times is 1.66 seconds, and the SD is 0.04 seconds. The helicopters were dropped from a height of approximately eight feet. There are much better helicopters; your job is to find one.

Investigations

Setup

The experiment is to be conducted by pairs of investigators. You may want to designate one partner to be the launcher and one to be the timer. However, both partners should try each of the jobs, so they have a sense of the work involved.

Record the flight times in a notebook. Be sure to annotate the recordings with observations about environmental effects, flight patterns, or any information that you think may be useful.

As a precursor to the main experiment, familiarize yourself with the product, the instruments, and the operation of the lab. Use the following list of questions as a guide in this process.

- How reproducible are the measurements of flight time? Try flying one or two helicopters several times, and examine summary statistics for the flight times for each helicopter.
- Do you notice any person-to-person differences, or other environmental effects? If possible, separate measurement error from "natural variability."
- Is there any variability between "identical" helicopters? That is, how similar are the flight times for two helicopters built according to the same specifications.
- Which wing dimensions lead to viable helicopters?
- How big do the changes in the dimensions of the helicopters need to be in order to easily observe a difference in the flight time?

It will be most effective to compare mean flight times for different helicopters. How many flight times should be recorded and averaged for each helicopter's mean response? What is an estimate for the standard error for the mean response? Too few flights per helicopter may result in indistinguishable results because the variability is too large. On the other hand, too many flights will be time consuming. Also, these helicopters have a very limited lifespan; they wear out quickly. The standard error that you determine here will be used to make design decisions for the rest of the experiment. It is important to get it right.

Of course, changes in your experimentation process can be made along the way to correct for earlier errors and new findings. Nonetheless, it is important to be careful and thorough in this part of the experimentation. Time spent here learning about the variability in the process could save time, limit confusion, and produce more reliable results later.

Begin the Search

Choose a set of wing dimensions to begin the search for the best helicopter. The wing width and length designate a point in what we call the design space of helicopters. The left plot in Figure 12.2 gives an example of such a design point; call it d_0.

Each helicopter has a "true" flight time. If it were plotted above the width and length of the helicopter, then the flight times might look like a surface map above the design space (Figure 12.2; right plot). The flight time is called the helicopter's response, and the map of flight times is called the ***response surface***. Our goal is to find the helicopter that corresponds to the maximum flight time—the location on the design space of the highest peak on the response surface.

To do this,. we search in small regions of the design space to get an idea of the local shape of the response surface. Then we use our rough, local picture of the surface to determine the general direction for continuing the search. We repeat this investigation in a new part of the design space until it looks as though we are in the neighborhood of the peak.

This search is described in three parts.

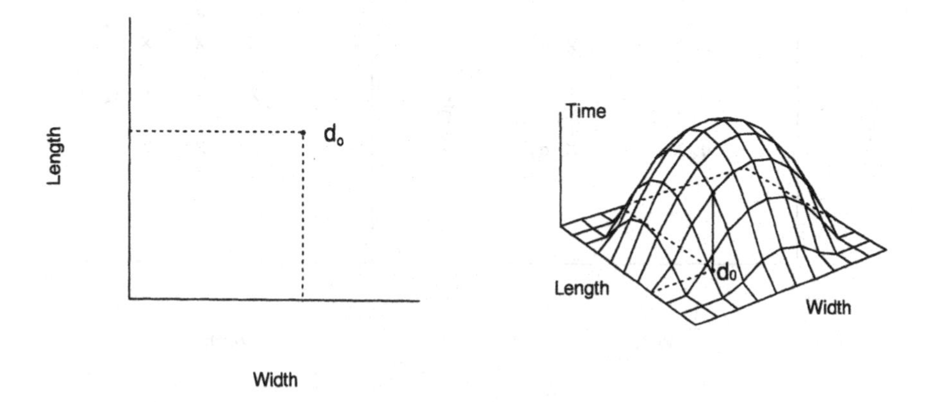

FIGURE 12.2. Examples of a design space (left) and response surface (right).

1. First-order Search

Form a rectangle around the design point d_0, with it as the center of the rectangle. Label each of the corners of the rectangle d_1, d_2, d_3, and d_4, as in the left plot in Figure 12.3. The earlier informal experimentation should give a sense of the appropriate size of the rectangle.

Build helicopters corresponding to the design points $d_1, \ldots, d_4$, fly each of them many times, and record the flight times in your notebook. Also draw a diagram of the design rectangle that is true to scale, and mark the average flight times on the diagram. (Why might it be a good idea to fly the helicopters in random order?)

If flight time is locally linear about the design point d_0, and if the rectangle is not too large, then the linear model

$$\text{time} = a + b \times \text{width} + c \times \text{length}$$

should provide a reasonable approximation to the response surface.

To determine estimates of these coefficients, fit the linear model using the method of least squares. You may use the computer to minimize the sum of squares or, as shown in the Theory section, the calculations are simple enough to be done on an ordinary calculator. Be sure to check whether the assumption of local linearity is reasonable. Examine the residuals from the fit. If nonlinearity is apparent, try correcting the problem with transformations of the design and response variables. This may not be possible if the rectangle is too big.

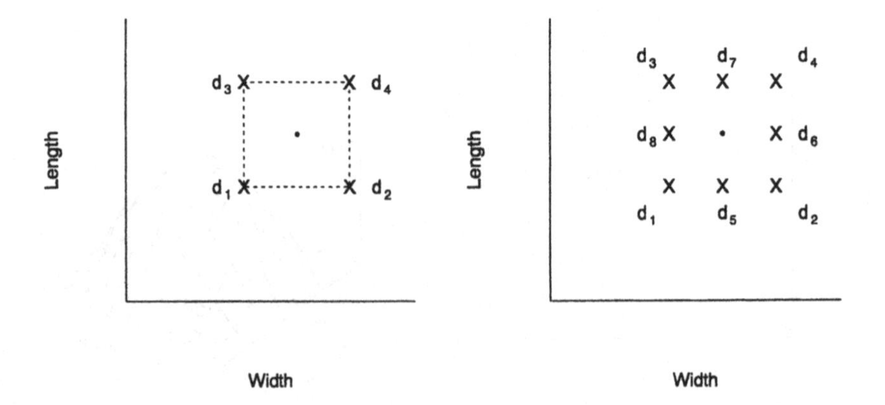

FIGURE 12.3. Examples of a rectangular design (left) and rectangular plus star design (right).

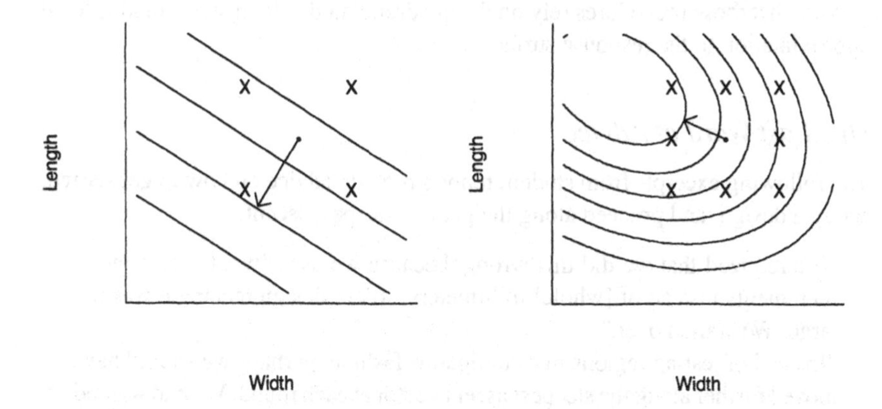

FIGURE 12.4. Examples of a linear contour plot (left) and quadratic contour plot (right).

2. The Path of Steepest Ascent

If the linear fit is adequate, then it can be used to determine the direction of future searches.

- Draw contour lines on the design region that correspond to the contours of the fitted plane.
- Draw a line from the center of d_0 perpendicular to the contour lines.
- Choose a design point along this line, build the corresponding helicopter, and measure the improvement in flight time. Continue to explore in this direction until you have settled on a new region to conduct your search. Be careful not to take steps that are too small or too large. Try moving far along the line and backing up halfway if the results indicate you have gone too far.
- Return to the first-order search above.

3. Second-order Search

If the linear model is inadequate, then a more complex design to fit a quadratic model may be required. The quadratic model is

$$\text{time} = d + e \times \text{width} + f \times \text{length} + \\ g \times \text{width}^2 + h \times \text{length}^2 + k \times \text{width} \times \text{length}.$$

To fit this model, more data must be collected. One approach is to augment the rectangular design by a star design as in the right plot of Figure 12.3. Once the additional data are collected at design points d_5, d_6, d_7, d_8, then the coefficients can be estimated via least squares. Again, these estimates can easily be obtained with the use of a calculator, as shown in the Theory section.

From the coefficients of the quadratic model, the contours of the quadratic can be sketched on the design space and the peak region can be determined.

Note that these procedures rely on the quadratic model being a reasonable local approximation to the response surface.

One Last Word of Advice

The following excerpts from student reports provide advice on how to get started, set up a design, and proceed along the path of steepest ascent.

> "We realized that we did this wrong, because we used [whole] centimeter increments instead of [whole] millimeters, so our design rectangle was too large. We started over."
>
> "Instead of testing regions in a contiguous fashion, perhaps we should have moved further along the steepest ascent vector at each round. We had wanted to have complete coverage of the test path, but our test method was too slow and used up too many helicopters along the way."
>
> "We figured out posthumously why our analyses were so bad is that we were using the wrong proportions for our path of steepest ascent. [Actually no one died.]"

Mistakes are bound to happen. The comments above will help you avoid some of the more serious ones.

Your notebook will serve as the final report. It should contain all data, statistical calculations, design diagrams, contour plots, and a narration of the experimental process. Make sure the narration includes the problems that you encountered, reasons for proceeding as you did, and final conclusions. All calculations can be done with a calculator. The only time a computer is needed is for drawing the quadratic contour plot.

Try to avoid making the following types of conclusions:

> "Our data are very suspicious."
>
> "We made an extremely vital mistake."
>
> "Since we are out of paper, we will just ..."
>
> "Great things were expected from our contour analysis, but unfortunately, fell far short of our goals."

Theory

Variability

In setting up the experiment, one of the first goals is to estimate the inherent variability in flight times. Table 12.1 shows the results of ten helicopter flights for two such helicopters. The average of the ten measurements taken on helicopter #1 is 1.67, and the standard deviation is 0.04.

In comparing the performance of various helicopters, the average flight time for each helicopter is used. The standard error for the average flight time can be estimated by $s/\sqrt{n}$, where s is an estimate of the standard deviation of flight time and n is the number of flights for that helicopter.

The results of flight times from many different helicopters can be combined to provide a more accurate estimate of the standard deviation. That is, of course, provided that all the helicopters have roughly the same variability in flight time. If so, the standard deviations for the helicopters can be pooled by forming their r.m.s. as follows:

$$s_p = \sqrt{\frac{s_1^2 + \cdots + s_m^2}{m}},$$

where

$$s_i^2 = \frac{1}{k-1}\sum_{j=1}^{k}(y_{i,j} - \bar{y}_i)^2$$

is the estimated flight time standard deviation for the ith helicopter, $y_{i,j}$ is the flight time for the jth flight of the ith helicopter, and $\bar{y}_i$ is the average flight time for the ith helicopter. If the individual standard deviations s_1, ..., s_m are based on a different number of observations, then a different weighting scheme for pooling standard deviations is required. (See the Exercises.)

Factorial Design

The rectangular set of design points in the left plot of Figure 12.3 are arranged in what is known as a 2×2 factorial. With this arrangement, the response (flight time) can be maximized over both dimensions in the design space (wing width and length) simultaneously. The *contour plot* to the left in Figure 12.4 is an example of the result of such a maximization. The arrow originating from the center of the rectangle points in the direction of the path of steepest ascent. The path is determined by the two coefficients b and c for wing width and wing length in the linear equation below:

$$\text{time} = a + b \times \text{width} + c \times \text{length}.$$

A contour line on the plot corresponds to a set of wing width and length values that yield the same estimated time. A simple rearrangement of the terms in the equation above gives the equation for these contour lines:

$$\text{length} = \frac{\text{time} - a}{c} - \frac{b}{c}\text{width}.$$

The slope of this line is $-b/c$. Hence the slope of a perpendicular to these contour lines is c/b. The path of steepest ascent from the center of the design rectangle can readily be sketched on the design space as in Figure 12.4.

Determining the path of steepest ascent via the factorial design is a much better search method than sequentially searching in one direction at a time: flight time is first maximized for helicopters with different wing widths but the same wing length, after which flight time is maximized over various wing lengths using the previously found "maximum" wing width. This search may end far from the true maximum.

The linear equation above serves as a local approximation to the response surface. If the response surface is smooth enough (i.e., if it has a continuous first derivative), then it can be approximated by a plane in a small region around d_0. The local linear approximation then provides a direction for continuing the search for the maximum flight time.

The linear model implies that, for a particular wing width, time is linear in wing length, and the slope of the line does not depend on the wing width. A more complex local approximation may be indicated if the linear approximation does not fit the data well. For example, there may be an interaction between wing width and length where, for a particular wing width, time is linear in length but the slope of the line depends on the width. Also, a quadratic function may yield a better local approximation. This may occur near the maximum, due to the curvature of the response surface. In this case, hopefully the maximum of the local quadratic approximation is near the true maximum of the response surface. However, the quadratic approximation may also be poor if, for example, the region over which the approximation is being made is too large.

Below is a quadratic model that includes an interaction term:

$$\begin{aligned} \text{time} &= d + e \times \text{width} + f \times \text{length} + g \times \text{width}^2 \\ &\quad + h \times \text{length}^2 + k \times \text{width} \times \text{length}. \end{aligned}$$

Here g and h are coefficients for the second-order terms for width and length, respectively. Ignoring these quadratic terms for the moment, notice that for a fixed wing width, the interaction term yields a model that is linear in length with coefficient $(f + k \times \text{width})$.

The 2×2 factorial is a very simple design. With only four design points, the six coefficients in the model above cannot be estimated. The rectangular design can be augmented by a star design as in the right plot of Figure 12.3. Then the flight times at the eight design points can be used to make a quadratic approximation to the response surface in the neighborhood of the design point d_0. With this approximation, it is not as easy to simply sketch the contours of the quadratic onto the design space. However, graphical software can be used to make the contour plots.

Fitting the Response Surface

To fit the local, linear approximation, the coefficients a, b, and c need to be estimated. For each design point, d_i, a helicopter has been flown several times and an average flight time reported, say $\bar{y}_i$. (Note that $d_i = (w_i, l_i)$, where w_i denotes the wing width and l_i the wing length.) We use the method of least squares to find the coefficients that minimize the sum of squared deviations for the averages $\bar{y}_i$. These estimated coefficients are called $\hat{a}$, $\hat{b}$, and $\hat{c}$.

Most statistical software can perform this minimization for you. However, the rectangular design of the experiment makes it very easy to determine these coefficients by simple arithmetic. The main advantage in doing it "by hand" is that the computer may not be available at the time of data collection. A second one is that you will get to exercise your math skills. A few simple calculations can quickly point out the path of steepest ascent, and the helicopter building and flying can continue uninterrupted.

The simple solution is based on a relabeling of the design axes to measure all helicopters relative to d_0, the center of the rectangle. Suppose helicopter d_1 has wings 3 mm narrower and 5 mm shorter than d_0. (Note that it is also the case that helicopter d_4 has wings that are both 3 mm wider and 5 mm longer than d_0's.) Change the scale so that 3 mm in width is one unit in width and 5 mm in length is one unit in length. Then d_1 has width -1 and length -1, relative to d_0. The rescaled widths and lengths for the rectangular design are displayed in Table 12.2. There is nothing special about the choice 3 mm and 5 mm; these numbers were only given as an example.

Now we can easily determine the least squares estimates of a, b, and c to be

$$\hat{a} = \frac{\bar{y}_1 + \bar{y}_2 + \bar{y}_3 + \bar{y}_4}{4}, \qquad \hat{b} = \frac{-\bar{y}_1 + \bar{y}_2 - \bar{y}_3 + \bar{y}_4}{4},$$

$$\hat{c} = \frac{-\bar{y}_1 - \bar{y}_2 + \bar{y}_3 + \bar{y}_4}{4}.$$

To draw the path of steepest ascent on your original design, convert back to the original units. For our example, where width is measured in increments of 3 mm and length in increments of 5 mm, the slope in original units will be $3c/5b$.

This same procedure can be used in estimating the coefficients in the quadratic model. Table 12.3 gives the measurements in the new units for each of the helicopters in the composite design in Figure 12.3 (right plot).

The least squares estimates are

TABLE 12.2. Design variables for the 2 × 2 factorial rectangular design.

Design point	Average time	Width	Length
d_1	$\bar{y}_1$	−1	−1
d_2	$\bar{y}_2$	+1	−1
d_3	$\bar{y}_3$	−1	+1
d_4	$\bar{y}_4$	+1	+1

TABLE 12.3. Design variables for the composite rectangular plus star design.

Design Point	Width	Length	Width^2	Length^2	Width × Length
d_1	−1	−1	1	1	1
d_2	1	−1	1	1	−1
d_3	−1	1	1	1	−1
d_4	1	1	1	1.	1
d_5	0	−1	0	1	0
d_6	1	0	1	0	0
d_7	0	1	0	1	0
d_8	−1	0	1	0	0

$$\hat{d} = \frac{-\bar{y}_1 - \bar{y}_2 - \bar{y}_3 - \bar{y}_4 + 2\bar{y}_5 + 2\bar{y}_6 + 2\bar{y}_7 + 2\bar{y}_8}{4},$$

$$\hat{e} = \frac{-\bar{y}_1 + \bar{y}_2 - \bar{y}_3 + \bar{y}_4 - \bar{y}_6 + \bar{y}_8}{6}, \quad \hat{f} = \frac{-\bar{y}_1 - \bar{y}_2 + \bar{y}_3 + \bar{y}_4 - \bar{y}_5 + \bar{y}_7}{6},$$

$$\hat{g} = \frac{\bar{y}_1 + \bar{y}_2 + \bar{y}_3 + \bar{y}_4 - 2\bar{y}_5 - 2\bar{y}_7}{4}, \quad \hat{h} = \frac{\bar{y}_1 + \bar{y}_2 + \bar{y}_3 + \bar{y}_4 - 2\bar{y}_6 - 2\bar{y}_8}{4},$$

$$\hat{k} = \frac{\bar{y}_1 - \bar{y}_2 - \bar{y}_3 + \bar{y}_4}{4}.$$

If the linear approximation in the 2×2 factorial design is too crude, it will show up in the residuals

$$\bar{y}_1 - (\hat{a} + \hat{b}w_1 + \hat{c}l_1) = \frac{\bar{y}_1 - \bar{y}_2 - \bar{y}_3 + \bar{y}_4}{4}.$$

In this case, the quantity above estimates the interaction between wing width and length and as such is a measure of model misfit. (To see this, notice that it matches the least squares estimate of k, the coefficient of the interaction term in the quadratic model.) On the other hand, if the linear approximation is good, then the "residual" $\hat{k}$ above should have expectation 0 and SE $\sigma_n/2$, where σ_n is the standard error of the average flight time. That is, $\sigma_n = \sigma/\sqrt{n}$, where n is the number of flights for each helicopter. To determine if there is significant model misfit, the "residual" can be compared against the replication-based estimate of the standard deviation.

Under the linear model, the standard errors of the coefficients in the least squares fit of the linear approximation are all $\sigma_n/2$. In the quadratic fit, the standard errors for $\hat{d}, \hat{e}, \hat{f}, \hat{g}, \hat{h}$, and $\hat{k}$ are, respectively, $\sqrt{5}\sigma_n/2, \sigma_n/\sqrt{6}, \sigma_n/\sqrt{6}, \sqrt{3}\sigma_n/2, \sqrt{3}\sigma_n/2$, and $\sigma_n/2$. These coefficients can also be compared against the replication-based estimate of the standard deviation.

Exercises

1. Derive the least squares estimates of $\hat{a}$, $\hat{b}$, and $\hat{c}$ from the minimization of

$$\sum_{i=1}^{4}[\bar{y}_i - (a + bw_i + cl_i)]^2$$

with w_i and l_i as in Table 12.2.

2. Show that the path of steepest ascent from d_0 for the fitted plane (Exercise 1) has slope is $\hat{c}/\hat{b}$.
3. Suppose m helicopters are labeled $i = 1, \ldots m$, and each is flown k_i times, where $y_{i,j}$ is the flight time for the jth flight of the ith helicopter. Show that

$$s_i^2 = \frac{1}{k_i - 1}\sum_{j=1}^{k_i}(y_{i,j} - \bar{y}_i)^2$$

is an unbiased estimate for the variance of the flight time for the ith helicopter.

4. Suppose the SD of flight time is the same for all m helicopters in Exercise 3. Find a pooled estimator s_p for the SD. Show that s_p^2 is unbiased.
5. Show that the residuals $\bar{y}_i - \hat{y}_i$ for the first-order model are such that $r_1 = -r_2 = -r_3 = r_4$. Explain why this is the case.
6. Show that the residual r_1 has expectation 0 and SE $\sigma_n/2$ under the linear model.
7. Show that the residual for the first-order model is an estimate of the interaction weight × length.
8. Prove that the SEs of $\hat{a}$, $\hat{b}$, and $\hat{c}$ are all $\sigma_n/2$.
9. Find the SE of $\hat{d}$ and $\hat{f}$ in the star design.
10. Suppose the contours of a response surface are elliptical and the response is the following function:

$$\exp\left[-\left(w^2 + \frac{1}{4}l^2 - \frac{1}{4}wl\right)\right].$$

Maximize this function with respect to l, holding w fixed at $1/2$. Call the maximizer l^*. Then holding l^* fixed, maximize over w, and show that the overall maximum is not achieved.

11. Show that the linear model is set up so that **w**, the vector of widths, **l**, the vector of lengths, and **1** are orthogonal.
12. Are the vectors in the quadratic model orthogonal? Use what you discover to derive the least squares estimators for d, e, f, g, h, and k in the quadratic model.

Notes

The idea for this lab came out of a discussion with Bill Kahn. At the time he was Chief Statistician for the Gore Company. One of Kahn's roles was to train the company's engineers in the design of experiments. For proprietary reasons, he could not share any of the data that these engineers collected in their experiments, but he explained that his most effective teaching technique was to let the engineers design and run a simple, fun experiment from scratch. He left us with a paper helicopter, from which we developed this lab. Andy Choy, a student in one of our earlier classes, suggested the use of a paper clip to help stabilize the helicopter's flight.

Later, we discovered Bisgaard's ([Bis91]) description of the use of a paper helicopter in a classroom setting to teach design of experiments to engineers. Bisgaard outlines how the helicopter can be used as a connecting theme throughout an entire introductory statistics course. We do not know whether or not that helicopter is the same as the one in Figure 12.1. Other examples of how to teach design of experiments to engineers appear in Hunter ([Hun77]), Kopas and McAllister ([KM92]), and Mead ([Mead90]). Hoadley and Kettenring ([HK90]) provide an interesting discussion on communications between statisticians and engineers/physical scientists.

The material on response surface analysis in the Theory section is adapted from Box, Hunter and Hunter ([BHH78]) Chapter 15. Other, more detailed treatments of the subject appear in Box and Draper ([BD87]) and Khuri and Cornell ([KC87]).

References

[BD87] G.P. Box and N.R. Draper. *Empirical Model Building and Response Surfaces*. John Wiley & Sons, New York, 1987.

[BHH78] G.P. Box, W.G. Hunter, and J.S. Hunter. *Statistics for Experimenters: An Introduction to Design, Data Analysis, and Model Building*. John Wiley & Sons, New York, 1978.

[Bis91] S. Bisgaard. Teaching statistics to engineers. *Am. Stat.*, **45**:274–283, 1991.

[HK90] A.B. Hoadley and J.R. Kettenring. Communications between statisticians and engineers/physical scientists. *Technometrics*, **32**:243–247, 1990.

[Hun77] W.G. Hunter. Some ideas about teaching design of experiments, with 2^5 examples of experiments conducted by students. *Am. Stat.*, **31**:12–17, 1977.

[KC87] A.I. Khuri and J.A. Cornell. *Response Surfaces: Designs and Analyses*. Marcel Dekker, New York, 1987.

[KM92] D.A. Kopas and P.R. McAllister. Process improvement exercises for the chemical industry. *Am. Stat.*, **46**:34–41, 1992.

[Mead90] R. Mead. Statistical games 1 – Tomato. *Teach. Stat.*, **12**:76–78, 1990.

Appendix A
Writing Lab Reports

To prepare your lab report, distill your investigations into a central argument. Then organize your findings into units of evidence for your argument. A good lab report is more than a compilation of your statistical analyses. A good report should be well organized, and it should demonstrate clear and sound reasoning, contain easily interpreted data displays, and use good grammar.

Organization[1]

Format your report into recognizable sections to clarify the structure of the paper. Two levels of headings are usually helpful to provide a general outline. A paper without section headings drags and is difficult to follow.

The choice of sections should match the reader's expectations. For example, a research article is generally divided into sections labeled: Introduction, Methodology, Results, and Discussion. When the sections are jumbled, such as when discussion or experimental detail is found in the statement of the results, the readers may become confused.

Use the Introduction to state the problem you are addressing and your findings. Without giving away all your points, let the reader know where your paper is headed.

---

[1]The first three paragraphs of this section are summarized from, S. Tollefson, "Encouraging Student Writing," S. Tollefson, University of California, p. 24.

- Catch the reader's attention. Start with an example, a quotation, a statistic, a question, or a complaint and use it as a theme that you refer to throughout the paper.
- The Introduction sets the tone for your report. Explain why the problem you are addressing is important. Appear to be interested in the topic.
- Break up a long Introduction into several paragraphs. One huge paragraph at the outset of a paper can put readers off.
- Avoid such phrases as "I will discuss" or "this report will examine." Better to just dive right in.

Use the Methodology section to describe your data and how they were collected. This information helps the reader assess the appropriateness of your analysis and the significance of your findings.

- Describe the subject(s) under study. Be as specific as possible. Make clear who was included in the study and who was not.
- Outline the procedures used for collecting the data. For example, if the data are from a sample survey, then the reader may need to know the sampling method, the interview process, and the exact wording of the questions asked. Also address any problems with the data such as nonresponse.
- Explain how the variables measured can be used to address the scientific question of interest. Clearly distinguish between the main outcome of the study and auxiliary information. Be sure to provide the units in which the responses were measured.

Use the Results section to present your findings. Limit the presentation to those results that are most relevant to your argument and most understandable to the reader.

- Be parsimonious in your use of supporting tables and graphs. Too much extraneous information overloads the reader and obscures the importance of your main thesis. The reader is often willing to accept a brief statement summarizing your additional findings, especially if the material presented is well displayed and to the point. When preparing your data displays, follow the guidelines that appear later in this appendix. Each display must be discussed in the prose.
- Limit the use of statistical jargon. Save the most technical material for an Appendix where you show the advanced reader your more sophisticated ideas and more complicated calculations.
- Include in your report any findings that point to a potential shortcoming in your argument. These problems should be considered in the Discussion section.
- If you include a figure from another paper, cite the original source in your figure caption.

Use the Discussion section to pull together your results in defense of your main thesis.

- Be honest. Address the limitations of your findings. Discuss, if possible, how your results can be generalized.

- Be careful not to overstate the importance of your findings. With statistical evidence, we can rarely prove a conjecture or definitively answer a question. More often than not, the analysis provides support for or against a theory, and it is your job to assess the strength of the evidence presented.
- Relate your results to the rest of the scientific literature. Remember to give credit to the ideas of others. Consider the following questions:
  — Do your results confirm earlier findings or contradict them?
  — What additional information does your study provide over past studies?
  — What are the unique aspects of your analysis?
  — If you could continue research into the area, what would you suggest for the next step?

Clarity and Structure of Prose[2]

Information in a passage of text is interpreted more easily and more uniformly if it is placed where readers expect to find it. Readers naturally emphasize the material that arrives at the end of a sentence. When the writer puts the emphatic material at the beginning or middle of a sentence, the reader is highly likely to emphasize the wrong material and to incorrectly interpret the message of the sentence. Readers also expect the material at the beginning of the sentence to provide them a link to previous material and a context for upcoming material. When old information consistently arrives at the beginning of the sentence, it helps readers to construct the logical flow of the argument.

Observe the following structural principles:

- Follow a grammatical subject as soon as possible with its verb.
- Place at the end of the sentence the new information you want the reader to emphasize.
- Place the person or thing whose story a sentence is telling at the beginning of the sentence.
- Place appropriate old information (material already stated in the discourse) at the beginning of the sentence for linkage backward and contextualization forward.
- Provide context for your reader before asking that reader to consider anything new.
- Try to ensure that the relative emphases of the substance coincide with the relative expectations raised by the structure.

Here is an example from Gopen and Swan of scientific prose that begins sentences with new information and ends with old information. After reading the paragraph, we have no clear sense of where we have been or where we are going.

[2]The material in this section is summarized from G.D. Gopen and J.A. Swan, "The Science of Scientific Writing," *Am. Sci.*, **78**:550–558.

> Large earthquakes along a given fault segment do not occur at random intervals because it takes time to accumulate the strain energy for the rupture. The rates at which tectonic plates move and accumulate strain at their boundaries are approximately uniform. Therefore, in first approximation, one may expect that large ruptures of the same fault segment will occur at approximately constant time intervals. If subsequent main shocks have different amounts of slip across the fault, then the recurrence time may vary, and the basic idea of periodic mainshocks must be modified. For great plate boundary ruptures the length and slip often vary by a factor of 2. Along the southern segment of the San Andreas fault the recurrence interval is 145 years with variations of several decades. The smaller the standard deviation of the average recurrence interval, the more specific could be the long term prediction of a future mainshock.

Gopen and Swan revised the paragraph to abide by the structural principles outlined above. The phrases in square brackets are suggestions for connections between sentences. These connections were left unarticulated in the original paragraph; they point out the problems that had existed with the logical flow of the argument.

> Large earthquakes along a given fault segment do not occur at random intervals because it takes time to accumulate the strain energy for the rupture. The rates at which tectonic plates move and accumulate strain at their boundaries are approximately uniform. Therefore, nearly constant time intervals (at first approximation) would be expected between large ruptures of the same fault segment. [However?], the recurrence time may vary; the basic idea of periodic mainshocks may need to be modified if subsequent main shocks have different amounts of slip across the fault. [Indeed?], the length and slip of great plate boundary ruptures often vary by a factor of 2. [For example?], the recurrence intervals along the southern segment of the San Andreas fault is 145 years with variations of several decades. The smaller the standard deviation of the average recurrence interval, the more specific could be the long term prediction of a future mainshock.

Data Displays[3]

The aim of good data graphics is to display data accurately and clearly, and the rules for good data display are quite simple. Examine data carefully enough to know what they have to say, and then let them say it with a minimum amount of adornment. Do this while following reasonable regularity practices in the depiction of scale, and label clearly and fully.

---

[3]The material in this section is summarized from "How to Display Data Badly," H. Wainer, *The American Statistician* **38**:137–147, 1984.

The following list provides guidelines for how to make good data displays.

- *Density*– Holding clarity and accuracy constant, the more information displayed the better. When a graph contains little information, the plot looks empty and raises suspicions that there is nothing to be communicated. However, avoid adding to the displays extraneous graphics such as three-dimensional bars, stripes, and logos. Chart junk does not increase the quantity of information conveyed in the display; it only hides it.
- *Scale*–
  - Graph data in context; show the scale of your axes.
  - Choose a scale that illuminates the variation in the data.
  - Do not change scale in mid-axis.
  - If two plots are to be compared, make their scales the same.
- *Labels*– Captions, titles, labels, and legends must be legible, complete, accurate, and clear.
- *Precision*– Too many decimal places can make a table hard to understand. The precision of the data should dictate the precision reported. For example, if weight is reported to the nearest 5 pounds then a table presenting average weights should not be reported to the nearest 1/100 of a pound.
- *Dimensions*– If the data are one-dimensional, then use a visual metaphor that is one-dimensional. Increasing the number of dimensions can make a graph more confusing. Additional dimensions can cause ambiguity: is it length, area, or volume that is being compared?
- *Color*– Adding color to a graph is similar to adding an extra dimension to the graph. The extra dimension should convey additional information. Using color in a graph can make us think that we are communicating more than we are.
- *Order*– Sometimes the data that are to be displayed have one important aspect and others that are trivial. Choose a display that makes it easy to make the comparison of greatest interest. For example: (a) ordering graphs and tables alphabetically can obscure structure in the data that would have been obvious had the display been ordered by some aspect of the data; (b) Stacking information graphically indicates the total but can obscure the changes in individual components, especially if one component both dominates and fluctuates greatly; (c) comparisons are most easily made by placing information all on one plot.

Grammar[4]

Bad grammar distracts the reader from what you are trying to say. Always assume that whoever reads your paper will be evaluating your grammar, sentence structure, and style, as well as content. Content cannot really be divorced from form.

[4]The material in this section is reprinted with permission from S. Tollefson, "Encouraging Student Writing," University of California, p. 27.

Examples of grammatical problems:

- *subject–verb or noun–pronoun agreement*
  Theories of cosmology suggests that the universe must have more than three dimensions. ("suggest")
  Everyone investigating mass extinctions by asteroids may bring their own prejudices to the investigations. ("Scientists investigating ... may bring their...")
- *faulty comparison—either incomplete or mixing apples and oranges*
  Thin metal strands deposited in a silicon wafer make it better. ("better" than what?)
  The stars in some galaxies are much more densely packed than the Milky Way. ("than those in the Milky Way")
- *sentence fragment*
  The HTLV virus, nearly unknown six years ago (although some evidence of a longer existence has been found) rapidly becoming the Black Plague of modern times. (add "is" before "rapidly")
  The virus may be handled in the laboratory. But only with care. (These should be one sentence.)
- *misuse of tenses and confusion of parts of speech*
  Some researchers feel badly about how lab animals are treated. ("feel bad")
  Ever since Quagmire Chemical Company requested a heat exchanger, we investigated several options. ("have investigated")
- *idiom—usually a misuse of prepositions*
  The Human Subjects Committee insisted to do it their way ("insisted on doing it...")
- *modification—usually a word or phrase in the wrong place*
  When applying an electric field, liquid crystal molecules align themselves and scattering of light is reduced. (Dangling—should be "When one applies...")
  Incinerating industrial wastes can produce compounds toxic to humans such as dioxins. (misplaced—should be "can produce compounds, such as dioxins, toxic to humans.")
- *parallel structure*
  In gel electrophoresis, large ions move slowly and small ones are traveling more quickly. ("...and small ones travel more quickly")
  The vaccinia virus may be used to vaccinate against small pox and as a carrier of genetic material from one organism to another. ("...to vaccinate...and to carry")
- *passive voice—not always bad, but overused*
  Short stature and low IQ can be caused by an extra chromosome. (more often than not, it's preferable to say "An extra chromosome causes...")
  Trisomy-21 is still often called "mongolism." ("People still often call...")
- *predication—illogical connection among subject/verb/complements*
  Viewing occultations of certain asteroids suggests that they have moons. (it's not the viewing, but the occultations themselves that suggest this)
  Exposure to intense x-rays is the reason these crystals glow. (exposure itself

is not the reason—what the x-rays do to the structure causes the glow; or you could say "Exposure to intense x-rays causes these crystals to glow.")

- *reference—faulty or vague*
  The deprojector and image stretcher allow us to examine tilted galaxies as if they were facing us head on. It's a great breakthrough. ("Development of these devices is a great...")
- *run-together sentence*
  The meteor impact site in the Indian Ocean is the best possibility at the moment, however, other sites do exist. (semicolon or period needed before "however")
  The layer of semenium is worldwide, it shows only a few gaps. (semicolon or period needed after "worldwide")

Revising and Proofreading

After you have completed a draft, look at the paper again. Learn to see where new material is needed, where material should be deleted, and where reorganization is required. Once you have spent enormous effort on the analysis and writing, proofread your manuscript two or three times. Keep looking for unclear passages, format errors, poor reasoning, etc., right down to the moment you submit the paper.

Use the following questions to appraise your manuscript:

- Is the problem clearly stated?
- Are the statistical statements correct?
- Are the data displays informative?
- Are the conclusions based on sound evidence?
- Are the grammar and sentence structure correct?
- Are the style and tone appropriate for the venue?

Appendix B
Probability

Random Variables

A *discrete* random variable is a random variable that can take on a finite or at most a countably infinite number of different values. If a discrete random variable X takes values $x_1, x_2, \ldots$, its distribution can be described by its *probability mass function*, or *frequency function*,

$$f(x_i) = \mathbb{P}(X = x_i).$$

These satisfy $0 \leq f(x_i) \leq 1$ and $\sum_i f(x_i) = 1$.

The *distribution function* of any random variable X is defined as

$$F(x) = \mathbb{P}(X \leq x).$$

$F(x)$ is a nondecreasing, left-continuous function which can take values between 0 and 1.

The *expected value* of a discrete random variable X with probability mass function f is defined as

$$\mathbb{E}(X) = \sum_i x_i f(x_i).$$

One can also compute the expected value of any function $g(X)$ by

$$\mathbb{E}\left[g(X)\right] = \sum_i g(x_i) f(x_i).$$

A *continuous* random variable is a random variable that can take on uncountably many values, for example, all real numbers, or all real numbers in the interval [−1,

1]. The analog of the probability mass function for continuous random variables is the *density function* $f(x)$, which is a nonnegative, piecewise continuous function such that $\int_{-\infty}^{\infty} f(x)dx = 1$. For any real numbers $a \leq b$,

$$\mathbb{P}(a \leq X \leq b) = \int_a^b f(x)dx.$$

It follows that for any single point a, $\mathbb{P}(X = a) = 0$. Additionally, we may write the distribution function as

$$F(x) = \mathbb{P}(X \leq x) = \int_{-\infty}^{x} f(t)dt.$$

Conversely, the density may be expressed through the distribution function by

$$f(x) = \frac{d}{dx}F(x).$$

The expected value of a continuous random variable X with density function f is

$$\mathbb{E}(X) = \int_{-\infty}^{\infty} xf(x)dx,$$

and for any function $g(X)$,

$$\mathbb{E}[g(X)] = \int_{-\infty}^{\infty} g(x)f(x)dx.$$

The *joint distribution function* of variables X and Y is

$$F(x, y) = \mathbb{P}(X \leq x, Y \leq y).$$

For two discrete random variables, one can define their joint probability mass function as

$$f(x, y) = \mathbb{P}(X = x, Y = y).$$

Then the *marginal* probability mass functions of X and Y can be computed as, respectively,

$$f_X(x) = \sum_y f(x, y) \quad \text{and} \quad f_Y(y) = \sum_x f(x, y).$$

One can similarly define the joint density function $f(x, y)$ of two continuous random variables X and Y. It is a nonnegative piecewise continuous function of two variables x and y such that $\int_{-\infty}^{\infty}\int_{-\infty}^{\infty} f(x, y)dxdy = 1$. The marginal densities of X and Y are then defined by

$$f_X(x) = \int_{-\infty}^{\infty} f(x, y)dy \quad \text{and} \quad f_Y(y) = \int_{-\infty}^{\infty} f(x, y)dx.$$

The conditional distribution of Y given X is a probability mass function (or a density function) written $f(y|x)$ such that

$$f(x, y) = f(y|x)f_X(x).$$

Two random variables X and Y are *independent* if their joint probability mass function or density $f(x, y)$ can be written as

$$f(x, y) = f_X(x) f_Y(y),$$

or equivalently $f(y|x) = f_Y(y)$.

Intuitively, independence means that knowing the value of one variable does not change the distribution of the other variable. Independent variables X and Y have the following important property:

$$\mathbb{E}(XY) = \mathbb{E}(X)\mathbb{E}(Y),$$

and in general, for any functions g and h,

$$\mathbb{E}\left[g(X)h(Y)\right] = \mathbb{E}\left[g(X)\right]\mathbb{E}\left[h(Y)\right].$$

Properties of Expectation, Variance and Covariance

The *variance* of a random variable X is defined as

$$\mathrm{Var}(X) = \mathbb{E}\left[X - \mathbb{E}(X)\right]^2 = \mathbb{E}(X^2) - [\mathbb{E}(X)]^2.$$

The *standard deviation* of X is defined as

$$\mathrm{SD}(X) = \sqrt{\mathrm{Var(X)}}.$$

The *covariance* of two random variables X and Y is

$$\mathrm{Cov}(X, Y) = \mathbb{E}[X - \mathbb{E}(X)][Y - \mathbb{E}(Y)] = \mathbb{E}(XY) - \mathbb{E}(X)\mathbb{E}(Y),$$

and the *correlation coefficient* is

$$\mathrm{corr}(X, Y) = \frac{\mathrm{Cov}(X, Y)}{\mathrm{SD}(X)\mathrm{SD}(Y)}.$$

X and Y are called *uncorrelated* if their correlation coefficient is equal to 0. Independent variables are always uncorrelated, but uncorrelated variables are not necessarily independent.

If a and b are real numbers, and X and Y are random variables, then

1. $\mathbb{E}(aX + b) = a\mathbb{E}(X) + b$.
2. $\mathbb{E}(X + Y) = \mathbb{E}(X) + \mathbb{E}(Y)$.
3. $\mathrm{Var}(X) \geq 0$.
4. $\mathrm{Var}(aX + b) = a^2\mathrm{Var}(X)$.
5. $\mathrm{Var}(X + Y) = \mathrm{Var}(X) + \mathrm{Var}(Y) + 2\mathrm{Cov}(X, Y)$.
6. If X and Y are uncorrelated, $\mathrm{Var}(X + Y) = \mathrm{Var}(X) + \mathrm{Var}(Y)$
7. $-1 \leq \mathrm{corr}(X, Y) \leq 1$, and if $\mathrm{corr}(X, Y) = \pm 1$, then there exist some constants a and b such that $Y = aX + b$.

Examples of Discrete Distributions

1. *Bernoulli* with parameter $p, 0 \le p \le 1$.

   - X can be thought of as an outcome of a coin toss ($X = 1$ if head; $X = 0$ if tail) for a coin with $\mathbb{P}(\text{head}) = p$.
   - X can take values 0 or 1.

$$f(1) = \mathbb{P}(X = 1) = p, \quad f(0) = \mathbb{P}(X = 0) = 1 - p;$$

$$\mathbb{E}(X) = p, \quad \mathrm{Var}(X) = p(1 - p).$$

2. *Binomial* with parameters n and p, n a positive integer and $0 \le p \le 1$.

   - X can be thought of as the number of heads in n independent coin tosses for a coin with $\mathbb{P}(\text{head}) = p$.
   - X can take values $k = 0, 1, \ldots, n$.

$$f(k) = \binom{n}{k} p^k (1 - p)^{n-k} = \frac{n!}{k!(n-k)!} p^k (1 - p)^{n-k}.$$

$$\mathbb{E}(X) = np, \quad \mathrm{Var}(X) = np(1 - p).$$

   - If $X_1, \ldots, X_n$ are independent Bernoulli random variables with common parameter p, then $X = X_1 + \ldots + X_n$ is binomial with parameters n and p.
   - A useful identity for binomial coefficients:

$$\sum_{k=0}^{n} \binom{n}{k} x^k y^{n-k} = (x + y)^n.$$

   - If we let $x = p$, $y = 1 - p$, we see $\sum_{k=0}^{n} f(k) = 1$. If we let $x = y = 1$, we get the identity

$$\sum_{k=0}^{n} \binom{n}{k} = 2^n.$$

3. *Geometric* with parameter $p, 0 \le p \le 1$.

   - X can be thought of as the number of coin tosses up to and including the first head for a coin with $\mathbb{P}(\text{head}) = p$.
   - X can take values $k = 1, 2, \ldots$.

$$f(k) = (1 - p)^{k-1} p.$$

$$\mathbb{E}(X) = \frac{1}{p}, \quad \mathrm{Var}(X) = \frac{1 - p}{p^2}.$$

4. *Negative binomial* with parameters r and p, r a positive integer and $0 \le p \le 1$.

   - X can be thought of as the number of coin tosses up to and including the time a head appears for the rth time for a coin with $\mathbb{P}(\text{head}) = p$.

- X can take values $k = r, r+1, \ldots$.

$$f(k) = \binom{k-1}{r-1}(1-p)^{k-r}p^r.$$

$$\mathbb{E}(X) = \frac{r}{p}, \quad \mathrm{Var}(X) = \frac{r(1-p)}{p^2}.$$

- If $X_1, \ldots, X_r$ are independent geometric random variables with parameter p, then $X = X_1 + \ldots + X_r$ is negative binomial with parameters r and p.

5. *Poisson* with parameter $\lambda > 0$.

- X can take values $k = 0, 1, \ldots$.

$$f(k) = \frac{e^{-\lambda}\lambda^k}{k!}.$$

$$\mathbb{E}(X) = \lambda, \quad \mathrm{Var}(X) = \lambda.$$

- If $X_1, \ldots, X_n$ are independent Poisson random variables with parameters $\lambda_1, \ldots, \lambda_n$, then their sum $X = X_1 + \cdots + X_n$ is a Poisson random variable with parameter $\lambda_1 + \cdots + \lambda_n$.

6. *Hypergeometric* with parameters n, M, and N all positive integers.

- Suppose an urn contains N balls: M black and $N - M$ white. If n balls are drawn without replacement, then X denotes the number of black balls among them.
- X can take values $k = \max(0, n - N + M), \ldots, \min(n, M)$.

$$f(k) = \frac{\binom{M}{k}\binom{N-M}{n-k}}{\binom{N}{n}},$$

$$\mathbb{E}(X) = n\frac{M}{N}, \quad \mathrm{Var}(X) = n\frac{M}{N}\left(1 - \frac{M}{N}\right)\frac{N-n}{N-1}.$$

- The hypergeometric distribution describes sampling without replacement from a finite population. If we keep n fixed and let M and N go to infinity in such a way that $M/N \to p$, then the hypergeometric probability mass function converges to the binomial probability mass function with parameters n and p. In other words, when the population is very large, there is essentially no difference between sampling with and without replacement.

7. *Multinomial* with parameters $n, p_1, \ldots, p_r$, n a positive integer, $0 \le p_i \le 1$, $\sum_{i=1}^r p_i = 1$.

- Suppose each of n independent trials can result in an outcome of one of r types, with probabilities $p_1, \ldots, p_r$, and let X_i be the total number of outcomes of type i in n trials.

- Each X_i can take values $n_i = 0, 1, \dots, n$, such that $\sum_i^n X_i = n$.

$$\mathbb{P}(X_1 = n_1, \dots, X_r = n_r) = \binom{n}{n_1, \dots, n_r} p_1^{n_1} p_2^{n_2} \dots p_r^{n_r},$$

provided that $n_1 + n_2 + \dots + n_r = n$, where

$$\binom{n}{n_1, \dots, n_r} = \frac{n!}{n_1! n_2! \cdots n_r!}.$$

- The binomial distribution is a special case of multinomial with $r = 2$.

Examples of Continuous Distributions

1. *Uniform* on the interval (0, 1).

$$f(x) = \begin{cases} 1 & \text{if } 0 \le x \le 1 \\ 0 & \text{otherwise} \end{cases},$$

$$F(x) = \begin{cases} 0 & \text{if } x < 0 \\ x & \text{if } 0 \le x \le 1 \\ 1 & \text{if } x > 1 \end{cases},$$

$$\mathbb{E}(X) = \frac{1}{2}, \quad \mathrm{Var}(X) = \frac{1}{12}.$$

Uniform on (0, 1) can be generalized to uniform on any interval (a, b) for a, b real numbers with $a < b$. If X is uniform on (0, 1), then $X' = a + (b - a)X$ is uniform on (a, b). It is easy to see that X' has density

$$f(x) = \begin{cases} \dfrac{1}{b-a} & \text{if } a \le x \le b \\ 0 & \text{otherwise} \end{cases}$$

and

$$\mathbb{E}(X') = \frac{a+b}{2}, \quad \mathrm{Var}(X') = \frac{(b-a)^2}{12}.$$

2. *Exponential* with parameter $\lambda > 0$.

$$f(x) = \begin{cases} \lambda e^{-\lambda x} & \text{if } x \ge 0 \\ 0 & \text{otherwise} \end{cases},$$

$$F(x) = \begin{cases} 1 - e^{-\lambda x} & \text{if } x \ge 0 \\ 0 & \text{otherwise} \end{cases},$$

$$\mathbb{E}(X) = \frac{1}{\lambda}, \quad \mathrm{Var}(X) = \frac{1}{\lambda^2}.$$

- The Poisson process provides an important connection between exponential and Poisson random variables. Suppose $X_1, X_2, \ldots$ are independent exponentials with parameter λ. Let $S_0 = 0$, and $S_n = X_1 + \cdots + X_n$ for all n. Fix some $t > 0$. Let $N(t)$ be the last index n such that $S_n < t$, i.e. $S_{N(t)} < t \leq S_{N(t)+1}$. Then $N(t)$ is a Poisson random variable with parameter λt. Moreover, if we think of $S_1, \ldots, S_n$ as points on a real line, then, conditional on $N(t) = n$, the locations of these n points are independently and uniformly distributed in the interval $(0, t)$.

3. *Gamma* with parameters $\lambda > 0$ and $\alpha > 0$.

$$f(x) = \begin{cases} \dfrac{\lambda^\alpha}{\Gamma(\alpha)} x^{\alpha-1} e^{-\lambda x} & \text{if } x \geq 0 \\ 0 & \text{otherwise} \end{cases},$$

where the gamma function $\Gamma(\alpha)$ is defined by $\Gamma(\alpha) = \int_0^\infty t^{\alpha-1} e^{-t} dt$ and for α a positive integer $\Gamma(\alpha) = (\alpha - 1)!$.

$$\mathbb{E}(X) = \frac{\alpha}{\lambda}, \quad \text{Var}(X) = \frac{\alpha}{\lambda^2}.$$

- Exponential with parameter λ is a special case of Gamma with $\alpha = 1$. Additionally, if $X_1, \ldots, X_n$ are independent exponential random variables with parameter λ, then $X = X_1 + \cdots + X_n$ has a Gamma distribution with parameters λ and n.

4. *Beta* with parameters $a > 0$ and $b > 0$.

$$f(x) = \begin{cases} \dfrac{1}{B(a, b)} x^{a-1}(1-x)^{b-1} & \text{if } 0 \leq x \leq 1 \\ 0 & \text{otherwise} \end{cases},$$

where the beta function $B(a, b)$ is defined by

$$B(a, b) = \int_0^1 t^{a-1}(1-t)^{b-1} dt = \frac{\Gamma(a)\Gamma(b)}{\Gamma(a+b)}.$$

$$\mathbb{E}(X) = \frac{a}{a+b}, \quad \text{Var}(X) = \frac{ab}{(a+b)^2(a+b+1)}.$$

5. *Standard normal*, written $\mathcal{N}(0, 1)$.

$$f(x) = \frac{1}{\sqrt{2\pi}} \exp\left(-\frac{1}{2}x^2\right),$$

$$\mathbb{E}(X) = 0, \quad \text{Var}(X) = 1.$$

6. *Normal* with parameters μ and $\sigma^2 > 0$, written $\mathcal{N}(\mu, \sigma^2)$.

$$f(x) = \frac{1}{\sqrt{2\pi\sigma^2}} \exp\left(-\frac{1}{2\sigma^2}(x-\mu)^2\right),$$

$$\mathbb{E}(X) = \mu, \quad \mathrm{Var}(X) = \sigma^2.$$

- If X is $\mathcal{N}(\mu, \sigma^2)$, then $(X - \mu)/\sigma$ is $\mathcal{N}(0, 1)$, a standard normal. If Y is a standard normal, then $\sigma Y + \mu$ is $\mathcal{N}(\mu, \sigma^2)$.
- If X_i's are independent $\mathcal{N}(\mu_i, \sigma_i^2)$, and $a_1, \ldots, a_n$ are real numbers, then $X = a_1 X_1 + \cdots + a_n X_n$ is $\mathcal{N}(\sum_{i=1}^n a_i \mu_i, \sum_{i=1}^n a_i^2 \sigma_i^2)$.

7. *Lognormal* with parameters μ and $\sigma^2 > 0$.

$$f(x) = \begin{cases} \dfrac{1}{x\sqrt{2\pi\sigma^2}} \exp\left(-\dfrac{1}{2\sigma^2}(\log y - \mu)^2\right) & \text{if } x \geq 0 \\ 0 & \text{otherwise} \end{cases},$$

$$\mathbb{E}(X) = \exp\left(\mu + \frac{1}{2}\sigma^2\right), \quad \mathrm{Var}(X) = \exp(2\mu + \sigma^2)\left(\exp(\sigma^2) - 1\right).$$

- X is called lognormal if the logarithm of X has a normal distribution with parameters μ and σ^2, or equivalently, if U is $\mathcal{N}(\mu, \sigma^2)$, then $X = e^U$.

8. *Chi-square* distribution (χ_n^2) with parameter n, a positive integer called *degrees of freedom*.

- The easiest way to define χ_n^2 is as follows: let $X_1, \ldots, X_n$ be independent normal random variables with mean 0 and variance 1. Then $X_1^2 + X_2^2 + \cdots + X_n^2$ has a χ_n^2 distribution.
- χ_n^2 is a special case of Gamma distribution with $\lambda = 1/2$ and $\alpha = n/2$.

$$\mathbb{E}(\chi_n^2) = n, \quad \mathrm{Var}(\chi_n^2) = 2n.$$

- If $X_1, \ldots, X_n$ are a sample from $\mathcal{N}(\mu, \sigma^2)$, with $\bar{x} = \frac{1}{n}\sum_{i=1}^n x_i$ the sample mean and $s^2 = \frac{1}{n-1}\sum_{i=1}^n (x_i - \bar{x})^2$ the sample variance, then

$$\frac{n-1}{\sigma^2} s^2$$

has a χ_{n-1}^2 distribution.

9. *t distribution*, or *Student's distribution* (t_n), with a positive integer parameter n (also called degrees of freedom).

- If Z is $\mathcal{N}(0, 1)$, Y is χ_n^2, and Z and Y are independent, then

$$\frac{Z}{\sqrt{Y/n}}$$

has a t_n distribution.
- When n is large, the t_n distribution becomes very similar to the standard normal distribution.

$$\mathbb{E}(t_n) = 0 \ \text{ for } n > 1; \quad \mathrm{Var}(t_n) = \frac{n}{n-2} \ \text{ for } n > 2;$$

otherwise $\mathbb{E}(t_n)$ and $\mathrm{Var}(t_n)$ are not defined.

- If $x_1, \dots, x_n$ are a sample from $\mathcal{N}(\mu, \sigma^2)$, with $\bar{x}$ the sample mean and s^2 the sample variance, then

$$\frac{\bar{x} - \mu}{s/\sqrt{n}}$$

has a t_{n-1} distribution. Here we are using another important property of normal sample statistics: $\bar{x}$ and s^2 are independent.

10. *F distribution* (F_{n_1,n_2}) with n_1 and n_2 degrees of freedom.

- If Y_1 and Y_2 are independent $\chi^2_{n_1}$ and $\chi^2_{n_2}$ random variables, then

$$\frac{Y_1/n_1}{Y_2/n_2}$$

has an F distribution with n_1 and n_2 degrees of freedom.

$$\mathbb{E}(F_{n_1,n_2}) = \frac{n_2}{n_2 - 2} \quad \text{for } n_2 > 2.$$

11. *Standard bivariate normal*, written $\mathcal{N}_2(0, 0, 1, 1, \rho)$, where $|\rho| \leq 1$.

$$f(x, y) = \frac{1}{2\pi\sqrt{1-\rho^2}} \exp\left(-\frac{1}{2(1-\rho^2)}(x^2 - 2\rho xy + y^2)\right).$$

12. *Bivariate normal* with parameters $\mu_X, \mu_Y, \sigma_X^2, \sigma_Y^2$ and ρ: μ_X and μ_Y are real numbers; $\sigma_X^2, \sigma_Y^2 > 0$; $|\rho| \leq 1$. Written $\mathcal{N}_2(\mu_X, \mu_Y, \sigma_X^2, \sigma_Y^2, \rho)$.

- If (X, Y) has a $\mathcal{N}_2(0, 0, 1, 1, \rho)$ standard bivariate normal distribution, then $(\sigma_X X + \mu_X, \sigma_Y Y + \mu_Y)$ has a bivariate normal distribution $\mathcal{N}_2(\mu_X, \mu_Y, \sigma_X^2, \sigma_Y^2, \rho)$. Similarly, if $(\tilde{X}, \tilde{Y})$ is bivariate normal distribution $\mathcal{N}_2(\mu_X, \mu_Y, \sigma_X^2, \sigma_Y^2, \rho)$, then $((\tilde{X} - \mu_X)/\sigma_X, (\tilde{Y} - \mu_Y)/\sigma_Y)$ is $\mathcal{N}_2(0, 0, 1, 1, \rho)$.
- Marginal distributions of X and Y are $\mathcal{N}(\mu_X, \sigma_X^2)$ and $\mathcal{N}(\mu_Y, \sigma_Y^2)$, respectively, and $\rho = \mathrm{corr}(X, Y)$. If $\rho = 0$, then X and Y are independent (note that this property does not hold in general).
- The conditional distribution of Y given $X = x$ is normal with mean $\mu_Y + \rho\frac{\sigma_Y}{\sigma_X}(x - \mu_X)$ and variance $\sigma_Y^2(1 - \rho^2)$.
- Note that if two variables have univariate normal distributions, their joint distribution is not necessarily bivariate normal.

Limit Theorems

1. *Law of Large Numbers:* Let $X_1, X_2, \dots$ be a sequence of independent random variables with mean μ and variance σ^2. Let $\bar{X}_n = \frac{1}{n}\sum_{i=1}^{n} X_i$. Then for any $t > 0$,

$$\mathbb{P}(|\bar{X}_n - \mu| > t) \to 0 \text{ as } n \to \infty.$$

2. *Central Limit Theorem:* Let $X_1, X_2, \ldots$ be a sequence of independent, identically distributed random variables with mean μ and variance σ^2. Let $\bar{X}_n = \frac{1}{n}\sum_{i=1}^n X_i$. Let $\Phi(z) = \mathbb{P}(\mathcal{N}(0,1) \leq z)$ be the distribution function of a standard normal. Then for any z,

$$\left|\mathbb{P}\left(\frac{\bar{X}_n - \mu}{\sigma/\sqrt{n}} \leq z\right) - \Phi(z)\right| \to 0 \text{ as n} \to \infty.$$

This mode of convergence is called convergence in distribution.

3. *Another Central Limit Theorem:* Let $x_1, \ldots, x_N$ be real numbers, and fix n, $0 < n < N$. Let S be a simple random sample of size n from $x_1, \ldots, x_N$, taken without replacement. Let $F(t)$ be the distribution function of the sample sum. Then for any t (Höglund [Hog78]),

$$\left|F(t) - \Phi\left(\frac{t - n\bar{x}}{s\sqrt{pq}}\right)\right| \leq \frac{C\sum_{k=1}^N (x_k - \bar{x})^3}{s^3\sqrt{pq}},$$

where $p = n/N, q = 1 - p$,

$$\bar{x} = \frac{1}{N}\sum_{k=1}^N x_k, \text{ and } s^2 = \frac{1}{N}\sum_{k=1}^N x_k^2.$$

4. *Poisson approximation to binomial:* Let $X_1, X_2, \ldots$ be a sequence of independent binomial random variables, where X_n is binomial$(n, \lambda/n)$ for $n = 1, 2, 3, \ldots$, and $0 < \lambda < 1$. Let Y have a Poisson(λ) distribution. Then, for $x = 0, 1, 2, \ldots$,

$$|\mathbb{P}(X_n \leq x) - \mathbb{P}(Y \leq x)| \to 0 \text{ as } n \to \infty.$$

References

[Hog78] T. Höglund. Sampling from a finite population: a remainder term estimate. *Scand. J. Stat.*, 5:69–71, 1978.

Appendix C
Tables

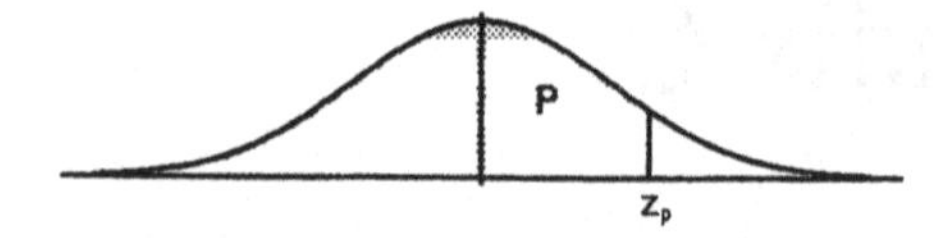

TABLE C.1. Cumulative normal distribution—values of P corresponding to z_p for the standard normal curve.

z_p	.00	.01	.02	.03	.04	.05	.06	.07	.08	.09
.0	.5000	.5040	.5080	.5120	.5160	.5199	.5239	.5279	.5319	.5359
.1	.5398	.5438	.5478	.5517	.5557	.5596	.5636	.5675	.5714	.5753
.2	.5793	.5832	.5871	.5910	.5948	.5987	.6026	.6064	.6103	.6141
.3	.6179	.6217	.6255	.6293	.6331	.6368	.6406	.6443	.6480	.6517
.4	.6554	.6591	.6628	.6664	.6700	.6736	.6772	.6808	.6844	.6879
.5	.6915	.6950	.6985	.7019	.7054	.7088	.7123	.7157	.719	.7224
.6	.7257	.7291	.7324	.7357	.7389	.7422	.7454	.7486	.7517	.7549
.7	.7580	.7611	.7642	.7673	.7704	.7734	.7764	.7794	.7823	.7852
.8	.7881	.7910	.7939	.7967	.7995	.8023	.8051	.8078	.8106	.8133
.9	.8159	.8186	.8212	.8238	.8264	.8289	.8315	.8340	.8365	.8389
1.0	.8413	.8438	.8461	.8485	.8508	.8531	.8554	.8577	.8599	.8621
1.1	.8643	.8665	.8686	.8708	.8729	.8749	.8707	.8790	.8810	.8830
1.2	.8849	.8869	.8888	.8907	.8925	.8944	.8962	.8980	.8997	.9015
1.3	.9032	.9049	.9066	.9082	.9099	.9115	.9131	.9147	.9162	.9177
1.4	.9192	.9207	.9222	.9236	.9251	.9265	.9279	.9292	.9306	.9319
1.5	.9332	.9345	.9357	.9370	.9382	.9394	.9406	.9418	.9429	.9441
1.6	.9452	.9463	.9474	.9484	.9495	.9505	.9515	.9525	.9535	.9545
1.7	.9554	.9564	.9573	.9582	.9591	.9599	.9608	.9616	.9625	.9633
1.8	.9641	.9649	.9656	.9664	.9671	.9678	.9686	.9693	.9699	.9706
1.9	.9713	.9719	.9726	.9732	.9738	.9744	.9750	.9756	.9761	.9767
2.0	.9772	.9778	.9783	.9788	.9793	.9798	.9803	.9808	.9812	.9817
2.1	.9821	.9826	.9830	.9834	.9838	.9842	.9846	.9850	.9854	.9857
2.2	.9861	.9864	.9868	.9871	.9875	.9878	.9881	.9884	.9887	.9890
2.3	.9893	.9896	.9898	.9901	.9904	.9906	.9909	.9911	.9913	.9916
2.4	.9918	.9920	.9922	.9925	.9927	.9929	.9931	.9932	.9934	.9936
2.5	.9938	.9940	.9941	.9943	.9945	.9946	.9948	.9949	.9951	.9952
2.6	.9953	.9955	.9956	.9957	.9959	.9960	.9961	.9962	.9963	.9964
2.7	.9965	.9966	.9967	.9968	.9969	.9970	.9971	.9972	.9973	.9974
2.8	.9974	.9975	.9976	.9977	.9977	.9978	.9979	.9979	.9980	.9981
2.9	.9981	.9982	.9982	.9983	.9984	.9984	.9985	.9985	.9986	.9986
3.0	.9987	.9987	.9987	.9988	.9988	.9989	.9989	.9989	.9990	.9990
3.1	.9990	.9991	.9991	.9991	.9992	.9992	.9992	.9992	.9993	.9993
3.2	.9993	.9993	.9994	.9994	.9994	.9994	.9994	.9995	.9995	.9995
3.3	.9995	.9995	.9995	.9996	.9996	.9996	.9996	.9996	.9996	.9997
3.4	.9997	.9997	.9997	.9997	.9997	.9997	.9997	.9997	.9997	.9998

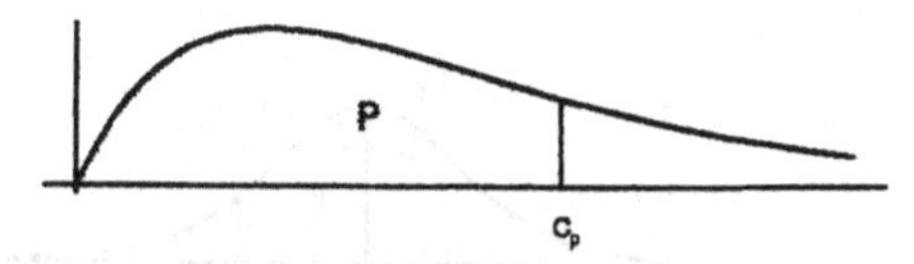

TABLE C.2. Percentiles of the Chisquare distribution—values of c_p corresponding to P.

df	$\chi^2_{.005}$	$\chi^2_{.01}$	$\chi^2_{.025}$	$\chi^2_{.05}$	$\chi^2_{.1}$	$\chi^2_{.9}$	$\chi^2_{.95}$	$\chi^2_{.975}$	$\chi^2_{.99}$	$\chi^2_{.995}$
1	0.000039	0.00016	0.00098	0.0039	0.0158	2.71	3.84	5.02	6.63	7.88
2	0.0100	0.0201	0.0506	0.1026	0.2107	4.61	5.99	7.38	9.21	10.60
3	0.0717	0.115	0.216	0.352	0.584	6.25	7.81	9.35	11.34	12.84
4	0.207	0.297	0.484	0.711	1.064	7.78	9.49	11.14	13.28	14.86
5	0.412	0.554	0.831	1.15	1.61	9.24	11.07	12.83	15.09	16.75
6	0.676	0.872	1.24	1.64	2.20	10.64	12.59	14.45	16.81	18.55
7	0.989	1.24	1.69	2.17	2.83	12.02	14.07	16.01	18.48	20.28
8	1.34	1.65	2.18	2.73	3.49	13.36	15.51	17.53	20.09	21.95
9	1.73	2.09	2.70	3.33	4.17	14.68	16.92	19.02	21.67	23.59
10	2.16	2.56	3.25	3.94	4.87	15.99	18.31	20.48	23.21	25.19
11	2.60	3.05	3.82	4.57	5.58	17.28	19.68	21.92	24.72	26.76
12	3.07	3.57	4.40	5.23	6.30	18.55	21.03	23.34	26.22	28.30
13	3.57	4.11	5.01	5.89	7.04	19.81	22.36	24.74	27.69	29.82
14	4.07	4.66	5.63	6.57	7.79	21.06	23.68	26.12	29.14	31.32
15	4.60	5.23	6.26	7.26	8.55	22.31	25.00	27.49	30.58	32.80
16	5.14	5.81	6.91	7.96	9.31	23.54	26.30	28.85	32.00	34.27
18	6.26	7.01	8.23	9.39	10.86	25.99	28.87	31.53	34.81	37.16
20	7.43	8.26	9.59	10.85	12.44	28.41	31.41	34.17	37.57	40.00
24	9.89	10.86	12.40	13.85	15.66	33.20	36.42	39.36	42.98	45.56
30	13.79	14.95	16.79	18.49	20.60	40.26	43.77	46.98	50.89	53.67
40	20.71	22.16	24.43	26.51	29.05	51.81	55.76	59.34	63.69	66.77
60	35.53	37.48	40.48	43.19	46.46	74.40	79.08	83.30	88.38	91.95
120	83.85	86.92	91.57	95.70	100.62	140.23	146.57	152.21	158.95	163.65

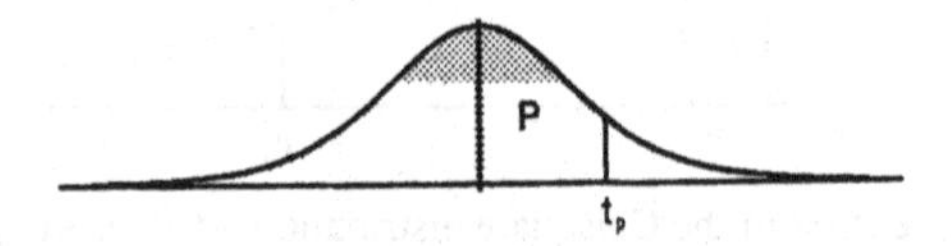

TABLE C.3. Percentiles of the t distribution—values of t_p corresponding to P.

df	$t_{.60}$	$t_{.70}$	$t_{.80}$	$t_{.90}$	$t_{.95}$	$t_{.975}$	$t_{.99}$	$t_{.995}$
1	0.325	0.727	1.376	3.078	6.314	12.706	31.821	63.657
2	0.289	0.617	1.061	1.886	2.920	4.303	6.965	9.925
3	0.277	0.584	0.978	1.638	2.353	3.182	4.541	5.841
4	0.271	0.569	0.941	1.533	2.132	2.776	3.747	4.604
5	0.267	0.559	0.920	1.476	2.015	2.571	3.365	4.032
6	0.265	0.553	0.906	1.440	1.943	2.447	3.143	3.707
7	0.263	0.549	0.896	1.415	1.895	2.365	2.998	3.499
8	0.262	0.546	0.889	1.397	1.860	2.306	2.896	3.355
9	0.261	0.543	0.883	1.383	1.833	2.262	2.821	3.250
10	0.260	0.542	0.879	1.372	1.812	2.228	2.764	3.169
11	0.260	0.540	0.876	1.363	1.796	2.201	2.718	3.106
12	0.259	0.539	0.873	1.356	1.782	2.179	2.681	3.055
13	0.259	0.538	0.870	1.350	1.771	2.160	2.650	3.012
14	0.258	0.537	0.868	1.345	1.761	2.145	2.624	2.977
15	0.258	0.536	0.866	1.341	1.753	2.131	2.602	2.947
16	0.258	0.535	0.865	1.337	1.746	2.120	2.583	2.921
17	0.257	0.534	0.863	1.333	1.740	2.110	2.567	2.898
18	0.257	0.534	0.862	1.330	1.734	2.101	2.552	2.878
19	0.257	0.533	0.861	1.328	1.729	2.093	2.539	2.861
20	0.257	0.533	0.860	1.325	1.725	2.086	2.528	2.845
21	0.257	0.532	0.859	1.323	1.721	2.080	2.518	2.831
22	0.256	0.532	0.858	1.321	1.717	2.074	2.508	2.819
23	0.256	0.532	0.858	1.319	1.714	2.069	2.500	2.807
24	0.256	0.531	0.857	1.318	1.711	2.064	2.492	2.797
25	0.256	0.531	0.856	1.316	1.708	2.060	2.485	2.787
26	0.256	0.531	0.856	1.315	1.706	2.056	2.479	2.779
27	0.256	0.531	0.855	1.314	1.703	2.052	2.473	2.771
28	0.256	0.530	0.855	1.313	1.701	2.048	2.467	2.763
29	0.256	0.530	0.854	1.311	1.699	2.045	2.462	2.756
30	0.256	0.530	0.854	1.310	1.697	2.042	2.457	2.750
40	0.255	0.529	0.851	1.303	1.684	2.021	2.423	2.704
60	0.254	0.527	0.848	1.296	1.671	2.000	2.390	2.660
120	0.254	0.526	0.845	1.289	1.658	1.980	2.358	2.617
∞	0.253	0.524	0.842	1.282	1.645	1.96	2.326	2.576

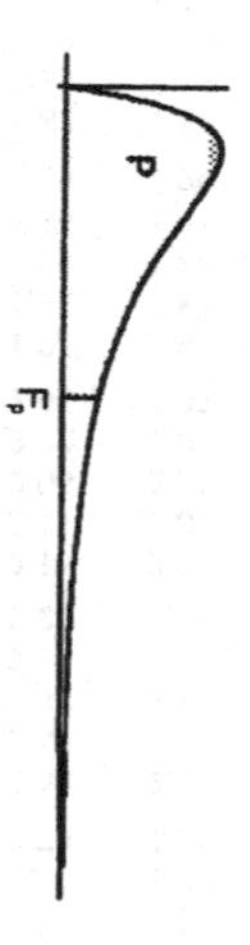

TABLE C.4. Percentiles of the F distribution—values of $F_{.90}$, for k degrees of freedom in the numerator and m degrees of freedom in the denominator.

$m\backslash k$	1	2	3	4	5	6	7	8	9	10	12	15	20	24	30	40	60	120
1	39.86	49.5	53.59	55.83	57.24	58.2	58.91	59.44	59.86	60.19	60.71	61.22	61.74	62.00	62.26	62.53	62.79	63.06
2	8.53	9.00	9.16	9.24	9.29	9.33	9.35	9.37	9.38	9.39	9.41	9.42	9.44	9.45	9.46	9.47	9.47	9.48
3	5.54	5.46	5.39	5.34	5.31	5.28	5.27	5.25	5.24	5.23	5.22	5.2	5.18	5.18	5.17	5.16	5.15	5.14
4	4.54	4.32	4.19	4.11	4.05	4.01	3.98	3.95	3.94	3.92	3.9	3.87	3.84	3.83	3.82	3.8	3.79	3.78
5	4.06	3.78	3.62	3.52	3.45	3.4	3.37	3.34	3.32	3.3	3.27	3.24	3.21	3.19	3.17	3.16	3.14	3.12
6	3.78	3.46	3.29	3.18	3.11	3.05	3.01	2.98	2.96	2.94	2.9	2.87	2.84	2.82	2.8	2.78	2.76	2.74
7	3.59	3.26	3.07	2.96	2.88	2.83	2.78	2.75	2.72	2.7	2.67	2.63	2.59	2.58	2.56	2.54	2.51	2.49
8	3.46	3.11	2.92	2.81	2.73	2.67	2.62	2.59	2.56	2.54	2.5	2.46	2.42	2.4	2.38	2.36	2.34	2.32
9	3.36	3.01	2.81	2.69	2.61	2.55	2.51	2.47	2.44	2.42	2.38	2.34	2.3	2.28	2.25	2.23	2.21	2.18
10	3.29	2.92	2.73	2.61	2.52	2.46	2.41	2.38	2.35	2.32	2.28	2.24	2.2	2.18	2.16	2.13	2.11	2.08
11	3.23	2.86	2.66	2.54	2.45	2.39	2.34	2.3	2.27	2.25	2.21	2.17	2.12	2.1	2.08	2.05	2.03	2.00
12	3.18	2.81	2.61	2.48	2.39	2.33	2.28	2.24	2.21	2.19	2.15	2.1	2.06	2.04	2.01	1.99	1.96	1.93
13	3.14	2.76	2.56	2.43	2.35	2.28	2.23	2.2	2.16	2.14	2.1	2.05	2.01	1.98	1.96	1.93	1.9	1.88
14	3.1	2.73	2.52	2.39	2.31	2.24	2.19	2.15	2.12	2.1	2.05	2.01	1.96	1.94	1.91	1.89	1.86	1.83
15	3.07	2.7	2.49	2.36	2.27	2.21	2.16	2.12	2.09	2.06	2.02	1.97	1.92	1.9	1.87	1.85	1.82	1.79
16	3.05	2.67	2.46	2.33	2.24	2.18	2.13	2.09	2.06	2.03	1.99	1.94	1.89	1.87	1.84	1.81	1.78	1.75
17	3.03	2.64	2.44	2.31	2.22	2.15	2.1	2.06	2.03	2.0	1.96	1.91	1.86	1.84	1.81	1.78	1.75	1.72
18	3.01	2.62	2.42	2.29	2.2	2.13	2.08	2.04	2.0	1.98	1.93	1.89	1.84	1.81	1.78	1.75	1.72	1.69
19	2.99	2.61	2.4	2.27	2.18	2.11	2.06	2.02	1.98	1.96	1.91	1.86	1.81	1.79	1.76	1.73	1.7	1.67
20	2.97	2.59	2.38	2.25	2.16	2.09	2.04	2.0	1.96	1.94	1.89	1.84	1.79	1.77	1.74	1.71	1.68	1.64
21	2.96	2.57	2.36	2.23	2.14	2.08	2.02	1.98	1.95	1.92	1.87	1.83	1.78	1.75	1.72	1.69	1.66	1.62
22	2.95	2.56	2.35	2.22	2.13	2.06	2.01	1.97	1.93	1.9	1.86	1.81	1.76	1.73	1.7	1.67	1.64	1.6
23	2.94	2.55	2.34	2.21	2.11	2.05	1.99	1.95	1.92	1.89	1.84	1.8	1.74	1.72	1.69	1.66	1.62	1.59
24	2.93	2.54	2.33	2.19	2.1	2.04	1.98	1.94	1.91	1.88	1.83	1.78	1.73	1.7	1.67	1.64	1.61	1.57
25	2.92	2.53	2.32	2.18	2.09	2.02	1.97	1.93	1.89	1.87	1.82	1.77	1.72	1.69	1.66	1.63	1.59	1.56
26	2.91	2.52	2.31	2.17	2.08	2.01	1.96	1.92	1.88	1.86	1.81	1.76	1.71	1.68	1.65	1.61	1.58	1.54
27	2.9	2.51	2.3	2.17	2.07	2.0	1.95	1.91	1.87	1.85	1.8	1.75	1.7	1.67	1.64	1.6	1.57	1.53
28	2.89	2.5	2.29	2.16	2.06	2.0	1.94	1.9	1.87	1.84	1.79	1.74	1.69	1.66	1.63	1.59	1.56	1.52
29	2.89	2.5	2.28	2.15	2.06	1.99	1.93	1.89	1.86	1.83	1.78	1.73	1.68	1.65	1.62	1.58	1.55	1.51
30	2.88	2.49	2.28	2.14	2.05	1.98	1.93	1.88	1.85	1.82	1.77	1.72	1.67	1.64	1.61	1.57	1.54	1.5
40	2.84	2.44	2.23	2.09	2.0	1.93	1.87	1.83	1.79	1.76	1.71	1.66	1.61	1.57	1.54	1.51	1.47	1.42
60	2.79	2.39	2.18	2.04	1.95	1.87	1.82	1.77	1.74	1.71	1.66	1.6	1.54	1.51	1.48	1.44	1.4	1.35
120	2.75	2.35	2.13	1.99	1.9	1.82	1.77	1.72	1.68	1.65	1.6	1.55	1.48	1.45	1.41	1.37	1.32	1.26

TABLE C.5. Percentiles of the F distribution—values of $F_{.95}$, for k degrees of freedom in the numerator and m degrees of freedom in the denominator.

$m\backslash k$	1	2	3	4	5	6	7	8	9	10	12	15	20	24	30	40	60	120
1	161.45	199.5	215.71	224.58	230.16	233.99	236.77	238.88	240.54	241.88	243.91	245.95	248.01	249.05	250.1	251.14	252.2	253.25
2	18.51	19.0	19.16	19.25	19.3	19.33	19.35	19.37	19.38	19.4	19.41	19.43	19.45	19.45	19.46	19.47	19.48	19.49
3	10.13	9.55	9.28	9.12	9.01	8.94	8.89	8.85	8.81	8.79	8.74	8.7	8.66	8.64	8.62	8.59	8.57	8.55
4	7.71	6.94	6.59	6.39	6.26	6.16	6.09	6.04	6.0	5.96	5.91	5.86	5.8	5.77	5.75	5.72	5.69	5.66
5	6.61	5.79	5.41	5.19	5.05	4.95	4.88	4.82	4.77	4.74	4.68	4.62	4.56	4.53	4.5	4.46	4.43	4.4
6	5.99	5.14	4.76	4.53	4.39	4.28	4.21	4.15	4.1	4.06	4.0	3.94	3.87	3.84	3.81	3.77	3.74	3.7
7	5.59	4.74	4.35	4.12	3.97	3.87	3.79	3.73	3.68	3.64	3.57	3.51	3.44	3.41	3.38	3.34	3.3	3.27
8	5.32	4.46	4.07	3.84	3.69	3.58	3.5	3.44	3.39	3.35	3.28	3.22	3.15	3.12	3.08	3.04	3.01	2.97
9	5.12	4.26	3.86	3.63	3.48	3.37	3.29	3.23	3.18	3.14	3.07	3.01	2.94	2.9	2.86	2.83	2.79	2.75
10	4.96	4.1	3.71	3.48	3.33	3.22	3.14	3.07	3.02	2.98	2.91	2.85	2.77	2.74	2.7	2.66	2.62	2.58
11	4.84	3.98	3.59	3.36	3.2	3.09	3.01	2.95	2.9	2.85	2.79	2.72	2.65	2.61	2.57	2.53	2.49	2.45
12	4.75	3.89	3.49	3.26	3.11	3.0	2.91	2.85	2.8	2.75	2.69	2.62	2.54	2.51	2.47	2.43	2.38	2.34
13	4.67	3.81	3.41	3.18	3.03	2.92	2.83	2.77	2.71	2.67	2.6	2.53	2.46	2.42	2.38	2.34	2.3	2.25
14	4.6	3.74	3.34	3.11	2.96	2.85	2.76	2.7	2.65	2.6	2.53	2.46	2.39	2.35	2.31	2.27	2.22	2.18
15	4.54	3.68	3.29	3.06	2.9	2.79	2.71	2.64	2.59	2.54	2.48	2.4	2.33	2.29	2.25	2.2	2.16	2.11
16	4.49	3.63	3.24	3.01	2.85	2.74	2.66	2.59	2.54	2.49	2.42	2.35	2.28	2.24	2.19	2.15	2.11	2.06
17	4.45	3.59	3.2	2.96	2.81	2.7	2.61	2.55	2.49	2.45	2.38	2.31	2.23	2.19	2.15	2.1	2.06	2.01
18	4.41	3.55	3.16	2.93	2.77	2.66	2.58	2.51	2.46	2.41	2.34	2.27	2.19	2.15	2.11	2.06	2.02	1.97
19	4.38	3.52	3.13	2.9	2.74	2.63	2.54	2.48	2.42	2.38	2.31	2.23	2.16	2.11	2.07	2.03	1.98	1.93
20	4.35	3.49	3.1	2.87	2.71	2.6	2.51	2.45	2.39	2.35	2.28	2.2	2.12	2.08	2.04	1.99	1.95	1.9
21	4.32	3.47	3.07	2.84	2.68	2.57	2.49	2.42	2.37	2.32	2.25	2.18	2.1	2.05	2.01	1.96	1.92	1.87
22	4.3	3.44	3.05	2.82	2.66	2.55	2.46	2.4	2.34	2.3	2.23	2.15	2.07	2.03	1.98	1.94	1.89	1.84
23	4.28	3.42	3.03	2.8	2.64	2.53	2.44	2.37	2.32	2.27	2.2	2.13	2.05	2.01	1.96	1.91	1.86	1.81
24	4.26	3.4	3.01	2.78	2.62	2.51	2.42	2.36	2.3	2.25	2.18	2.11	2.03	1.98	1.94	1.89	1.84	1.79
25	4.24	3.39	2.99	2.76	2.6	2.49	2.4	2.34	2.28	2.24	2.16	2.09	2.01	1.96	1.92	1.87	1.82	1.77
26	4.23	3.37	2.98	2.74	2.59	2.47	2.39	2.32	2.27	2.22	2.15	2.07	1.99	1.95	1.9	1.85	1.8	1.75
27	4.21	3.35	2.96	2.73	2.57	2.46	2.37	2.31	2.25	2.2	2.13	2.06	1.97	1.93	1.88	1.84	1.79	1.73
28	4.2	3.34	2.95	2.71	2.56	2.45	2.36	2.29	2.24	2.19	2.12	2.04	1.96	1.91	1.87	1.82	1.77	1.71
29	4.18	3.33	2.93	2.7	2.55	2.43	2.35	2.28	2.22	2.18	2.1	2.03	1.94	1.9	1.85	1.81	1.75	1.7
30	4.17	3.32	2.92	2.69	2.53	2.42	2.33	2.27	2.21	2.16	2.09	2.01	1.93	1.89	1.84	1.79	1.74	1.68
40	4.08	3.23	2.84	2.61	2.45	2.34	2.25	2.18	2.12	2.08	2.0	1.92	1.84	1.79	1.74	1.69	1.64	1.58
60	4.0	3.15	2.76	2.53	2.37	2.25	2.17	2.1	2.04	1.99	1.92	1.84	1.75	1.7	1.65	1.59	1.53	1.47
120	3.92	3.07	2.68	2.45	2.29	2.18	2.09	2.02	1.96	1.91	1.83	1.75	1.66	1.61	1.55	1.5	1.43	1.35

TABLE C.6. Percentiles of the F distribution—values of $F_{.975}$, for k degrees of freedom in the numerator and m degrees of freedom in the denominator.

m\k	1	2	3	4	5	6	7	8	9	10	12	15	20	24	30	40	60	120
1	647.79	799.5	864.16	899.58	921.85	937.11	948.22	956.66	963.28	968.63	976.71	984.87	993.1	997.25	1001.41	1005.6	1009.8	1014.02
2	38.51	39.0	39.17	39.25	39.3	39.33	39.36	39.37	39.39	39.4	39.41	39.43	39.45	39.46	39.46	39.47	39.48	39.49
3	17.44	16.04	15.44	15.1	14.88	14.73	14.62	14.54	14.47	14.42	14.34	14.25	14.17	14.12	14.08	14.04	13.99	13.95
4	12.22	10.65	9.98	9.6	9.36	9.2	9.07	8.98	8.9	8.84	8.75	8.66	8.56	8.51	8.46	8.41	8.36	8.31
5	10.01	8.43	7.76	7.39	7.15	6.98	6.85	6.76	6.68	6.62	6.52	6.43	6.33	6.28	6.23	6.18	6.12	6.07
6	8.81	7.26	6.6	6.23	5.99	5.82	5.7	5.6	5.52	5.46	5.37	5.27	5.17	5.12	5.07	5.01	4.96	4.9
7	8.07	6.54	5.89	5.52	5.29	5.12	4.99	4.9	4.82	4.76	4.67	4.57	4.47	4.41	4.36	4.31	4.25	4.2
8	7.57	6.06	5.42	5.05	4.82	4.65	4.53	4.43	4.36	4.3	4.2	4.1	4.0	3.95	3.89	3.84	3.78	3.73
9	7.21	5.71	5.08	4.72	4.48	4.32	4.2	4.1	4.03	3.96	3.87	3.77	3.67	3.61	3.56	3.51	3.45	3.39
10	6.94	5.46	4.83	4.47	4.24	4.07	3.95	3.85	3.78	3.72	3.62	3.52	3.42	3.37	3.31	3.26	3.2	3.14
11	6.72	5.26	4.63	4.28	4.04	3.88	3.76	3.66	3.59	3.53	3.43	3.33	3.23	3.17	3.12	3.06	3.0	2.94
12	6.55	5.1	4.47	4.12	3.89	3.73	3.61	3.51	3.44	3.37	3.28	3.18	3.07	3.02	2.96	2.91	2.85	2.79
13	6.41	4.97	4.35	4.0	3.77	3.6	3.48	3.39	3.31	3.25	3.15	3.05	2.95	2.89	2.84	2.78	2.72	2.66
14	6.3	4.86	4.24	3.89	3.66	3.5	3.38	3.29	3.21	3.15	3.05	2.95	2.84	2.79	2.73	2.67	2.61	2.55
15	6.2	4.77	4.15	3.8	3.58	3.41	3.29	3.2	3.12	3.06	2.96	2.86	2.76	2.7	2.64	2.59	2.52	2.46
16	6.12	4.69	4.08	3.73	3.5	3.34	3.22	3.12	3.05	2.99	2.89	2.79	2.68	2.63	2.57	2.51	2.45	2.38
17	6.04	4.62	4.01	3.66	3.44	3.28	3.16	3.06	2.98	2.92	2.82	2.72	2.62	2.56	2.5	2.44	2.38	2.32
18	5.98	4.56	3.95	3.61	3.38	3.22	3.1	3.01	2.93	2.87	2.77	2.67	2.56	2.5	2.44	2.38	2.32	2.26
19	5.92	4.51	3.9	3.56	3.33	3.17	3.05	2.96	2.88	2.82	2.72	2.62	2.51	2.45	2.39	2.33	2.27	2.2
20	5.87	4.46	3.86	3.51	3.29	3.13	3.01	2.91	2.84	2.77	2.68	2.57	2.46	2.41	2.35	2.29	2.22	2.16
21	5.83	4.42	3.82	3.48	3.25	3.09	2.97	2.87	2.8	2.73	2.64	2.53	2.42	2.37	2.31	2.25	2.18	2.11
22	5.79	4.38	3.78	3.44	3.22	3.05	2.93	2.84	2.76	2.7	2.6	2.5	2.39	2.33	2.27	2.21	2.14	2.08
23	5.75	4.35	3.75	3.41	3.18	3.02	2.9	2.81	2.73	2.67	2.57	2.47	2.36	2.3	2.24	2.18	2.11	2.04
24	5.72	4.32	3.72	3.38	3.15	2.99	2.87	2.78	2.7	2.64	2.54	2.44	2.33	2.27	2.21	2.15	2.08	2.01
25	5.69	4.29	3.69	3.35	3.13	2.97	2.85	2.75	2.68	2.61	2.51	2.41	2.3	2.24	2.18	2.12	2.05	1.98
26	5.66	4.27	3.67	3.33	3.1	2.94	2.82	2.73	2.65	2.59	2.49	2.39	2.28	2.22	2.16	2.09	2.03	1.95
27	5.63	4.24	3.65	3.31	3.08	2.92	2.8	2.71	2.63	2.57	2.47	2.36	2.25	2.19	2.13	2.07	2.0	1.93
28	5.61	4.22	3.63	3.29	3.06	2.9	2.78	2.69	2.61	2.55	2.45	2.34	2.23	2.17	2.11	2.05	1.98	1.91
29	5.59	4.2	3.61	3.27	3.04	2.88	2.76	2.67	2.59	2.53	2.43	2.32	2.21	2.15	2.09	2.03	1.96	1.89
30	5.57	4.18	3.59	3.25	3.03	2.87	2.75	2.65	2.57	2.51	2.41	2.31	2.2	2.14	2.07	2.01	1.94	1.87
40	5.42	4.05	3.46	3.13	2.9	2.74	2.62	2.53	2.45	2.39	2.29	2.18	2.07	2.01	1.94	1.88	1.8	1.72
60	5.29	3.93	3.34	3.01	2.79	2.63	2.51	2.41	2.33	2.27	2.17	2.06	1.94	1.88	1.82	1.74	1.67	1.58
120	5.15	3.8	3.23	2.89	2.67	2.52	2.39	2.3	2.22	2.16	2.05	1.94	1.82	1.76	1.69	1.61	1.53	1.43

TABLE C.7. Percentiles of the F distribution—values of $F_{.99}$, for k degrees of freedom in the numerator and m degrees of freedom in the denominator.

$m \backslash k$	1	2	3	4	5	6	7	8	9	10	12	15	20	24	30	40	60	120
1	4052.18	4999.5	5403.35	5624.58	5763.65	5858.99	5928.36	5981.07	6022.47	6055.85	6106.32	6157.28	6208.73	6234.63	6260.65	6286.78	6313.03	6339.39
2	98.5	99.0	99.17	99.25	99.3	99.33	99.36	99.37	99.39	99.4	99.42	99.43	99.45	99.46	99.47	99.47	99.48	99.49
3	34.12	30.82	29.46	28.71	28.24	27.91	27.67	27.49	27.35	27.23	27.05	26.87	26.69	26.6	26.5	26.41	26.32	26.22
4	21.2	18.0	16.69	15.98	15.52	15.21	14.98	14.8	14.66	14.55	14.37	14.2	14.02	13.93	13.84	13.75	13.65	13.56
5	16.26	13.27	12.06	11.39	10.97	10.67	10.46	10.29	10.16	10.05	9.89	9.72	9.55	9.47	9.38	9.29	9.2	9.11
6	13.75	10.92	9.78	9.15	8.75	8.47	8.26	8.1	7.98	7.87	7.72	7.56	7.4	7.31	7.23	7.14	7.06	6.97
7	12.25	9.55	8.45	7.85	7.46	7.19	6.99	6.84	6.72	6.62	6.47	6.31	6.16	6.07	5.99	5.91	5.82	5.74
8	11.26	8.65	7.59	7.01	6.63	6.37	6.18	6.03	5.91	5.81	5.67	5.52	5.36	5.28	5.2	5.12	5.03	4.95
9	10.56	8.02	6.99	6.42	6.06	5.8	5.61	5.47	5.35	5.26	5.11	4.96	4.81	4.73	4.65	4.57	4.48	4.4
10	10.04	7.56	6.55	5.99	5.64	5.39	5.2	5.06	4.94	4.85	4.71	4.56	4.41	4.33	4.25	4.17	4.08	4.0
11	9.65	7.21	6.22	5.67	5.32	5.07	4.89	4.74	4.63	4.54	4.4	4.25	4.1	4.02	3.94	3.86	3.78	3.69
12	9.33	6.93	5.95	5.41	5.06	4.82	4.64	4.5	4.39	4.3	4.16	4.01	3.86	3.78	3.7	3.62	3.54	3.45
13	9.07	6.7	5.74	5.21	4.86	4.62	4.44	4.3	4.19	4.1	3.96	3.82	3.66	3.59	3.51	3.43	3.34	3.25
14	8.86	6.51	5.56	5.04	4.69	4.46	4.28	4.14	4.03	3.94	3.8	3.66	3.51	3.43	3.35	3.27	3.18	3.09
15	8.68	6.36	5.42	4.89	4.56	4.32	4.14	4.0	3.89	3.8	3.67	3.52	3.37	3.29	3.21	3.13	3.05	2.96
16	8.53	6.23	5.29	4.77	4.44	4.2	4.03	3.89	3.78	3.69	3.55	3.41	3.26	3.18	3.1	3.02	2.93	2.84
17	8.4	6.11	5.18	4.67	4.34	4.1	3.93	3.79	3.68	3.59	3.46	3.31	3.16	3.08	3.0	2.92	2.83	2.75
18	8.29	6.01	5.09	4.58	4.25	4.01	3.84	3.71	3.6	3.51	3.37	3.23	3.08	3.0	2.92	2.84	2.75	2.66
19	8.18	5.93	5.01	4.5	4.17	3.94	3.77	3.63	3.52	3.43	3.3	3.15	3.0	2.92	2.84	2.76	2.67	2.58
20	8.1	5.85	4.94	4.43	4.1	3.87	3.7	3.56	3.46	3.37	3.23	3.09	2.94	2.86	2.78	2.69	2.61	2.52
21	8.02	5.78	4.87	4.37	4.04	3.81	3.64	3.51	3.4	3.31	3.17	3.03	2.88	2.8	2.72	2.64	2.55	2.46
22	7.95	5.72	4.82	4.31	3.99	3.76	3.59	3.45	3.35	3.26	3.12	2.98	2.83	2.75	2.67	2.58	2.5	2.4
23	7.88	5.66	4.76	4.26	3.94	3.71	3.54	3.41	3.3	3.21	3.07	2.93	2.78	2.7	2.62	2.54	2.45	2.35
24	7.82	5.61	4.72	4.22	3.9	3.67	3.5	3.36	3.26	3.17	3.03	2.89	2.74	2.66	2.58	2.49	2.4	2.31
25	7.77	5.57	4.68	4.18	3.85	3.63	3.46	3.32	3.22	3.13	2.99	2.85	2.7	2.62	2.54	2.45	2.36	2.27
26	7.72	5.53	4.64	4.14	3.82	3.59	3.42	3.29	3.18	3.09	2.96	2.81	2.66	2.58	2.5	2.42	2.33	2.23
27	7.68	5.49	4.6	4.11	3.78	3.56	3.39	3.26	3.15	3.06	2.93	2.78	2.63	2.55	2.47	2.38	2.29	2.2
28	7.64	5.45	4.57	4.07	3.75	3.53	3.36	3.23	3.12	3.03	2.9	2.75	2.6	2.52	2.44	2.35	2.26	2.17
29	7.6	5.42	4.54	4.04	3.73	3.5	3.33	3.2	3.09	3.0	2.87	2.73	2.57	2.49	2.41	2.33	2.23	2.14
30	7.56	5.39	4.51	4.02	3.7	3.47	3.3	3.17	3.07	2.98	2.84	2.7	2.55	2.47	2.39	2.3	2.21	2.11
40	7.31	5.18	4.31	3.83	3.51	3.29	3.12	2.99	2.89	2.8	2.66	2.52	2.37	2.29	2.2	2.11	2.02	1.92
60	7.08	4.98	4.13	3.65	3.34	3.12	2.95	2.82	2.72	2.63	2.5	2.35	2.2	2.12	2.03	1.94	1.84	1.73
120	6.85	4.79	3.95	3.48	3.17	2.96	2.79	2.66	2.56	2.47	2.34	2.19	2.03	1.95	1.86	1.76	1.66	1.53

Index

Springer Texts in Statistics *(continued from page ii)*

Lehmann: Testing Statistical Hypotheses, Second Edition
Lehmann and Casella: Theory of Point Estimation, Second Edition
Lindman: Analysis of Variance in Experimental Design
Lindsey: Applying Generalized Linear Models
Madansky: Prescriptions for Working Statisticians
McPherson: Applying and Interpreting Statistics: A Comprehensive Guide, Second Edition
Mueller: Basic Principles of Structural Equation Modeling: An Introduction to LISREL and EQS
Nguyen and Rogers: Fundamentals of Mathematical Statistics: Volume I: Probability for Statistics
Nguyen and Rogers: Fundamentals of Mathematical Statistics: Volume II: Statistical Inference
Noether: Introduction to Statistics: The Nonparametric Way
Nolan and Speed: Stat Labs: Mathematical Statistics Through Applications
Peters: Counting for Something: Statistical Principles and Personalities
Pfeiffer: Probability for Applications
Pitman: Probability
Rawlings, Pantula and Dickey: Applied Regression Analysis
Robert: The Bayesian Choice: From Decision-Theoretic Foundations to Computational Implementation, Second Edition
Robert and Casella: Monte Carlo Statistical Methods, Second Edition
Rose and Smith: Mathematical Statistics with *Mathematica*
Ruppert: Statistics and Finance: An Introduction
Santner and Duffy: The Statistical Analysis of Discrete Data
Saville and Wood: Statistical Methods: The Geometric Approach
Sen and Srivastava: Regression Analysis: Theory, Methods, and Applications
Shao: Mathematical Statistics, Second Edition
Shorack: Probability for Statisticians
Shumway and Stoffer: Time Series Analysis and Its Applications
Simonoff: Analyzing Categorical Data
Terrell: Mathematical Statistics: A Unified Introduction
Timm: Applied Multivariate Analysis
Toutenburg: Statistical Analysis of Designed Experiments, Second Edition
Wasserman: All of Statistics: A Concise Course in Statistical Inference
Whittle: Probability via Expectation, Fourth Edition
Zacks: Introduction to Reliability Analysis: Probability Models and Statistical Methods